Plasmonics in Chemistry and Biology

Plasmonics in Chemistry and Biology

edited by

Marc Lamy de la Chapelle
Nordin Felidj

JENNY STANFORD
PUBLISHING

Published by

Jenny Stanford Publishing Pte. Ltd.
Level 34, Centennial Tower
3 Temasek Avenue
Singapore 039190

Email: editorial@jennystanford.com
Web: www.jennystanford.com

British Library Cataloguing-in-Publication Data
A catalogue record for this book is available from the British Library.

Plasmonics in Chemistry and Biology

ISBN 978-981-4800-03-7 (Hardcover)
ISBN 978-0-429-45875-0 (eBook)

Contents

Preface xiii

1. Plasmon-Driven Surface Functionalization of Gold Nanoparticles **1**

Mai Nguyen, Inga Tijunelyte, Marc Lamy de la Chapelle, Claire Mangeney, and Nordin Felidj

1.1 Plasmon-Induced Surface Functionalization by Diazonium Salts 2

1.1.1 Grafting by Laser Heating and Threshold Energy Dose E_{th} 3

1.1.2 Plasmon-Induced Grafting on 1D Structure Arrays of Gold Nanostripes 6

1.1.3 Plasmon-Driven Grafting on 2D Structure Array of Gold Nanorods 11

1.1.3.1 Description of the gold nanorod array 11

1.1.3.2 Plasmon-driven grafting on gold nanorod array 12

1.1.4 Plasmon-Driven Multi-Functionalization of Gold Nanodisks Array 15

1.1.5 Conclusion 21

1.2 Plasmon-Initiated Surface Functionalization by Thiol-Ene "Click" Chemistry 22

1.2.1 Fabrication of Substrates 22

1.2.2 In situ Thiol-Ene Click Reaction 23

1.2.3 Conclusion 28

2. Concept and Development of Multi-Functional Hybrid Systems: Photoswitchable and Thermotunable Plasmonic Materials **33**

Mai Nguyen, Leila Boubekeur-Lecaque, Claire Mangeney, Stéphanie Lau-Truong, Alexandre Chevillot-Biraud, François Maurel, Nordin Felidj, and Jean Aubard

2.1 Introduction 33
2.2 Elaboration and Properties of the Multifunctional Hybrid System 37
2.2.1 Preparation of Gold Nanoparticle Arrays 37
2.2.2 Preparation of GNPs Arrays Covered by PNIPAM with AB Chromophore End Groups (GNPA-PNIPAM-AB) 37
2.2.3 AFM and Optical (Extinction) Characterization of the Thermosensitive Properties of the GNPA-PNIPAM-AB System 39
2.3 Reversible Changes of the LSP Resonance of GNPA-PNIPAM-AB Upon *cis/trans* Isomerization of Azobenzene 42
2.4 SERS Experiments of GNPA-PNIPAM-AB at Various Temperatures and upon AB *cis/trans* Isomerization 44
2.4.1 ThermoInduced Reversible Changes of Azobenzene SERS Intensity 44
2.4.2 SERS Intensity Changes upon *cis/trans* Isomerization of Azobenzene 48
2.5 Conclusion 52

3. Reversible Adsorption of Biomolecules on Thermosensitive Polymer-Coated Plasmonic Nanostructures **57**

Nguyen Thi Tuyet Mai, Jean Aubard, Claire Mangeney, and Nordin Felidj

3.1 Introduction 57
3.2 Experimental 59

3.2.1 Materials 59
3.2.2 Elaboration of Gold Nanostructure Arrays 59
3.2.3 Functionalization of Gold Nanostructures by PNIPAM Brushes 59
3.2.3.1 Synthesis of diazonium salt 59
3.2.3.2 Initiator-modified gold surfaces 60
3.2.3.3 Atomic Transfer Radical Polymerization (ATRP) of NIPAM 60
3.3 Results and Discussion 61
3.3.1 Characterization of PNIPAM-Coated Gold Nanodots 61
3.3.2 Adsorption of Proteins on the PNIPAM-Grafted Gold Nanostructured Surface 64
3.4 Conclusion 67

4. Reactivity and Bio Samples Probed by Tip-Enhanced Raman Spectroscopy 71

Zhenglong Zhang, Robert Meyer, and Volker Deckert

4.1 Introduction—an Explanation of Tip-Enhanced Raman Spectroscopy 71
4.2 Plasmon-Driven Chemical Reactions 72
4.2.1 Hot Electron–Induced Chemical Reactions 73
4.2.2 Plasmon-Driven Chemical Reactions in SERS 75
4.2.3 Plasmon-Driven Chemical Reaction at the Tip of a Probe 77
4.3 Probing Biological Samples 82
4.3.1 Human Cells and Its Components 83
4.3.2 Virus and Bacteria 86
4.3.3 From Amino Acids to Peptides and Fibrils 89
4.3.4 DNA and RNA 94
4.4 Conclusion 97

5. Surface-Enhanced Spectro-Electrochemistry of Biological and Molecular Catalysts on Plasmonic Electrodes **109**

Patrycja Kielb and Inez M. Weidinger

5.1 Principles of Electrocatalysis 110
5.1.1 Why Do We Need to Understand Electrocatalytic Reactions? 110
5.1.2 Metal–Porphyrin Complexes in Biology and Chemistry 113
5.2 Methods to Probe Structure and Function of Catalysts on Electrodes 115
5.2.1 Electrochemistry 115
5.2.2 Infrared and Resonance Raman Spectroscopy of Porphyrins 117
5.2.3 Surface-Enhanced Vibrational Spectroscopy 119
5.2.4 Surface-Enhanced Spectro-Electrochemistry on Porphyrin Systems 121
5.2.5 Time Resolved SER Spectroscopy 123
5.3 Examples 123
5.3.1 Cellobiose Dehydrogenase 124
5.3.2 Cytochrome c Oxidase 127
5.3.3 Hangman Complexes 130
5.4 Conclusions 134

6. Fluorescence Spectroscopy Enhancement on Photonic Nanoantennas **139**

Jérôme Wenger

6.1 Introduction and Motivation 139
6.2 Brief Theoretical Background: The Physics of Fluorescence Enhancement 141
6.3 Experimental Approaches to Enhance Fluorescence 144
6.3.1 Top-Down Milling 145
6.3.2 Bottom-Up Self-Assembly 147

6.3.3 Dielectric Alternatives to Plasmonic Metals 148
6.4 Biochemical Applications of Enhanced Fluorescence 148
6.4.1 Real-Time DNA Sequencing 149
6.4.2 Nanoscale Organization of Lipid Membranes 150
6.4.3 Förster Resonance Energy Transfer 150
6.5 Conclusion 151

7. Plasmonic-Based SERS-Traceable Drug Nanocarriers in Cancer Theranostics 159

Monica Potara, Timea Nagy-Simon, Sorina Suarasan, and Simion Astilean

7.1 Introduction 160
7.2 SERS Encoded Plasmonic Nanoparticles for Cancer Detection and Imaging 163
7.3 Combining SERS Imaging with Therapy for Cancer Theranostics 169
7.3.1 SERS-Traceable Plasmonic Nanoparticles in Chemotherapeutic Drug Delivery Applications 171
7.3.2 SERS-Traceable Plasmonic Nanoparticles in Photosensitizer Delivery Applications 179
7.4 Conclusions 188

8. Label-Free SERS Detection of Heme-Proteins with Porous Silver Nanocubes 199

Maximilien Cottat, Marella de Angelis, Elizaveta Panfilova, Nikolai Khlebtsov, Roberto Pini, and Paolo Matteini

8.1 Introduction 199
8.2 Protein Detection Using Standard and Porous Nanocubes 201
8.2.1 Experimental Section 201
8.2.1.1 Fabrication of the nanoparticles 201
8.2.1.2 SERS analysis 202

8.2.2 Results and Discussion 203
8.2.2.1 Characterization of the nanoparticles 203
8.2.2.2 Simulation of the E-field distribution 204
8.2.2.3 SERS detection 206
8.2.2.4 SERS sensitivity 210
8.3 Conclusions 212

9. Observation of Biomolecules and Their Dynamics in SERS 219

Jean Emmanuel Clément, Thibault Brulé, Aymeric Leray, and Eric Finot

9.1 Introduction 219
9.2 Raman Spectroscopy of Proteins 221
9.3 Detection of Single Structures 222
9.3.1 Complexity: Sorting of Molecules 223
9.3.2 Submolecular Resolution: Spectral Pointillism 225
9.4 Dynamics 227
9.4.1 Molecular Counting 227
9.4.2 Strong Volatility 228
9.5 Conclusion 229

10. Intracellular Surface-Enhanced Raman Spectroscopy 233

Jack Taylor, Anna Huefner, Jonathan Wingfield, and Sumeet Mahajan

10.1 Intracellular Applications of SERS 233
10.2 Nanoparticle–Cell Interactions 235
10.2.1 Cellular Internalisation Methods 235
10.2.2 The Endocytotic Pathway 236
10.2.3 Manipulating Interactions 239
10.2.4 Toxicity 242
10.3 Intracellular SERS 244
10.3.1 Advances in SERS-Reporter Research 246

10.3.2 Advances in Reporter-Free SERS 253
10.4 Conclusions and Outlook 264

11. SERS-Based Nanotechnology for Imaging of Cellular Properties 277

Ewelina Wiercigroch and Kamilla Malek

11.1 Cellular Interaction and Uptake of SERS Nanosensors 278
11.2 Designing of Cellular SERS Nanoprobes 284
11.2.1 Simple Functionalization of the Metal Surface 284
11.2.2 Antibody- and Peptide-Functionalized SERS Tags 289
11.2.3 Other Solutions for Cellular SERS-Sensing 294
11.3 Conclusion 295

Index 303

Preface

Over the past decade, plasmonic nanoparticles have been the subject of extensive research, owing to their remarkable optical properties. These properties arise from a collective oscillation of the conductive electrons at the nanoparticle surface, under light irradiation. These resonant oscillations, so-called localized surface plasmon (LSP), are mostly efficient for noble metals. Such optical phenomenon is characterized by: (i) a strong absorption and scattering in the far field for specific wavelengths depending on the geometrical parameters of the nanoparticles and (ii) a strong amplification of the local electric field at the proximity of the nanoparticles. These outstanding properties have stimulated a huge enthusiasm for the use of plasmonic nanoparticles in surface-enhanced spectroscopies (Raman, fluorescence) or in nano-optics (optical switches and modulators, plasmonic waveguides, etc.). Quite recently, it has been shown that the activation and the initiation of chemical reactions or physical processes can be provided using the LSP excitation. Such exploitation presents two main advantages: an enhanced yield and a fine control of the chemical reactions at the nanoscale. These topics have become very active and are in line with molecular plasmonics. The objective of this book, thus, is to explore this new field and to provide a wide view on the exploitation of plasmonics in the chemical and biological fields. In particular, it will be seen that LSP excitation can be employed as a tool to promote and probe specific physical, chemical, or biological effects at the nanoscale.

The book is divided in two main sections concerning (i) the LSP induction of physical or chemical processes (Chapters 1 to 6) and (ii) the probe of biological processes by plasmonics (Chapters 7 to 11).

The first section is focused on the use of plasmonic properties to induce chemical reactions at the nanometer scale. The first chapter, by N. T. Nguyen et al., illustrates two examples of LSP-induced chemical reactions: (i) a chemical surface functionalization of aryl films derived from diazonium salts and (ii) a thiol-ene click-chemistry reaction. Chapter 2 discusses the control of energy transfer from a photochromic molecule and a plasmonic nanoparticle through a thermoresponsive polymer linker. In Chapters 3, 4, and 5 (authored by N. T. Nguyen et al., Z. Zhang et al., and P. Kielb et al., respectively), the reactivity and adhesion of biological systems are investigated using tip-enhanced Raman spectroscopy (TERS) and surface-enhanced Raman scattering (SERS) techniques. Chapter 6, by J. Wenger, focuses on fluorescence enhancement using plasmonic nanoantennas.

The second section concerns the observation of biological structures and processes, essentially using the SERS technique based on the plasmonic properties of metallic nanoparticles. Thus, several applications are described in Chapters 7 to 11. In Chapter 7, M. Potara et al. explain how to design some SERS traceable nanoparticles for tumor detection and cancer diagnosis. Such objects can also be used as nanocarriers to transport drug to the tumor and as a consequence to deliver cancer treatment by using their plasmonic properties. In Chapters 8 and 9, M. Cottat et al. and J.-E. Clément et al., respectively, use the SERS technique to detect proteins and to probe their structures and their potential modifications induced by chemical or biological processes. Multivariate statistical methods are notably exploited to observe the various protein conformations and their dynamics. Chapters 10 and 11, by J. Taylor et al. and E. Wiercigorch et al., respectively, discuss the investigation of cell structures, mechanisms, and compartments. These chapters demonstrate that the SERS method can be used to image cells and to study some specific parts of cells, such as lysosomes and endosomes. The second section of the book illustrates how SERS can be powerful to describe the study of biological processes and objects at the nano- or microscopic scales.

Thus, the different chapters of this book pave the way for new approaches and new insights on the wide applications of plasmonics in chemistry and biology.

Marc Lamy de la Chapelle
Nordin Felidj

Chapter 1

Plasmon-Driven Surface Functionalization of Gold Nanoparticles

Mai Nguyen,[a,b] Inga Tijunelyte,[a,c] Marc Lamy de la Chapelle,[c] Claire Mangeney,[a,d] and Nordin Felidj[a]

[a]*Laboratoire ITODYS, Université Paris Diderot, Sorbonne Paris Cité, CNRS UMR 7086, 15 rue Jean-Antoine de Baïf, 75205 Paris Cedex 13, France*
[b]*School of Chemical Engineering, Hanoi University of Science and Technology, 1 Dai Co Viet road, Hanoi, Vietnam*
[c]*Laboratoire CSPBAT, Université Paris 13, Sorbonne Paris Cité, CNRS, UMR7244, 74 rue Marcel Cachin, 93017 Bobigny, France*
[d]*UFR Biomédicale, UMR 8601 Université Paris Descartes Sorbonnne Paris Cité, 45 Rue des Saint Pères, 70005 Paris, France*

Nguyenthituyetmai87@gmail.com

Over the past decade, plasmonic nanoparticles (NPs) have been the subject of extensive research, due to their remarkable optical properties [1, 2]. These properties arise from a collective oscillation of the conductive electrons at the NP surface, under light irradiation [3]. These resonant oscillations, so-called localized surface plasmon (LSP), are mostly efficient for noble metals (gold, silver, copper), with a LSP frequency strongly dependent on the geometrical features of the NPs and the chemical nature of the metal [4, 5]. These LSP modes are characterized in the far field by an enhanced extinction mainly in the visible and near-infrared

Plasmonics in Chemistry and Biology
Edited by Marc Lamy de la Chapelle and Nordin Felidj

ISBN 978-981-4800-03-7 (Hardcover), 978-0-429-45875-0 (eBook)
www.jennystanford.com

spectral region, and a strong electric field enhancement in the near-field of the NPs [6]. Therefore, plasmonic NPs can act as a light source, as well as a heat generator and electron reservoir at the nanoscale, controllable by the light energy [7]. These outstanding properties have stimulated huge enthusiasm for the use of plasmonic NPs in medicine (hyperthermia, drug delivery, bio-imaging), surface-enhanced spectroscopy or nanooptics (optical switches and modulators, plasmonic waveguides, etc.) [8–12].

In chemistry, plasmonic NPs can offer a unique platform to selectively boost the chemical reaction yields with a spatial control at the nanoscale [13, 14]. The mechanisms involved in plasmon-mediated chemical reactions result mainly from three processes: a local field enhancement, heat generation, and hot-electron transfer [7]. These processes have recently been mentioned for triggering chemical reactions. Although all these examples have demonstrated the unique potential of plasmonic NPs to activate various kinds of chemical reactions, their use for plasmon-mediated chemical surface functionalization at the nanoscale has been poorly investigated so far. To address this issue and afford a selective plasmon-mediated chemical surface functionalization on gold nanoparticles, we propose two novel approaches. The first one is related with combining plasmonic AuNPs as electron reservoirs and aryl diazonium salts as electron-induced surface-functionalization agents [15–17] and will be overviewed in Section 1.1 of this chapter. The second approach, discussed in Section 1.2, involves local surface functionalization by thiols grafted via plasmon-initiated thiol-ene "click" chemistry [18].

1.1 Plasmon-Induced Surface Functionalization by Diazonium Salts

Regarding literature, aryl diazonium salts are able to bind covalently to gold surfaces via electron transfer and to form grafted polymers with thicknesses ranging from 1 nm to above 300 nm [19–21]. In first part of this chapter, we demonstrate that plasmonic gold NPs can act as a booster to promote the grafting of diazonium derived aryl films at the gold NPs surface, in a confined nanoscale area. The aryl films derived from the diazonium salt are grafted on the gold NPs, exclusively in areas corresponding to a maximum of the near-field intensity enhancement of the gold NPs. Moreover, the grafting yield and the aryl film thickness can be monitored by the LSP wavelength as well as the energy dose produced by the incident light. Interestingly, we monitored the incident light polarization on gold nanoparticles, to trigger plasmon excitation and obtain regiospecific local surface double-functionalization by diazonium salts. By playing on the polarization of the incident excitation light, the plasmon-driven functionalization of these isotropic NPs should result in a regioselective surface anisotropy. The plasmon-driven regioselective grafting was studied using scanning electron microscopy (SEM), atomic force microscopy (AFM), extinction micro-spectroscopy, and surface-enhanced Raman spectroscopy (SERS). Calculations based on the discrete dipole approximation (DDA) method were also performed to compare the spatial localization of the functional groups around the gold nanoparticles with the electromagnetic field distribution. These approaches open way for a vast field of study on the nanoscale confinement of organic layers at the plasmonic nanostructure surface.

1.1.1 Grafting by Laser Heating and Threshold Energy Dose E_{th}

The laser irradiation is expected to generate a local heating, which can be sufficiently high to activate the thermal decomposition of diazonium salts and the resulting grafting of aryl layers, as described by M. Busson et al. [22] This laser-induced thermal grafting could occur simultaneously with a plasmon-induced electron transfer and these two mechanisms should therefore be investigated separately to evaluate their respective efficiencies. In order to first investigate a photothermal mechanism due to laser heating at the nanostripe surface, we measured the threshold energy dose E_{th}, corresponding to a threshold temperature T_{th} above which the grafting of aryl layers could be induced photothermally, in the absence of any plasmonic excitation. The grafting by laser heating of aryl films derived from the 4-(2-hydroxyethyl)-benzene diazonium tetrafluoroborate (HEBDT) salt is demonstrated through the use of lithographically designed gold nanostripes (width W = 125 ± 5 nm, height H = 50 ± 5 nm, length $L \sim$ 100 μm, and interparticle distance $\Lambda \sim$ 5 μm) lying on an ITO coated-glass substrate [23]. Similarly, to metal NP (a stripe can be understood as a "nanoparticle" with one infinitely extending dimension), their LSP resonance and extinction peak can be switched *on* and *off* by changing the incident polarization [24]. When the polarization of the incident illumination is perpendicular to the stripe length (transverse polarization), the observed enhanced extinction peak is assigned to a dipolar LSP resonance, consistent with previous results [25]. In contrast, for an incident electric field parallel to the stripes (longitudinal polarization), no LSP resonance is excited, leading to an extinction spectrum with a low signal, similarly to a gold flat film. The LSP wavelength is located at $\lambda_{LSP} \sim$ 640 nm for a transverse polarization. The gold nanostripes (GNTs) were homogeneously immersed in a solution of HEBDT (3 mM).

The threshold energy dose E_{th} was estimated using AFM, by comparing the height of a single GNT immersed in HEBDT, before and after laser irradiation with two different laser powers (P_0 = 0.089 mWμm^{-2} and P_1 = 0.86 mW μm^{-2}) and by varying the irradiation time. The experiments were carried out with a longitudinal polarization, for which no LSP resonance was excited. As an example, Fig. 1.1a shows the profile of the cross section of a single stripe measured by AFM, before and after laser irradiation under longitudinal polarization, after 240 seconds using a power P_0 = 0.089 mWμm^{-2}. The height difference between the two profiles points out the grafting with a homogeneous thickness of 3 ± 0.5 nm. The aryl grafting is attributed to the thermal decomposition of the diazonium salts induced by the laser heating generating aryl cations able to attach to the metallic surface. For the two laser powers, the measured thicknesses of the aryl films were reported versus the irradiation time (not shown). The aryl film thickness appears to increase with the incubation time, reaching around H_{aryl} = 4 nm after t_1 = 40 s of irradiation for a laser power of P_1 = 0.86 mWμm^{-2} or after t_0 = 300 s of irradiation for a laser power of P_0 = 0.089 mWμm^{-2}. Interestingly, the aryl film thickness appears to depend exclusively on the energy dose (which represents the product of the laser power by the incubation time), deduced from the two sets of irradiation conditions (see Fig. 1.1b). For instance, an aryl thickness of ~4 nm is obtained for an energy dose of E = 26.5 mJμm^{-2}. From these curves, a threshold energy dose of E_{th} = 5.5 mJμm^{-2} could be deduced, above which the laser heating is sufficient to induce the grafting of the diazonium-derived aryl film. This value is also the threshold below which the laser heating is not sufficient to induce the grafting of the aryl.

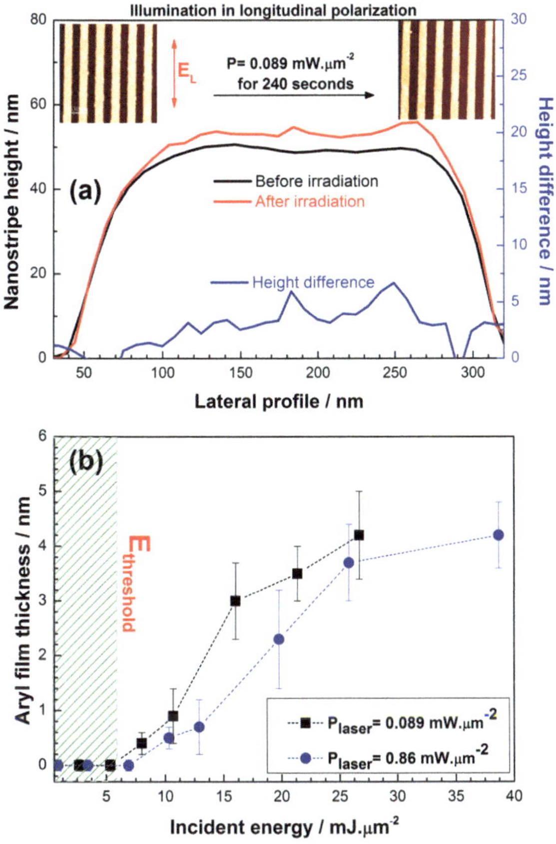

Figure 1.1 (a) Profiles of the lateral cross sections of a GNT, deduced from AFM measurements, in air: (black curve) before laser irradiation; (red curve) after irradiation in longitudinal polarization at 633 nm (P_0 = 0.089 mWμm^{-2}, 240 s); (blue curve) differential cross sections deduced from the two previous profiles, evidencing the deposition of an aryl film with a thickness of ~3.5 nm. Inset: AFM image of the stripe array before and after the irradiation. (b) Aryl film thickness (deduced from AFM measurements) versus the energy dose E (mJ μm^{-2}).

1.1.2 Plasmon-Induced Grafting on 1D Structure Arrays of Gold Nanostripes

To study exclusively the plasmon-induced grafting and be sure that the laser heating is not involved in the grafting mechanism, the next step consisted of working below this threshold energy dose, but under plasmon excitation. The nanostripe array was illuminated in its LSP resonance and in transverse polarization, thus under plasmon excitation. The incident laser power was kept constant at P_0 = 0.89 mWμm^{-2}, while varying the irradiation time. A GNTs array was first considered, with a LSP resonance at λ_{LSP} at 640 nm (in water), coinciding with the laser excitation (λ_{inc} = 633 nm). The sample was immersed in a HEBDT solution (3 mM) at an energy dose of 0.5 × E_{th}. Figure 1.2a represents a topographic AFM image of a single stripe in air, before and after irradiation in transverse polarization in the HEBDT solution. The profile of the lateral cross section, after laser irradiation, measured at the same position of the stripe before and after irradiation, and with the same non-contact tip under identical scanning conditions, clearly exhibits a larger width of the stripe (~25 nm larger), while its height slightly increases (~5 nm thicker). To confirm the assumption of a LSP-mediated grafting, the GNTs were also illuminated with a longitudinal polarization, for which no LSP excitation occurs. Under these conditions, no grafting was observed. Figure 1.2b confirms this assumption as the profile of the lateral cross section of the stripe remains identical after laser irradiation (Fig. 1.2b). To highlight the localization of the aryl film grafted on the stripe, under transverse polarization, the profiles of the lateral cross section of the stripes, after and before irradiation, were subtracted. The resulting differential profile of the cross sections, deduced from the AFM images in Fig. 1.2a, is displayed in Fig. 1.3a, revealing two lobes of 35 ± 5 nm oriented along the borders of the stripe. In contrast, in the case of a longitudinal polarization, the differential profile is almost flat and equal to zero (Fig. 1.2b and 1.3a).

In order to elucidate the origin of the lobes appearing under LSP excitation with transverse polarization, the electric field intensity distribution around a single stripe was mapped, using the discrete dipole approximation (DDA) method [26, 27].

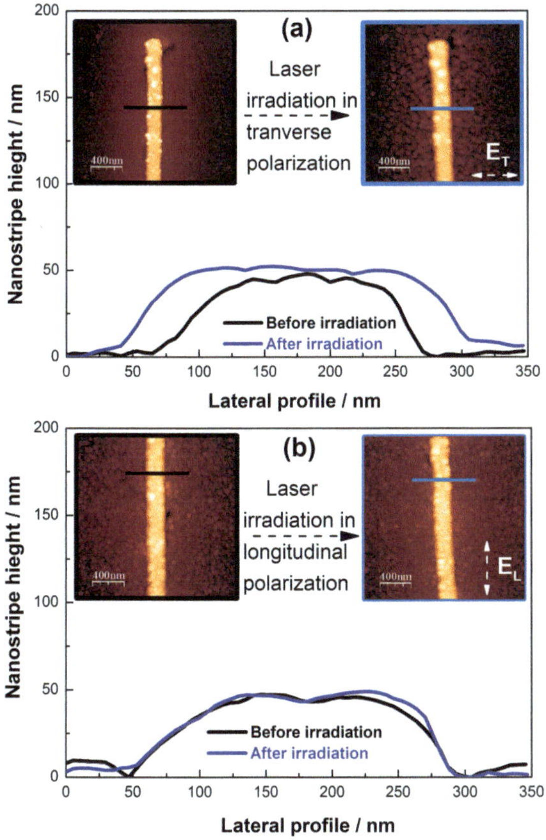

Figure 1.2 (a) Profiles of the lateral cross section of a single stripe (width $W = 125 \pm 5$ nm, height $H = 50 \pm 2$ nm, length $L \sim 100$ μm) deduced from the AFM images displayed in the inset: (black curve) before laser irradiation; (blue curve) after laser irradiation under transverse polarization (energy dose: $E = 0.5 \times E_{th}$). (b) Profiles of the lateral cross section of a single stripe deduced from the AFM images displayed in the inset: (black curve) before laser irradiation; (blue curve) after laser irradiation under longitudinal polarization (energy dose $E = 0.5 \times E_{th}$).

For calculation, we considered a gold stripe target with a width of 125 nm and height of 50 nm. The incident polarization was set perpendicular to the stripe. The interaction between the stripe and the substrate was taken into account, enabling us to make reliable comparisons between the experimental and calculated spectra. A good match was obtained between the experimental and calculated extinction spectra for the LSP wavelength (located ca. 640 nm in water) and the band shape. The optical near-field enhancement of the stripe was then calculated, using an incident wavelength of 633 nm. Figure 1.3b displays the distribution of the near-field intensity enhancement on a unit cell length of a smooth stripe target excited with a transverse polarization. The three-dimensional mapping of the optical near-field shows variations only in the cross-sectional area perpendicular to the stripe, with the stripe length remaining constant. Interestingly, the enhancement is mainly located on the side of the stripe at the ITO/glass interface, where a maximum of enhancement factor (EF) of ~1.5×10^2 (in intensity) is evidenced. In contrast, almost no enhancement is observed on top of the stripe. The remarkable agreement between the spatial confinement of the aryl grafted film (Fig. 1.3a) and the spatial distribution of the electromagnetic field enhancement (Fig. 1.3b) implies that the enhanced near-field intensity is responsible for the grafting of the diazonium-derived aryl film, under LSP excitation [28]. The aryl film grafting can be explained by the capacity of the gold nanostripes under LSP excitation to act as electron-rich reservoirs, in the area of maximum EF. This could stimulate the promotion of an electron transfer from the gold nanostripes to the diazonium salts, under LSP excitation, resulting in the formation of aryl radical species, able to bind the gold surface through covalent bonds. The aryl film grafting (~5 nm thick) observed on top of the stripe might be explained by a thermal effect, due to an increase of the temperature at the stripe surface, upon strong light absorption under LSP excitation. This increase of temperature may approach or exceed locally the threshold energy measured in the absence of LSP excitation, and can lead to the presence of the aryl film grafting on top of the gold nanostripes. It could also originate from the nanoscale surface roughness features located on top of the stripes, resulting in additional near-field enhancements randomly distributed on the stripe.

To further confirm the key role of the plasmonic excitation for the grafting of aryl films, a GNTs array presenting a LSP resonance λ_{LSP} = 530 nm (in water), far from the excitation line (λ_{inc} = 633 nm), was subjected to the same grafting experiment. After laser irradiation in transverse polarization, but without any LSP excitation (the laser is at 633 nm while the LSP is at

530 nm), the profile of the lateral cross section remains identical to the one before irradiation (data not shown). These results demonstrate the crucial role of LSP excitation in the grafting mechanism and confirm a plasmon-mediated grafting of the aryl film on the GNTs.

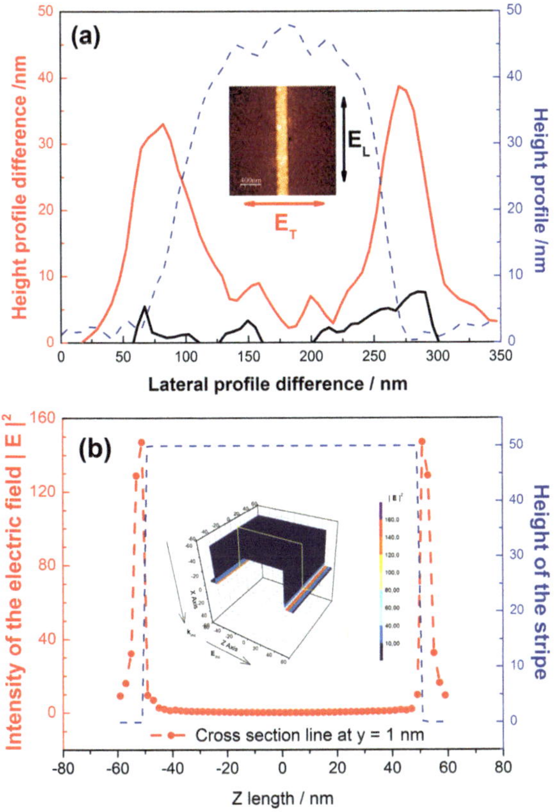

Figure 1.3 (a) Height profile difference of a single stripe deduced from the profile of the lateral cross section before and after irradiation displayed in Fig. 1.2 (λ_{exc} = 633 nm, energy dose E = 0.5 × E_{th}): (red curve) for a transverse polarization; (black curve) for a longitudinal polarization. The blue dashed curve indicates the profile of the lateral cross section of the stripe before irradiation. (b) Mapping of the near-field intensity enhancement $|E|^2$ on a unit cell with a W = 125 nm wide stripe, and H = 50 nm, using the DDA method. The transverse section profile corresponding to the 3D mapping of the NF enhancement is indicated in red. The dashed blue line represents the lateral cross section of the target.

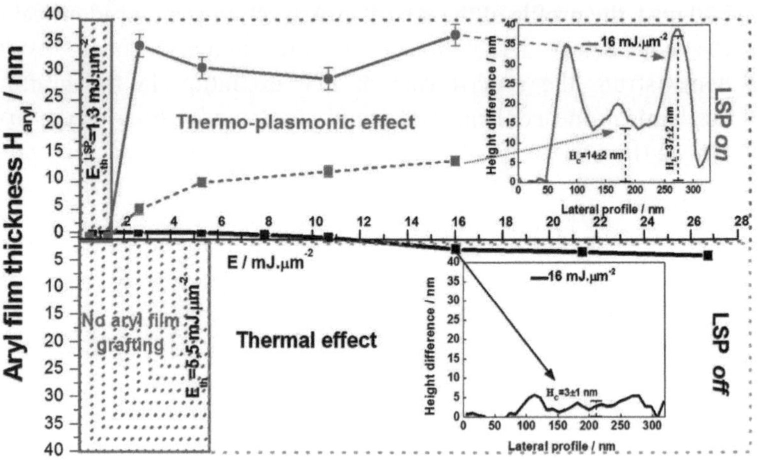

Figure 1.4 Aryl layer thickness grafted on a GNT (measured by AFM) versus the incident energy dose E (mJµm^{-2}). The black squares represent the aryl layer thickness after laser irradiation at various energies with a longitudinal polarization (no LSP excitation takes place). The red squares and red circles represent the aryl layer heights H_C and H_L measured at the center and the borders of a single stripe, respectively, after laser irradiation at various energies with a transverse polarization (under LSP excitation). The threshold energy below which aryl film grafting does not take place is lowered, from E_{th} = 5.5 mJµm^{-2} under off LSP excitation, to a new threshold value of E_{th}^{LSP} = 1.3 mJµm^{-2} under LSP excitation.

The influence of the energy dose upon the thickness of the aryl film and its confinement on the nanostripes was studied by varying the energy dose from 0.5 × E_{th} to 3 × E_{th}, under LSP excitation (see Fig. 1.4). Above E_{th}, the differential AFM images point out again a grafting of the aryl film mainly at the borders of the stripes, attributed to LSP excitation. In addition, an aryl film grafting significantly takes place at the center of the stripe, as shown in the inset of Fig. 1.4 (upper part). The higher the energy dose, the thicker the aryl film is at the center (up to $H_C \sim$ 15 nm at the center of the stripe for E = 3 × E_{th}). This grafting at the center of the stripe can be explained by a thermo-plasmonic effect, due to a strong light absorption under LSP excitation. Interestingly, the aryl film appears always thicker in transverse polarization (under LSP excitation) than in longitudinal polarization (no LSP excitation), for a fixed energy dose. The aryl film thickness at the center of the stripe is $H_C \sim 14$ nm for a transverse polarization (the inset of Fig. 1.4, upper part), while in longitudinal polarization, H_C is only ~3 nm at the center of the stripe (the inset of Fig. 1.4, lower part). These results highlight two grafting mechanisms occurring at the same time. When the energy dose is superior to the threshold energy, the thermally induced grafting, under laser heating, occurs all along the gold nanostripes, regardless of the polarization, transverse or longitudinal. In addition to the deposition of aryl layers within the area of high field enhancement under LSP excitation (leading to the formation of the lobes described above), the increase of temperature at the stripe surface caused by LSP excitation, may significantly contribute to the aryl film grafting at the center of the stripe, above the threshold energy dose [29]. As a consequence of LSP excitation, the threshold energy above which the aryl film grafting takes place has been lowered to a new threshold value of $E_{th}^{LSP} = 1.3$ mJμm^{-2}, as illustrated in Fig. 1.4.

In summary, a plasmon-mediated grafting of aryl films derived from diazonium salts on regular arrays of gold nanostripes was demonstrated. This grafting occurs specifically in the regions of maximum of field enhancement.

1.1.3 Plasmon-Driven Grafting on 2D Structure Array of Gold Nanorods

1.1.3.1 Description of the gold nanorod array

Lithographic gold nanorod (GNR) arrays (diameter $D = 100 \pm 5$ nm, length $L = 165 \pm 5$ nm, height $H = 47 \pm 3$ nm, interparticle distance $\Lambda \sim 300 \pm 5$ nm) were elaborated on an indium–tin-oxide (ITO) coated-glass substrate by electron beam lithography (EBL) [30]. A representative SEM image of the array of GNRs is displayed in Fig. 1.5a. Due to their anisotropic shape, they display two separate surface plasmon resonance (SPR) bands corresponding to their width and length, known as the transverse and longitudinal plasmon bands. Indeed, the longitudinal LSPR mode is associated with the electron oscillations along the length axis for an incident polarization parallel to this axis. The transverse LSPR mode is associated with the electron oscillations along the short axis for a polarization parallel to the short axis of the GNRs [12]. Therefore, the transverse LSPR is located at about 560 nm while the longitudinal LSPR is at approximately 780 nm as shown in Fig. 1.5b (the spectra are recorded in water).

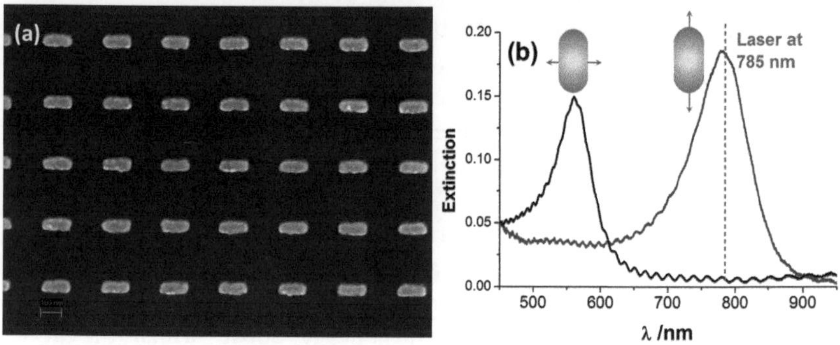

Figure 1.5 Characterization of the GNRs array. (a) SEM image; (b) Extinction spectra under transversal (black curve) and longitudinal polarization (blue curve) in water.

1.1.3.2 Plasmon-driven grafting on gold nanorod array

The GNRs arrays were then homogeneously immersed in the solution of the HEBDT salt (3 mM) and illuminated by a laser, during 10 s. A Krypton laser at 785 nm was used since its wavelength matches the maximum of extinction of the longitudinal LSPR band. This excitation wavelength has been selected for the plasmon-induced grafting of poly(aryl) films, since a strong field enhancement at the extremities of the rods is expected, provided that the incident polarization is set parallel to the long axis of the rod. The aryl film grafting is thus expected to be particularly efficient under such experimental conditions. The incident polarization was set parallel to the AuNRs long axis. It is worth mentioning that the laser heating induces the grafting reaction by itself, even in the absence of any plasmon excitation, when the exposure energy dose exceeds a given threshold E_{th}, as shown in a previous part. In the case of thermal grafting, the layers are observed all around the nanoparticles, with a uniform distribution. Therefore, an exposure dose below this threshold value, corresponding to $E = 50\%\ E_{th}$, was used here to avoid side reactions by laser heating and guarantee that the grafting of

electron mediated reduction of aryl diazonium salts, depending on the polarization of the incident illumination.

1.1.4 Plasmon-Driven Multi-Functionalization of Gold Nanodisks Array

The strategies developed so far provided a single functionalization of the surface with only one type of organic layer grafted in nanoscale region. Extension of these strategies to the multi-functionalization of surfaces represents a major breakthrough in plasmon-mediated chemistry, in order to achieve the grafting of various chemical groups in distinct nanoscale regions. We address this issue in this part by monitoring the incident light polarization on gold nanodisks, to trigger plasmon excitation and obtain regiospecific local surface double-functionalization by diazonium salts (see Fig. 1.7).

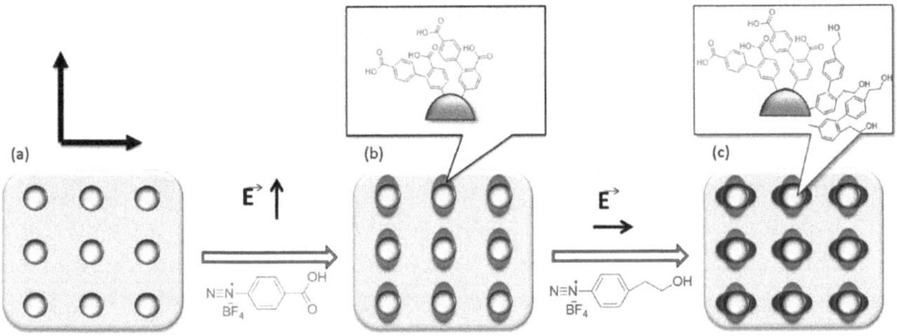

Figure 1.7 Scheme of the regioselective multifunctionalization of GNDs under plasmon excitation with polarized light. The nanostructures were first incubated for 180 s in an aqueous solution of carboxy-phenyl diazonium salt (3 mM) under laser irradiation polarized in *Y* direction. Then, the nanostructures were incubated for 180 s in an aqueous solution containing a hydroxy-ethyl phenyl diazonium salt (3 mM) under laser irradiation, polarized in *X* direction.

Lithographic gold nanodisks (GNDs), with no capping agents on their surface, were chosen for the proof of concept, as they display a symmetric geometry in the *XY* plane and a uniform surface chemistry. GNDs arrays (diameter $D = 100 \pm 5$ nm, height $H = 50 \pm 5$ nm, interparticle distance $\Lambda = 300$ nm) were elaborated ITO coated-glass substrates by EBL. The optical response of the nanodisks demonstrates the main peak at 655 nm, which represents the LSPR dipolar mode in aqueous solution. By playing on the polarization of the incident excitation light, the plasmon-driven functionalization of these isotropic NPs should result in a regioselective surface anisotropy. Functionalization agents based on aryl diazonium salts were selected as they were shown in the previous part to graft on Au nanostripes through plasmon-induced electron transfer. In this part, two types of aryl diazonium salts were used, bearing either hydroxyl (4-hydroxyethyl benzene diazonium tetrafluoroborate), or carboxyl-terminated groups (4-carboxy-benzene diazonium tetrafluoroborate). This plasmon-mediated multi-functionalization approach opens promising perspectives for the elaboration of nanostructured surfaces with chemical anisotropy.

In order to demonstrate that the LSPR excitation can locally trigger regiospecific surface functionalization, the GNDs arrays were homogeneously immersed in the solution of carboxy-terminated diazonium salts (3 mM) and illuminated by a laser polarized along the Y axis, during 180 s. The optical exposure was performed under normal incidence, and the He–Ne laser (λ_{inc} = 633 nm) was focused on the nanoarray through a microscope objective (×10 numerical aperture N.A. #0.25), resulting in a circular laser spot of ~ 5 μm diameter at the surface. The incident wavelength λ_{inc} = 633 nm of the laser excitation matches well the LSPR of the gold nanodisk (λ_{max} = 655 nm). It is worth mentioning that laser heating induces the grafting reaction, probably through a cationic pathway, when the exposure energy dose exceeds a given threshold E_{th}, even in the absence of any plasmon excitation, as shown in a previous part. In the case of thermal grafting, the layers are observed all around the nanoparticles, with a uniform distribution. Therefore, an exposure dose below this threshold value, corresponding to E = 50% E_{th}, was used here to avoid side reactions by laser heating and guarantees that the grafting of poly(aryl) layers on the nanostructures was solely induced by LSPR excitation. The obtained hybrid nanostructures were characterized by SEM as the metallic NPs and the poly(aryl) layers show different contrasts (Fig. 1.8a). Polymer lobes can be visualized easily along the Y-axis in the vicinity of the nanodisks. In order to verify that the plasmon-induced grafting occurs specifically in the regions of maximum of field enhancement, the mapping of the intensity of the electric field around the nanodisks was calculated by the DDA (see Fig. 1.8b). It evidences a strong amplification on each side of the nanodisk along the excitation axis with a progressive attenuation away from the axis. From the comparison of the spatial extent of the grafted layers in the *XY* plane, observed by SEM, it is clear that the regioselective grafting of the poly(aryl) layers is a polymer replica of the dipolar near-field intensity.

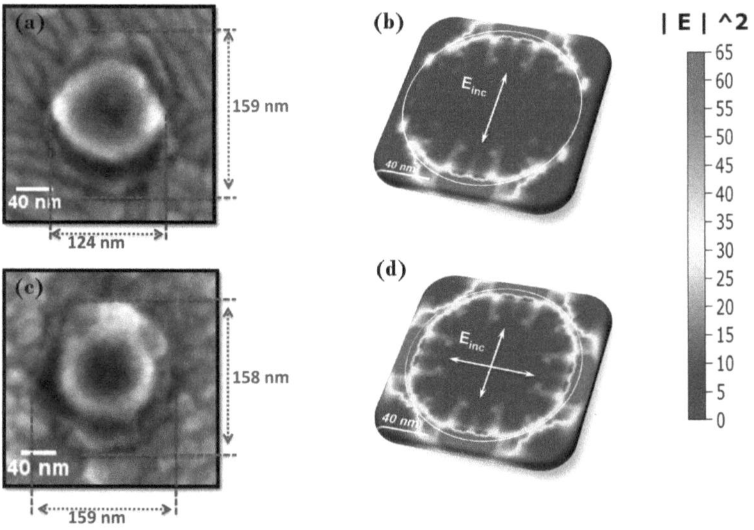

Figure 1.8 SEM images of a single nanoparticle after plasmon-induced grafting of (a) carboxyphenyl layers along the *Y* direction and (c) additional hydroxyethyl phenyl layers along the *X* axis. Irradiation conditions: λ_{inc} = 633 nm, 180 s with a power of P = 0.8 mW μm^{-2}. Mapping of the near-field intensity enhancement $|E|^2$ upon irradiation along the *Y* axis (b) and both *X* and *Y* axis (d), on a unit cell with a nanodisk of 100 nm diameter, H = 50 nm, using the DDA method.

The extension of the poly(aryl) layers along the *Z* direction (perpendicular to the nanoarray plane) was studied by AFM. The differential AFM profiles of the lateral cross sections (corresponding to the subtraction of the profiles obtained before and after plasmon-induced grafting) revealed no grafting on top of the disk and confirmed the confinement of the grafted layers only in the *X*–*Y* plane, on each sides of the GNDs (Fig. 1.9a). The SERS spectra, recorded after plasmon-induced grafting, looked very similar to those obtained after spontaneous reaction apart from an enhancement of the shoulder at 1614 cm^{-1}, at the higher frequency side of the aromatic ortho-meta C=C bond vibration band (not shown). The main Raman band at 1586 cm^{-1} can be assigned to the aromatic ring deformations in the first layer grafted onto the gold surface, whereas the shoulder at 1614 cm^{-1} corresponds to the same vibration in the poly(aryl) layer above, covalently linked to the former. The intensity increase of this shoulder thus confirms the polymeric nature of the grafted organic layer. Interestingly, strong variations of the SERS signal intensity were observed depending on the incident polarization. The spectra recorded in the *Y* direction, matching the polarization of plasmon-induced grafting, gave the highest SERS signal intensities. Nevertheless, the SERS signal is not fully switched off along the *X* direction, perpendicular to the polarization of plasmon-induced grafting. This is due to the spatial extension of the electromagnetic field enhancement away from the irradiation polarization axis leading to a poly(aryl) film, which is not strictly confined along the *Y* axis, but slightly spread out towards the *X*-axis (Fig 1.8b).

In a second step, the samples were thoroughly cleaned with ethanol and dried before a new immersion in an aqueous solution of the other diazonium salt, hydroxyethyl benzene diazonium. A second exposure with the same energy dose but an incident polarization oriented along the *X*-axis was then performed. Figure 1.8c shows the SEM image of the nanodisks after the second grafting step along the *X*-axis. New polymer nodules are observed on each side of the gold nanodisks along the *X* direction. On the basis of the SEM images recorded at both steps, it turns out that the two different poly(aryl) layers, bearing either carboxyl or hydroxyethyl substituents, were selectively integrated in different orientations around the gold nanodisks via near-field plasmonic excitation.

The AFM differential profile of the single disk before and after the two-step grafting (Fig. 1.9b) evidences the bi-directional anchoring of the poly(aryl) nanolayers, which perfectly matches the electric field distribution around the nanodisks, according to DDA calculations (Fig. 1.8d). Note that the corresponding optical and SERS spectra also confirm this bi-directional anchoring (data not shown).

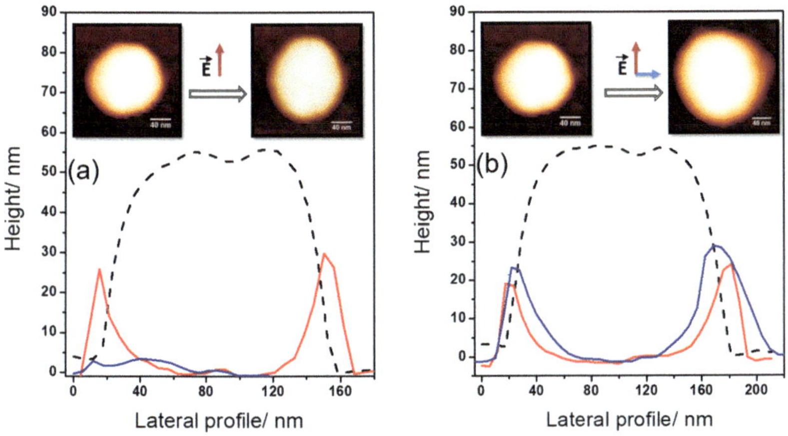

Figure 1.9 (a) Plasmon-induced grafting of the first carboxy-aryl layer, using an incident polarization along *Y* axis. Top: AFM image of a single disk before (left) and after (right) plasmon-induced grafting along *Y* axis. Bottom: Lateral cross section profile along *Y* axis after grafting (dashed line) and differential lateral cross section along *Y* axis (red line) revealing two main lobes along the *Y* direction and along *X* (blue line) confirming no grafting along *X* direction. (b) Plasmon-induced grafting of the second hydroxyethyl-aryl layer, using an incident polarization along *X* axis. Top: AFM image of a single disk before grafting (left) and after double plasmon-induced grafting along *Y* and *X* axis (right). Bottom: Lateral cross section profile along *X* axis after grafting (dashed line) and differential lateral cross section along *Y* axis (red curve) and *X* axis (blue curve) revealing two main lobes along the *Y* and *X* directions.

The influence of the energy dose upon the spatial extent of the poly(aryl) nodules along the *Y* and *X* direction was further estimated by SEM. Figure 1.10 displays the elongation observed in *Y* and *X* direction. It reveals two important features:

- The growth of the poly(aryl) layer with the energy dose is directional along the excitation polarization axis. Indeed, the elongation along *Y* increases with the energy dose when the excitation is polarized along the *Y* axis, while the elongation along the *X* axis remains constant, and inversely for the *X* polarization.
- The thickness of the grafted layers increases with the energy dose, until reaching ca. 40–60 nm, depending on the diazonium salt: The carboxyphenyl-derived layers appear slightly thicker than the hydroxyethyl ones.

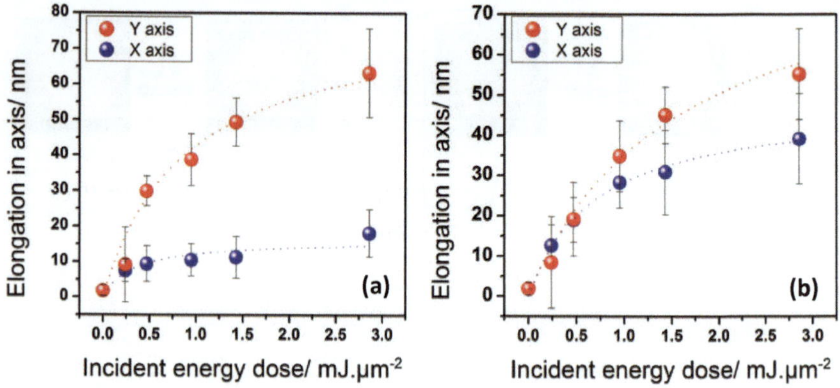

Figure 1.10 Plot of the thickness (measured by SEM) of the grafted poly(aryl) layer along the *Y* and *X* direction, as a function of the incident energy dose from 0 to 3 mJµm^{-2}, (a) after plasmon-induced first carboxy-aryl layer grafting along *Y* and (b) after second hydroxyethyl-aryl layer grafting along *X*.

The influence of the incident energy dose upon the growth of the carboxy-terminated poly(aryl) layer was also controlled by SERS. The ratio of the relative intensities I_{1614}/I_{1586} of the two bands at 1614 and 1586 cm^{-1}, assigned respectively to the aromatic ring deformations in the poly(aryl) layer and in the first layer directly grafted on gold via Au-C bonds was plotted against the dose (see Fig. 1.11).

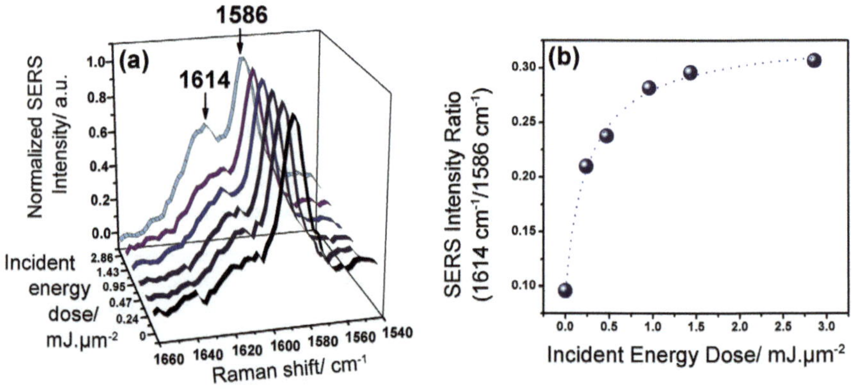

Figure 1.11 (a) SERS spectra in the 1540–1660 cm^{-1} region, recorded on the nanodisk arrays after plasmon-induced grafting of carboxy-phenyl layers along the *Y* direction; (b) Plot of the I_{1614}/I_{1586} SERS intensity ratio as a function of the incident energy dose from 0 to 3 mJµm^{-2}.

An increase in the energy dose leads to a clear enhancement of the shoulder at 1614 cm^{-1}, revealing the growth of the poly(aryl) layer. Although the SEM analysis indicates a continuous thickening of the poly(aryl) layers upon an increase of the incident energy dose, the I_{1614}/I_{1586} SERS ratio reaches a plateau value at around 1 $mJ\mu m^{-2}$, revealing the maximum distance from the nanostructure's surface probed by SERS. This energy dose corresponds to a poly(aryl) thickness of ca. 40 nm, as shown in Fig. 1.10, above which the additional poly(aryl) layers grafted on the gold nanodisks are no longer detected by SERS. Note that this distance correlates perfectly well with the spatial extent of the electromagnetic field, calculated by DDA. It is therefore possible to use this plasmon-mediated functionalization approach to probe the distance-dependence enhancement effect of SERS in hot spot regions. Moreover, it is an efficient way to modify only the regions of high electromagnetic field while leaving the other area of the nanoparticle surface chemically passive.

In summary, we have shown that it is possible to pattern the surface chemical properties of plasmonic gold nanodisks by two different types of functional poly(aryl) layers, bearing either carboxyl or hydroxyethyl pendant groups. This double-functionalization strategy is triggered by the polarization of the incident illumination leading to a site-selective hot-electrons mediated reduction of aryl diazonium salts.

1.1.5 Conclusion

Compared to conventional grafting methods, the approach described in this chapter offers several advantages: (i) It is very fast (it occurs in a few seconds upon laser irradiation, while the usual spontaneous methods require several hours); (ii) it gives thick poly(aryl) layers (up to 40–60 nm) using low diazonium salt concentrations; (iii) the grafting takes place selectively in the regions of maximum of field enhancement, in contrast to the homogeneous coating obtained under spontaneous grafting. This plasmon-mediated functionalization approach therefore opens promising prospects for the nanoscale confinement of organic layers at the gold nanoparticles surface. It could also provide a general strategy to attach molecules to hot spot regions and further improve their detection for (bio)sensing applications, using surface-enhanced Raman spectroscopy.

1.2 Plasmon-Initiated Surface Functionalization by Thiol-Ene "Click" Chemistry

Thiol-ene coupling reaction is one of the "click" chemistry processes and described as a simple method that is modular, widely scoped, and generating high yields. This reaction consists in grafting thiol-bearing molecules to a carbon double bond. It had a significant impact in polymer synthesis and material science [31, 32] and recently has attracted considerable interest in surface functionalization and modifications [33, 34].

By means of conventional approach thiol-ene reaction is performed through UV light or thermal activation, with the presence of a radical initiator. In this part of the chapter, we will demonstrate that the thiol-ene coupling can be locally triggered using the visible range light source and by exploiting nanoplasmonic effects. The kinetics of this reaction was monitored in situ by surface-enhanced Raman scattering (SERS). After investigating the reaction rates on different diameters of gold nanocylinders (NCs) made by EBL, we show that the reaction can be tuned by controlling the NCs plasmonic properties.

1.2.1 Fabrication of Substrates

Arrays of gold nano-cylinder (NCs) on glass were produced by electron beam lithography and lift-off techniques [35, 36] and were designed to contain the variable diameters of NCs: 110, 140, and 200 nm (electron microscope micrograph can be seen in Fig. 1.12a). EBL technique is known to allow the good control of the size and shape of the nanostructures. Thus, the substrates ensure the reproducibility of LSPR position and SERS signal. The height of the NCs was set to 60 nm where gold was evaporated on 3 nm of chromium (Fig. 1.12b) for better adhesion. The gap between two NCs was set at 200 nm (side to side) in order to avoid any near-field coupling effects.

Figure 1.12c demonstrates the extinction spectra of fabricated NCs in aqueous environment. LSPR position of the 110 nm diameter NCs array was observed close to 660 nm (as excitation laser wavelength used to initiate the reaction), whereas 140 nm and 200 nm NCs had a LSPR position red-shifted or nearly out of resonance, respectively.

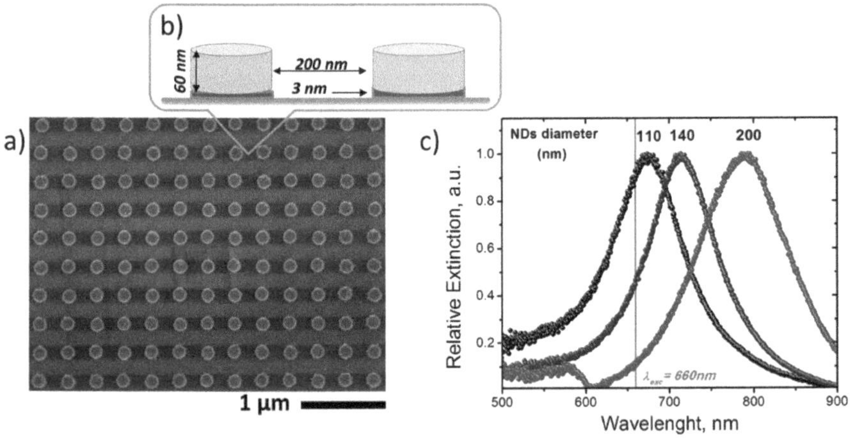

Figure 1.12 (a) Scanning electron microscopy image of an array of NCs with a diameter of 200 nm. (b) Scheme of the nanocylinders (NCs) used in the experiments. (c) Extinction spectroscopy measurements in water on 110 nm (black curve), 140 nm (dark blue curve) and 200 nm (blue curve) NCs arrays. The red line indicates the laser wavelength used for thiol-ene reaction initiation and SERS investigation.

1.2.2 In situ Thiol-Ene Click Reaction

Thiol-ene "click" reaction implicates the addition of a thiol to an alkene function via a free-radical mechanism [37] (Fig. 1.13).

Briefly, first thiol group is converted to a thiyl radical thanks to the interaction with a free radical initiator exposed to UV light or to a specific temperature [38]. Active thiyl radical react then with the alkene group and forms a carbon centered radical. From here, the reaction turns to a propagation step resulting in the formation of thio ether as a reaction product and a new thiyl radical, which then subsequently is involved in the reaction chain.

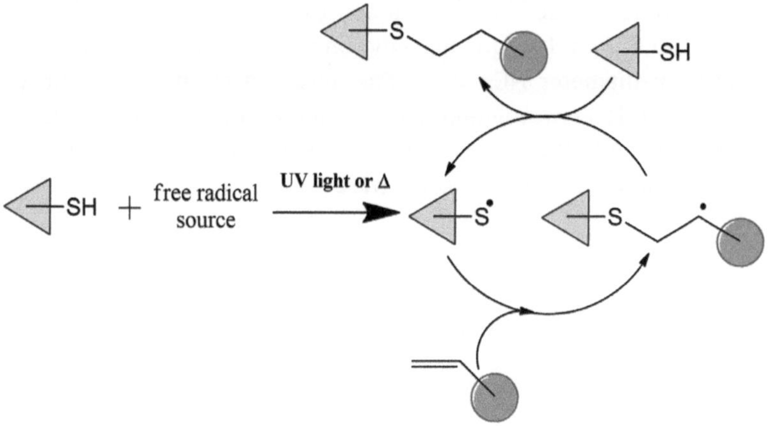

Figure 1.13 Schematic illustration of the free-radical thiol–ene coupling mechanism.

Herein, thiol-ene "click" reaction was carried on plasmonic substrates pre-functionalized by allyl mercaptan using 2,2′-Azobis(2-methylpropionamidine) dihydrochloride (AAPH) as free radical initiator and thiophenol as reactant (scheme of reaction can be seen in Fig. 1.14a). Regarding literature, the AAPH can be decomposed into cationic radicals under exposition to UV light (365 nm) or temperature above 60°C. In our case, the local reaction at the nanoscale level was initiated on the surface of the gold NCs in aqueous environment using a laser excitation wavelength of 660 nm. The laser beam was focused on the sample with a 60× objective (N.A. 0.7).

The reaction kinetics was monitored by SERS in real time using the same laser beam. SERS spectra were first recorded on the 110 nm NCs for 30 min and the process was then reproduced on the NCs of 140 nm and finally on the NCs of 200 nm.

Obtained SERS spectra for different arrays of NCs can be seen in Fig. 1.14b. The peaks observed in the SERS spectra correspond to the thiophenol signal. In order to follow the kinetics of reaction, the intensity of the peak at 1572 cm^{-1} (assigned to an aromatic C=C vibration) was plotted versus reaction time (Fig. 1.15a) for each NCs array. One can notice that the observed SERS intensity of the thiophenol peak increases with the time and saturates in less than 10 min for the 110 and the

140 nm diameters NCs. Time needed to complete local thiol-ene coupling reaction was very fast, while in literature it has been estimated to be in the range of several hours using normal conditions [39] (i.e., without NCs and using UV excitation). It is worth mentioning that the saturation of thiol-ene process even after 40 min was not reached for the 200 nm diameter NCs.

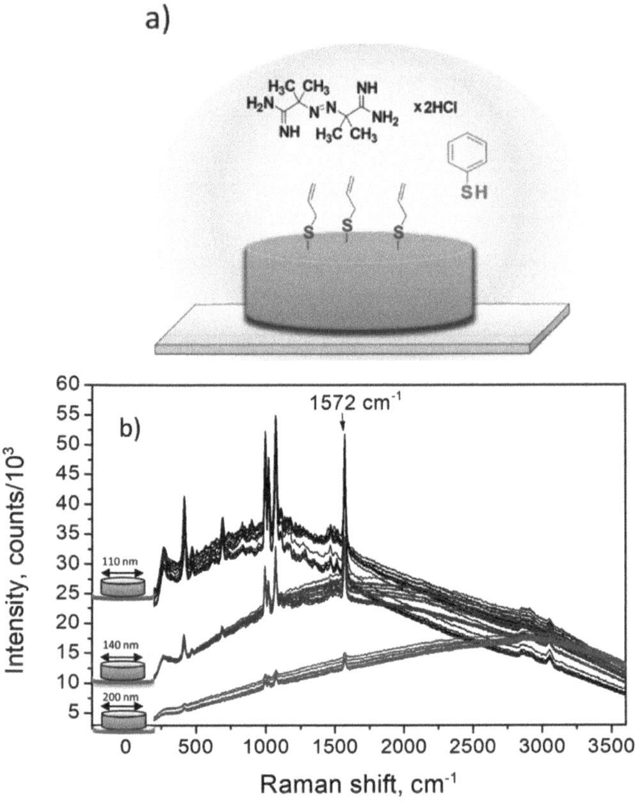

Figure 1.14 (a) Scheme of the thiol-ene reaction configuration. (b) SERS spectra of grafted thiophenol onto gold NCs of 110 nm (black spectra), 140 nm (dark blue spectra) and 200 nm (blue spectra) diameter.

To determine precisely the reaction time, the obtained experimental data were fitted with a first order Langmuir isotherm:

$$\theta = \theta_{\text{sat}}\left[1 - \exp^{-\frac{t}{\tau}}\right], \tag{1.1}$$

where θ is the surface coverage, θ_{sat} is the saturation of the coverage, t is the time and τ is the reaction time constant.

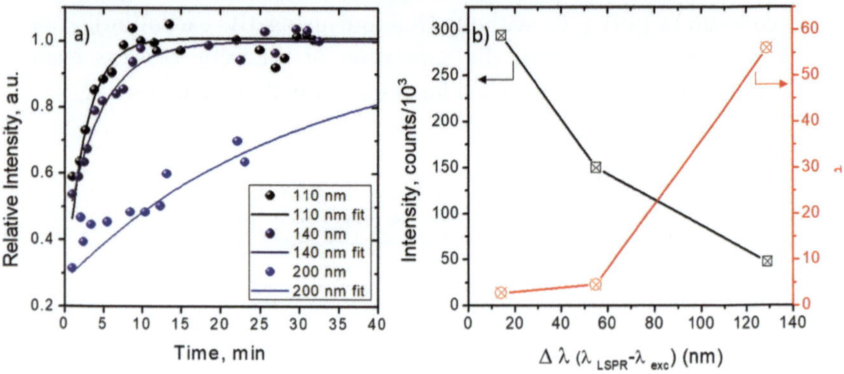

Figure 1.15 (a) Relative SERS intensities of the thiophenol peak at 1572 cm^{-1} versus reaction time for the 110 nm (black dot), 140 nm (dark blue dot) and 200 nm (blue dot) diameter NCs. Solid lines are the fitting using the equation (1) (110 nm: black line, 140 nm: dark blue line and 200 nm: blue line). (b) Absolute SERS intensity of the peak at 1572 cm^{-1} at saturation point and reaction time constant plotted versus the mismatch between the LSPR position of NCs and the used laser excitation wavelength.

Figure 1.15b demonstrates the absolute SERS intensity and the reaction time constant, τ (time required to reach 63% of SERS signal saturation) versus the mismatch between the LSPR position of NCs and the laser excitation wavelength, $\Delta\lambda = \lambda_{LSPR} - \lambda_{exc}$. One can notice that the reaction time is strongly dependent on the LSPR position of different diameters NCs. A twenty times faster reaction was witnessed on highly resonant 110 nm NCs compared to 200 nm NCs. Unexpected result was achieved for 140 nm diameter NCs which have a 54 nm difference between the LSPR position and the excitation wavelength. In this case, just slightly lower reaction efficiency was obtained compared to the one measured on 110 nm NCs.

The obtained results are not straightforward because grafting of thiophenol on the NCs can be governed by two other parallel mechanisms. First, from the literature it is known that short thiol bearing molecules (like allyl mercaptan in this case) form non-homogeneous self-assembled monolayer (SAM) due to the lack of intermolecular interactions [40]. Thus, non-blocked active gold sites can be available for spontaneous thiophenol grafting. Second, a competitive molecular replacement of grafted allyl mercaptan by thiophenol could take a place directly on the gold surface [41]. Therefore, the SERS signal achieved during the thiol-ene experiments could then also include probable contributions from these two competitive thiophenol direct interactions with the NC surface.

To address the latter concern and to prove that achieved experimental results actually correspond to the thiol-ene "click" process, three different experimental conditions were tested as negative controls, including: (i) duplication of the thiol-ene reaction excluding a radical initiator, (ii) testing reaction on mercaptoethanol functionalized NCs to monitor the competitive molecular displacement by thiophenol and (iii) observing the rate of direct thiophenol grafting to non-functionalized gold NC surface.

The first two negative control samples indicated low-rate replacement of short thiols previously deposited on the NCs by thiophenol. Moreover, the obtained SERS signal was nearly constant with time and considerably lower than the one recorded during the thiol-ene reaction [18]. On the other hand, the results obtained from the third negative control sample, where the kinetics of thiophenol SAM formation was monitored, gave a base for useful comparison with the thiol-ene reaction sample. First, it was revealed that the rate of thiophenol attachment by thiol-ene process is faster than the one of spontaneous grafting, suggesting the strong impact of the "click" chemistry process. Second, the SERS signal for 110 nm diameter NCs on both samples was equivalent at the saturation point, meaning that the same NC coverage is reached using either the thiol-ene reaction or the simple SAM formation.

Thiophenol SAM formation is not dependent on the plasmonic properties of the NCs since the thiol interaction with gold is not thermally or optically activated. Thus, to demonstrate the plasmon contribution to the thiol-ene reaction initiation we compared the absolute SERS intensities at the saturation points for both samples (Fig. 1.16). In the case of thiophenol SAM formation, the SERS signal on 110 nm diameter NCs is 2.5 higher than the one measured on the 200 nm NCs and can be attributed to plasmonic effects on the SERS signal enhancement. However, for the thiol-ene "click" reaction, the intensity ratio was close to 6. Such large ratio gives a clear evidence that the radical reaction is plasmon-induced on highly resonant NCs but less efficient for the NCs with the LSPR out of resonant.

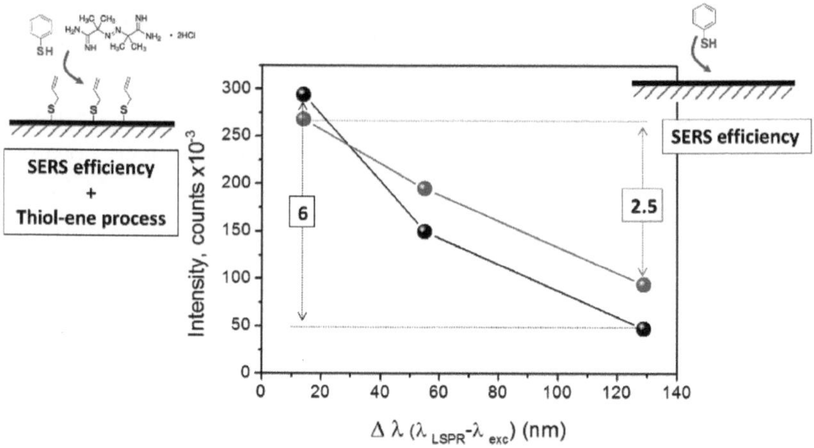

Figure 1.16 Absolute SERS intensity of the peak at 1572 cm^{-1} at saturation point plotted versus the mismatch between the LSPR position of NCs and the used laser excitation wavelength for thiol-ene reaction (black spheres) and thiophenol spontaneous SAM formation (red spheres).

1.2.3 Conclusion

The results discussed in this part of the chapter demonstrate that the thiol-ene "click" reaction, which by conventional means takes several hours to be completed under specific conditions (UV light or temperature), can be successfully triggered at the nanoscale range of the gold surface by exploiting the plasmonic properties of the nanostructures. Moreover, it was shown that local surface functionalization by plasmon-initiated thiol-ene reaction is performed in a few minutes on highly resonant nanoparticles and thus can be modulated and controlled by tuning the LSPR position.

References

1. Kawata, S. (2007). *Nanophotonics with Surface Plasmons.* (Elsevier, Amsterdam).

2. Lal, S., Clare, S. E., and Halas, N. J. (2008). Nanoshell-enabled photothermal cancer therapy: Impending clinical impact, *Acc. Chem. Res.*, **41**, 1842–1851.
3. Maier, S. A. (2007). *Plasmonics: Fundamentals and Applications.* (Springer US).
4. Felidj, N. (2008). Multipolar surface plasmon peaks on gold nanotriangles, *J. Chem. Phys.*, **128**, 094702–094705.
5. Becker, J. (2010). The optimal aspect ratio of gold nanorods for plasmonic bio-sensing, *Plasmonics*, **5**, 161–167.
6. Encai Hao, G. C. S. (2004). Electromagnetic fields around silver nanoparticles and dimers, *J. Chem. Phys.*, **120**, 357.
7. Baffou, G., and Quidant, R. (2014). Nanoplasmonics for chemistry, *Chem. Soc. Rev.*, **43**, 3898–3907.
8. Urban, A. S. (2011). Single-step injection of gold nanoparticles through phospholipid membranes, *ACS Nano*, **5**, 3585–3590.
9. Moskovits, M. (1985). Surface-enhanced spectroscopy, *Rev. Modern Phys.*, **57**, 783–826.
10. Häfele, V. (2015). Local refractive index sensitivity of gold nanodisks, *Opt. Express*, **23**, 10293–10300.
11. Pelton, M., Aizpurua, J., and Bryant, G. (2008). Metal-nanoparticle plasmonics, *Laser Photonics Rev.*, **2**, 136–159.
12. Pérez-Juste, J. (2005). Gold nanorods: Synthesis, characterization and applications, *Coordination Chem. Rev.*, **249**, 1870–1901.
13. Volpe, G. (2012). Near-field mapping of plasmonic antennas by multiphoton absorption in poly(methyl methacrylate), *Nano Lett.*, **12**, 4864–4868.
14. Gruber, C. (2015). Imaging nanowire plasmon modes with two-photon polymerization, *Appl. Phys. Lett.*, **106**, 081101.
15. Tijunelyte, I. (2018). Multi-functionalization of lithographically designed gold nanodisks by plasmon-mediated reduction of aryl diazonium salts, *Nanoscale Horiz.*, **3**, 53–57.
16. Nguyen, M. (2017). Regioselective surface functionalization of lithographically designed gold nanorods by plasmon-mediated reduction of aryl diazonium salts, *Chem. Commun.*, **53**, 11364–11367.
17. Nguyen, M. (2016). Plasmon-mediated chemical surface functionalization at the nanoscale, *Nanoscale*, **8**, 8633–8640.
18. Tijunelyte, I. (2016). Nanoplasmonics tuned "click chemistry", *Nanoscale*, **8**, 7105–7112.

19. Gehan, H. (2011). Design and optical properties of active polymer-coated plasmonic nanostructures, *J. Phys. Chem. Lett.*, **2**, 926–931.

20. Gehan, H. (2010). A general approach combining diazonium salts and click chemistries for gold surface functionalization by nanoparticle assemblies, *Langmuir*, **26,** 3975–3980.

21. Nguyen, M. (2015). Engineering thermoswitchable lithographic hybrid gold nanorods as plasmonic devices for sensing and active plasmonics applications, *ACS Photonics*, **2**, 1199–1208.

22. Ahmad, R. (2014). Tailoring the surface chemistry of gold nanorods through Au–C/Ag–C covalent bonds using aryl diazonium salts, *J. Phys. Chem. C*, **118**, 19098–19105.

23. Hohenau, A. (2006). Electron beam lithography, a helpful tool for nanooptics, *Microelectronic Eng.*, **83**, 1464–1467.

24. Sow, I. (2013). Revisiting surface-enhanced Raman scattering on realistic lithographic gold nanostripes, *J. Phys. Chem. C*, **117**, 25650–25658.

25. Schider, G., Krenn, J. R., Gotschy, W., Lamprecht, B., Ditlbacher, H., Leitner, A., and Aussenegg, F. R. (2001). Optical properties of Ag and Au nanowire gratings, *J. Appl. Phys.*, **90**, 3825.

26. Flatau, B. T. D. A. P. J. *User Guide for the Discrete Dipole Approximation Code DDSCAT7.0*, http://arxiv.org/pdf/0809.0337.pdf.

27. Purcell, E. M. P. C. R. (1973). Scattering and absorption of light by nonspherical dielectric grains, *Astrophys. J.*, **186**, 705–714.

28. Baffou, G., Girard, C., and Quidant, R. (2010). Mapping heat origin in plasmonic structures, *Phys. Rev. Lett.*, **104**, 136805.

29. Baffou, G., and Quidant, R. (2013). Thermo-plasmonics: Using metallic nanostructures as nano-sources of heat, *Laser Photonics Rev.*, **7**, 171–187.

30. Gotschy, W. (1996). Thin films by regular patterns of metal nanoparticles: Tailoring the optical properties by nanodesign, *Appl. Phys. B*, **63**, 381–384.

31. Lowe, A. B. (2014). Thiol-ene "click" reactions and recent applications in polymer and materials synthesis: A first update, *Polymer Chem.*, **5**, 4820–4870.

32. Lowe, A. B. (2010). Thiol-ene "click" reactions and recent applications in polymer and materials synthesis, *Polymer Chem.*, **1**, 17–36.

33. Han, X., Wu, C., and Sun, S. (2012). Photochemical reactions of thiol-terminated self-assembled monolayers (SAMs) for micropatterning

of gold nanoparticles and controlled surface functionality, *Appl. Surf. Sci.*, **258**, 5153–5156.

34. Lallana, E. (2012). Click chemistry for drug delivery nanosystems, *Pharm. Res.*, **29**, 1–34.
35. Grand, J. (2005). Role of localized surface plasmons in surface-enhanced Raman scattering of shape-controlled metallic particles in regular arrays, *Phys. Rev. B*, **72**, 033407.
36. Grand, J. (2003). Optimization of SERS-active substrates for near-field Raman spectroscopy, *Synthetic Metals*, **139**, 621–624.
37. Griesbaum, K. (1970). Problems and possibilities of the free-radical addition of thiols to unsaturated compounds, *Angew. Chem. Int. Ed. Engl.*, **9**, 273–287.
38. Dupin, D. (2007). Efficient synthesis of poly(2-vinylpyridine)–silica colloidal nanocomposite particles using a cationic azo initiator, *Langmuir*, **23**, 11812–11818.
39. Northrop, B. H., and Coffey, R. N. (2012). Thiol–ene click chemistry: Computational and kinetic analysis of the influence of alkene functionality, *J. Am. Chem. Soc.*, **134**, 13804–13817.
40. Chi, Zhang, J., and Ulstrup, J. (2006). Surface microscopic structure and electrochemical rectification of a branched alkanethiol self-assembled monolayer, *J. Phys. Chem. B*, **110**, 1102–1106.
41. Kakiuchi, T. (2000). Phase separation of alkanethiol self-assembled monolayers during the replacement of adsorbed thiolates on Au(111) with thiols in solution, *Langmuir*, **16**, 7238–7244.

Chapter 2

Concept and Development of Multi-Functional Hybrid Systems: Photoswitchable and Thermotunable Plasmonic Materials

Mai Nguyen,[a,b] Leila Boubekeur-Lecaque,[a] Claire Mangeney,[a,c] Stéphanie Lau-Truong,[a] Alexandre Chevillot-Biraud,[a] François Maurel,[a] Nordin Felidj,[a] and Jean Aubard[a]

[a]*Laboratoire ITODYS, Université Paris Diderot, Sorbonne Paris Cité, CNRS UMR 7086, 15 rue Jean–Antoine de Baïf, 75205 Paris Cedex 13, France*
[b]*Present address: University of Science and Technology of Hanoi, Vietnam Academy of Science and Technology, 18 Hoang Quoc Viet, Cau Giay, Ha Noi, Viet Nam*
[c]*Université Paris Descartes, Lab Chim & Biochim Pharmacolog & Toxicol, UMR 8601, F-75006 Paris, France*

jean.aubard@univ-paris-diderot.fr

2.1 Introduction

In Chapter 3 of this book [1], it is demonstrated how to obtain thermoresponsive plasmonic nanostructures by grafting poly (*N*-isopropylacrylamide) (PNIPAM) on gold nanoparticles (GNPs)

Plasmonics in Chemistry and Biology
Edited by Marc Lamy de la Chapelle and Nordin Felidj

ISBN 978-981-4800-03-7 (Hardcover), 978-0-429-45875-0 (eBook)
www.jennystanford.com

arrays.* Various applications are proposed to show the interest of this approach for developing new plasmonic systems for (bio) molecular sensing and enhanced spectroscopies (mainly SERS). To go further in designing multifunctional hybrid systems based on gold NPs arrays covered by PNIPAM, recent works have suggested to attach organic molecules with specific properties at the end of this polymer chains. Thus, plasmonic systems based on GNPs substrates covered by thermoresponsive polymer brushes labeled with photochromic units could lead, using suitable LSPR excitation, to (new) enhanced molecular (optical) properties [2]. Last but not least, these hybrid systems are of great interest because they could exhibit both photoswitchable and thermotunable LSPR optical changes. The choice of light-sensitive molecules attached to plasmonic nanostructures comes from the great interest in the design, functioning and applications of optical switches and electronic devices at the nanoscale [3]. In this context, organic photochromes are good candidates since specific changes in their electronic and structural properties occur under light irradiation [4].

Photochromism of organic compounds is defined as a reversible transformation induced in one or both direction by light between two isomeric states with different properties, such as change in their geometry, polarity, UV-Vis absorption bands, etc. Among various series of organic photochromes, diarylethenes (Scheme 2.1) have received increasing attention during the past 20 years thanks to their interesting optical and photochromic properties, which make them promising candidates for optical switches and memories [5].

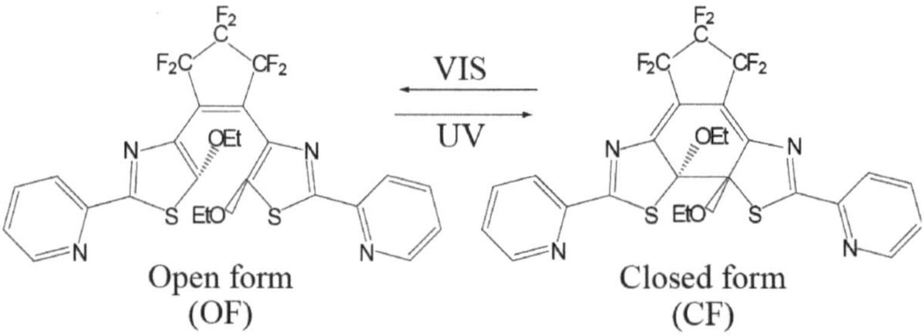

Scheme 2.1 Reversible photoisomerization in diarylethenes family upon UV-Vis light irradiation.

*See the abbreviations at the end of the text.

Unfortunately, due to the lack of suitable commercially available diarylethene derivatives that can be used for the attachment process (through an efficient click chemistry approach), azobenzene (AB) molecules have been preferred for practical reasons: Azobenzenes form one of the largest and most studied classes of photochromic molecules [6]; they have been used as light-triggered switches in a variety of polymers, surface-modified materials, protein probes, molecular machines, holographic recording devices, and metal ion chelators. AB exhibits remarkable photostability as negligible decomposition occurs even after prolonged irradiation. Moreover, large amounts of various azobenzene derivatives are commercially available, particularly the hydroxy derivative, which leads, in a one-step synthesis, to the alkyne derivative that can be easily used in a click chemistry synthesis (vide infra). AB molecules exist in two isomeric states, *trans* and *cis*, with the former being ~10 kcal/mol more stable than the latter. The *trans* to *cis* photoisomerization of azobenzenes is induced by UV irradiation (e.g., 365 nm), while visible light (>460 nm) reversibly transforms *cis* to *trans* (Scheme 2.2). These isomers have a different spatial arrangement of the aromatic moieties (Scheme 2.2) and show significantly different physical and chemical properties; thus, the *trans* to *cis* isomerization is accompanied by a considerable change in polarity, from zero dipole moment of the *trans* isomer to ~3 D for the dipole moment of the *cis* form. However, the optical changes on going from *trans* to *cis* isomer are weak (i.e., UV-Vis absorption band, Raman spectra, IR absorption) and its photochromic efficiency is quite low. Moreover, depending on the nature of the substituents on the aromatic groups, azobenzenes can undergo thermal *cis* to *trans* isomerization at room temperature [6].

In this chapter, we report on the procedure to attach onto lithographically designed gold NPs arrays a thermoresponsive polymer, PNIPAM, functionalized with an azobenzene chromophore at the end of the polymer chains (Scheme 2.3). The optical properties of the gold NPs (namely LSPR) and SERS spectra of attached AB molecules were investigated after photoirradiation (UV and visible) in air at room temperature and in water upon heating. It was observed that efficient *cis* to *trans* photoisomerization takes place due to LSPR excitation and probably originated from a coupling mechanism between GNPs and azobenzenes. Moreover, since the distance between the photoactive AB moiety and the gold NPs decreases upon heating above the critical temperature (LCST) of PNIPAM, it is expected that the coupling efficiency between them should increase as well [7, 8]. Indeed, it was shown in previous papers dedicated to optical properties of PNIPAM-GNPs plasmonic systems [1, 7–8] that the distance between the plasmonic GNPs arrays and the polymer end groups can be modulated upon increasing/decreasing the temperature around the LCST of PNIPAM. Thus, taking into account that the distance between the GNPs and the photochromic unit plays a major role in the coupling strength, such a thermal modulation of the distance should lead to observable modifications of optical properties of the hybrid system, i.e., LSPR spectral changes and photochromic efficiency.

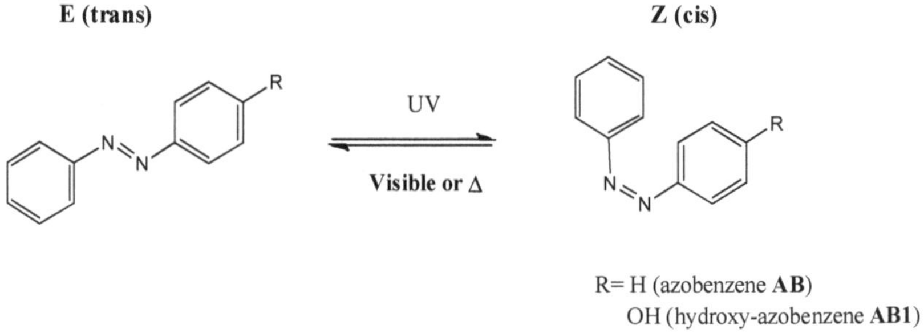

Scheme 2.2 Reversible isomerization of azobenzene moieties upon UV-Vis irradiation. The *trans* (*E*) isomer is planar and the *cis* (*Z*) isomer is bent and more compact.

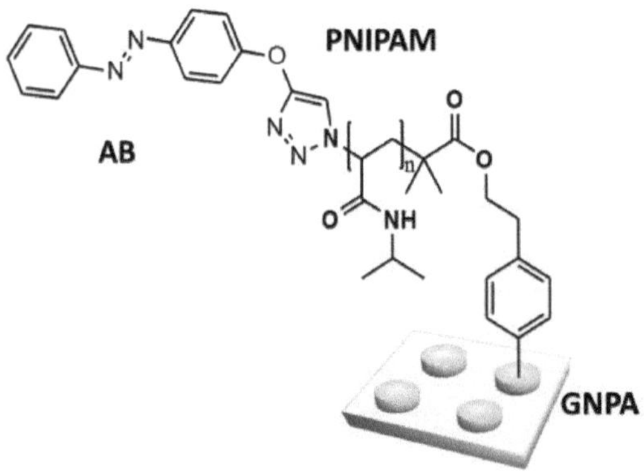

Scheme 2.3 Illustration of the attachment of an azobenzene (AB) derivative on a PNIPAM brush grafted on a GNP array (GNPA).

In the following, we will detail the fabrication and implementation of a three-component system coupling GNPs of 2D lithographic arrays to photochromic units through a PNIPAM linker. Optical properties of this multifunctional hybrid system were obtained thanks to UV-visible absorption and surface-enhanced Raman scattering (SERS) spectroscopies.

2.2 Elaboration and Properties of the Multifunctional Hybrid System

The photochrome–PNIPAM-Au hybrid system was built through a bottom-up approach (see Scheme 2.3) involving (i) the fabrication of plasmonic substrates consisting of gold nanoparticle arrays, (ii) the chemical surface modifications of GNPs for the covalent bonding of a thermoresponsive organic linker (PNIPAM), and (iii) the selective covalent attachment of azobenzene photochromes at the end of the polymer chains.

2.2.1 Preparation of Gold Nanoparticle Arrays

The electron beam lithography (EBL) method to prepare GNP arrays has been detailed in Chapter 3 [1]. The extinction spectra of four GNP arrays along with the AFM image of one selected array (D140) are shown in Fig. 2.1.

2.2.2 Preparation of GNPs Arrays Covered by PNIPAM with AB Chromophore End Groups (GNPA-PNIPAM-AB)

The stepwise strategy for the preparation of the multifunctional hybrid plasmonic system consists in two major steps: (1) the grafting of PNIPAM chains on the GNPs arrays, and (2) the attachment of the alkyne azobenzene derivative (AB2, obtained from 4-phenylazophenol, AB1, Scheme 2.4) at the end of the polymer chain.

This multistep functionalization is summarized in Fig. 2.2. It involves (i) a spontaneous covalent grafting of hydroxyl (–OH) terminated aryl diazonium moieties on the surface of the GNPs; (ii) the terminal hydroxyl groups are treated with 2-bromoisobutyryl bromide leading to bromo-terminated ester groups able to initiate the polymerization of NIPAM. Using this approach, the grafting of polymerization initiators could be confined exclusively on the gold nanoparticle surface. (iii) PNIPAM brushes are grown from the initiators using standard aqueous surface-initiated atom transfer radical polymerization (SI–ATRP) conditions. In step (iv), substitution of the bromo-end groups (of PNIPAM brushes) by azide provides the "click-active" substrates. Then, the azide-terminated substrates are reacted with the alkyne azobenzene derivative AB2 in typical click-chemistry conditions (azide-alkyne cycloaddition, step (v)) to yield azobenzene-terminated gold NPs array, abbreviated as **GNPA-PNIPAM-AB.**

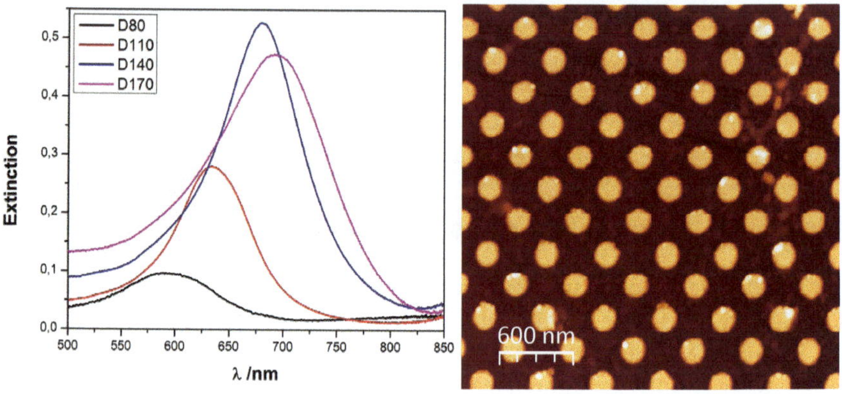

Figure 2.1 (Left) Extinction spectra of four GNP arrays (disk diameter, D = 80, 110, 140, and 170 nm) recorded in air at room temperature. These spectra were recorded with a spectrometer (LOT ORIEL, 74050 model, 400–1000 nm spectral range) equipped with a CCD camera (ANDOR, CCD-8855) under an upright optical microscope (Olympus, BX51TF) through a 100× objective (Olympus, NA: 0.8). (Right) AFM image of one array (D140) obtained in air at room temperature.

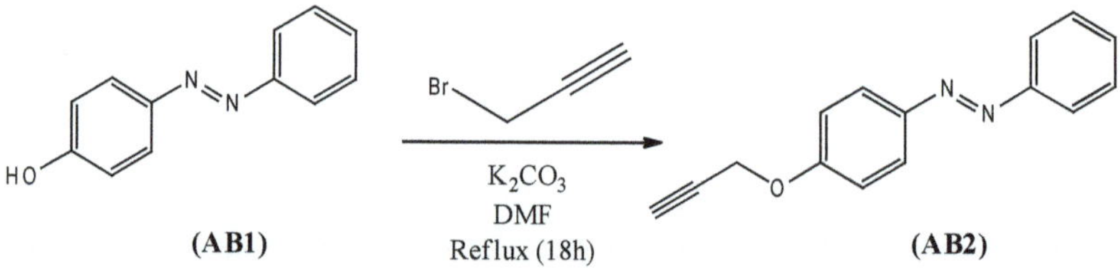

Scheme 2.4 From 4-phenylazophenol (AB1) a one-step synthesis (nucleophilic substitution by propargyl bromide) leads to the alkyne azobenzene derivative (AB2).

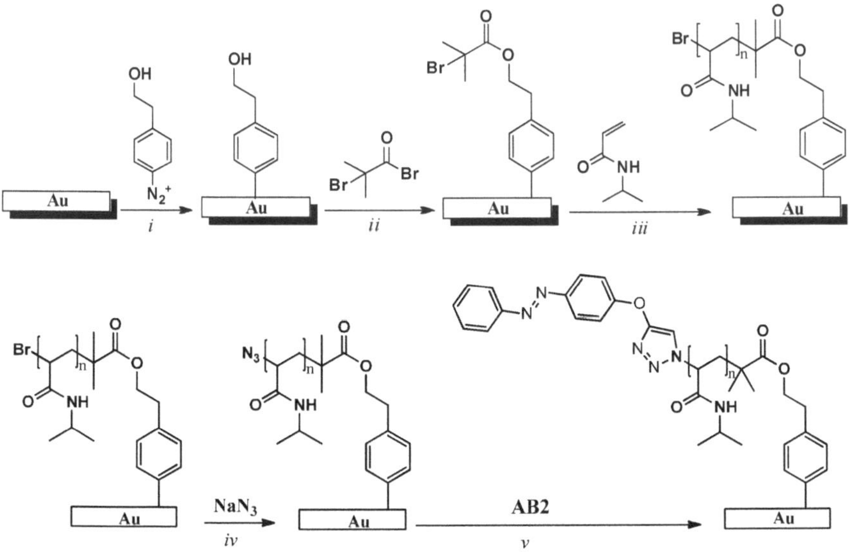

Figure 2.2 General scheme summarizing the stepwise strategy for the preparation of the multifunctional hybrid plasmonic system (GNPA-PNIPAM-AB). (i) spontaneous covalent grafting of hydroxyl (–OH) terminated aryl diazonium on the surface of the GNPs; (ii) 2-bromoisobutyryl bromide (0.1 M, toluene), trimethylamine (0.12 M), 5 min. (iii) NIPAM in surface-initiated atom transfer radical polymerization (SI–ATRP) conditions. (iv) NaN_3 (0.05 M, DMF) for 24 h, room temperature (v) β-cyclodextrin (1.25 mM), AB2 (0.05 M), Cu(II) sulfate pentahydrate (2.5 mM), L-ascorbic acid sodium salt (7.5 mM) in water under Ar at room temperature, 24 h.

2.2.3 AFM and Optical (Extinction) Characterization of the Thermosensitive Properties of the GNPA-PNIPAM-AB System

The thickness of PNIPAM grafted on GNP arrays was measured by AFM in air at room temperature. The AFM images of GNP arrays recorded before and after grafting of PNIPAM allow estimating the dry PNIPAM thickness which is around 5 nm (in air at room temperature, PNIPAM adopts a collapsed conformation). It should be noted that the height of the dried organic coating estimated to *ca.* 15 ± 2 nm (Fig. 2.3) reflects the total thickness including the initiator layer and the polymer brushes. This value, which represents the distance between the AB chromophore and the GNP surface (see, Scheme 2.3), is very close to the near-field area (*ca.* 10 nm) and it is expected that a quite strong coupling strength should take place between AB and gold NPs.

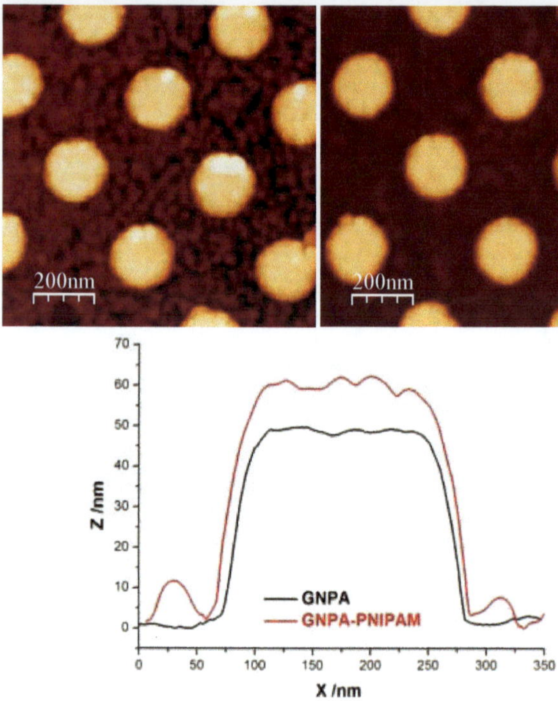

Figure 2.3 AFM images and height profiles in air at room temperature of a GNP array (GNPA; left and black curve) and GNPA-PNIPAM (right and red curve).

In order to characterize the thermosensitive properties of the system, we considered its optical extinction changes with temperature in water. At room temperature in water (below the LCST) the PNIPAM adopts a swollen (extended) conformation. From the AFM measurements in air at room temperature (in dry condition the polymer is collapsed, see Fig. 2.3) and taking into account that the swelling ratio, α, defined as $\alpha = h_{swollen}/h_{dry}$ (where $h_{swollen}$ and h_{dry} corresponds to the swollen and dry brush thickness), was estimated to *ca.* 2 [9], the PNIPAM thickness in water at room temperature was estimated to *ca.* 10 ± 2 nm. Thus, in these conditions, the total length of the spacer can be estimated to 20 ± 2 nm (at room temperature in water).

Increasing the external temperature up to 50°C in water (above the LCST of PNIPAM) induces a dramatic change in the polymer state which adopts a collapsed conformation of which the thickness significantly decreases upon heating to, 5 ± 2 nm. Thus for **GNPA-PNIPAM-AB** samples, under controlled conditions in water, the spacer length was reduced from 20 nm (at room temperature) to 15 nm after heating above the LCST (at 50°C).

The extinction spectra, for one selected array (D140), recorded in water at 20 and 50°C, are shown in Fig. 2.4. From these spectra, it appears that LSPR bands are quite sensitive to the conformation change of PNIPAM around LCST since, depending on the arrays, LSPR is red-shifted by 8–10 nm upon heating. Moreover, cycling experiments (not shown) evidenced that LSPR thermal switching of the **GNPA-PNIPAM-AB system** is reversible and reproducible [9]. This bathochromic shift with temperature is due to a change of the medium index in the vicinity of GNPs by conformational change of the polymer. The medium index becomes larger by heating above the LCST, due to the collapse of PNIPAM.

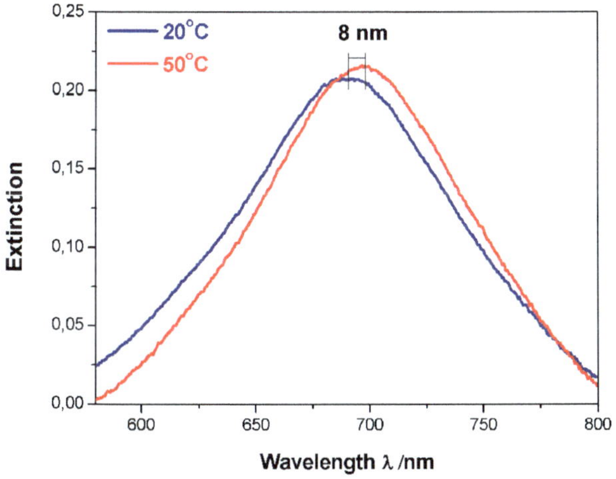

Figure 2.4 Changes with temperature of the extinction spectra of a ***GNPA-PNIPAM-AB*** sample (array D140) in water at 20°C (blue line) and 50°C (red line). Extinction spectra of GNPAs in water at various temperatures were recorded using a 100× immersion objective (Olympus, N.A: 1).

2.3 Reversible Changes of the LSP Resonance of GNPA-PNIPAM-AB Upon *cis/trans* Isomerization of Azobenzene

In order to carry out the isomerization of azobenzene (AB) units, the samples were photo-irradiated in air at room temperature and in water upon heating. Light irradiations (UV or visible) were performed using a spot light source (Hamamatsu, Lightningcure, LC8, equipped with a 200 W Xe-Hg lamp) through a NIR long pass filter, to protect samples from thermal effects, and appropriate band pass filters to select either UV (365 nm) or visible (540 nm) light, before reaching the sample.

Figure 2.5a shows extinction spectra recorded in air, of one selected **GNPA-PNIPAM-AB** sample (array D140) after UV (365 nm) and visible (540 nm) light irradiations for 90 min. For all samples, a small but reproducible blue shift of the LSPR band, depending on the arrays, is observed when going from *trans* to *cis* isomer upon 365 nm irradiation. Green light irradiation induced a reversible switching with a bathochromic shift of the LSPR band, which returns to the initial *trans* position value (see cycling experiments in the inset of Fig. 2.5a). These blue/red reproducible shifts, recorded upon cycling UV/Vis light irradiations, which are related to changes in the refractive index of AB in the environment of GNPs, could be assigned for the blue shift to a less conjugated bent structure of the *cis* isomer, while the *trans* isomer, which is endowed with a highly conjugated planar structure (see Scheme 2.2) leads to, as expected, a bathochromic (red) shift.

Extinction spectra for all sample arrays were recorded in water at room temperature after 365/540 nm cycling irradiation (Fig. 2.5b). For all samples, blue/red shifts were measured in the range, *ca.* 5–20 nm, but with a poor reproducibility. Surprisingly, these shifts are slightly more important than those measured in air at the same temperature where the dry polymer is shrunk (collapsed). On going from air to water, the LSPR band makes a large move toward the red for more than 20–40 nm, depending on the arrays, and as well known the LSPR sensitivity to the medium increases in the red/IR part of the spectrum. Thus, although the spacer length increases in water at 20°C (20 nm vs. 15 nm in air), the spectral shifts induced upon UV/Vis light irradiation are (slightly) greater than those measured in air.

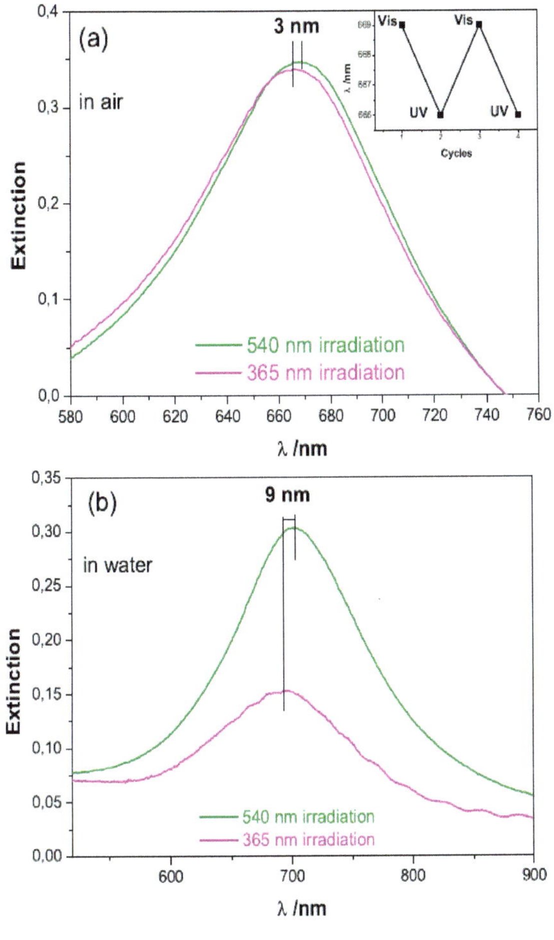

Figure 2.5 (a) Extinction spectra of a **GNPA-PNIPAM-AB** sample (array D140) upon repetitive cycling irradiations to UV (365 nm, pink line) and visible (540 nm, green line) each of 90 min duration. (a) Recorded in air at room temperature (20°C); the reproducible cycles are shown in the inset. (b) Recorded in water at room temperature (20°C); poor reproducibility of the cycles in this case (not shown). Either in air or in water, exposure to UV light led to *cis* isomer with a blue shift of the LSPR band while 540 nm green light led to *trans* isomer with a red shift of similar amount.

Upon heating up to 50°C in water, the distance between the photoactive AB moiety and the GNP is reduced from 20 to 15 nm (*vide supra*)—a value very close to the near-field area—and it is expected that the concomitant increase in the coupling strength should lead to more important changes of LSPR band of **GNPA-PNIPAM-AB**, upon 365/540 nm light irradiations, than those observed at room temperature. Surprisingly, the extinction spectra recorded at 50°C in water showed no significant (convincing) difference with the LSPR changes at 20°C in water upon 365/540 nm cycling irradiations for all samples. The reason for this failure is unclear but could be due to the damage of the samples after long UV irradiations along with hundred UV/Vis cycles.

Anyway, it is unexpected to not observe larger changes in the LSPR band on going from 20 to 50°C in water after 365/540 nm light irradiations for all samples. It is known that EM field enhancement increases as the distance to the GNP core decreases and in the present case it is more or less expected an increase in the E field intensity, felt by the AB unit at the end of the PNIPAM chain, of about $(70/65)^3$, i.e., 1.25 (see further on in the text, Eq. 2.1). Thus, the related expected "stronger" coupling strength should be difficult to observe with the low LSPR sensitivity of the **GNPA-PNIPAM-AB** system, upon 365/540 nm light irradiations.

Therefore, to follow reversible *trans*/*cis* photoisomerization of AB in our hybrid system, we turn toward SERS detection, which is expected to probe the reversible photoswitching of a very small amount of AB attached to the PNIPAM linker.

2.4 SERS Experiments of GNPA-PNIPAM-AB at Various Temperatures and upon AB *cis*/*trans* Isomerization

2.4.1 ThermoInduced Reversible Changes of Azobenzene SERS Intensity

SERS experiments with **GNPA-PNIPAM-AB** samples were carried out in air at room temperature and in water upon heating below and above the LCST of PNIPAM. Azobenzene molecules used in this study, i.e., AB1, AB2 and the AB unit in **GNPA-PNIPAM-AB** samples, display fluorescence properties and it was "difficult" in these conditions to record Raman spectra using 633 nm laser excitation. On the contrary, NIR excitation at 785 nm provides satisfactory Raman spectra (see for example the Raman spectrum of AB2 powder in Fig. 2.6b). Unfortunately, for **GNPA-PNIPAM-AB** samples, excitation at 785 nm led to SERS spectra showing huge and wide Raman bands in the 1300–1600 cm^{-1} spectral range, arising from amorphous carbon layers which develop at the sample surface due to thermal degradation, thus preventing good-quality SERS spectra to be recorded. Therefore, all SERS experiments with **GNPA-PNIPAM-AB** samples were conducted with 633 nm laser excitation. In these conditions, SERS spectra appear with a strong background attributed to surface-enhanced fluorescence (SEF) which is spectrally modified by LSPR excitation [10]; however, as discussed further on in the text, 633 nm excitation is able to induce some *cis* to *trans* photoisomerization, depending both on the laser power and on the acquisition time. Figure 2.6a shows SERS spectra in water at temperatures below and above the LCST of PNIPAM (resp. 20 and 54°C) for one selected sample (array D140). All SERS spectra display the characteristic Raman lines of similar azobenzene derivatives [11] and, as evidenced in Fig. 2.6b, there is a one to one correspondence with the six strong lines of the Raman spectrum of AB2 powder, at *ca.*, 1142 and 1186 cm^{-1} (CN stretching modes), 1414 and 1441 cm^{-1} (ring modes coupled to N=N stretching), 1464 cm^{-1} (N=N stretching) and 1595 cm^{-1} (ring modes, C=C stretching).

Despite the small amount of AB units (estimated to less than 5 in a 10 nm^2 area, [9]), located at the end of each polymer chain, SERS spectra recorded in the 700–1700 cm^{-1} range are of enough good quality allowing some quantitative analysis. Thus, it appears clearly that the SERS intensities increase with temperature on going from 20°C (below the LCST) to 54°C (above the LCST of PNIPAM). Cooling down the substrate to room temperature leads to a decrease of the SERS signal reversibly to the initial intensity recorded below the LCST. A twofold increase of the Raman band intensities is then detected, which is in perfect agreement with the theoretical enhancement, *ca.* 2.4, calculated from a simple EM model, which is now detailed just below.

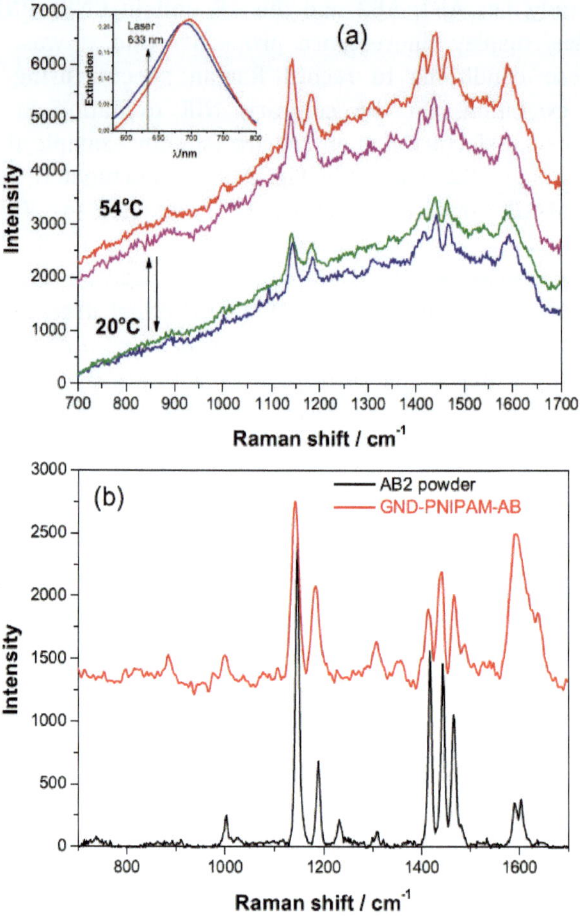

Figure 2.6 (a) SERS spectra excited at 633 nm, for one selected **GNPA-PNIPAM-AB** sample (array D140) in water after several 54/20°C heating/cooling cycles. The LSPR bands (in water at 20 and 54°C) are shown in the inset. (b) Raman spectrum of AB2 powder (black curve) and the SERS spectrum of a **GNPA-PNIPAM-AB** sample in water at 54°C (red curve). The SERS spectrum, excited at 633 nm, is corrected from the strong fluorescence background (see main text).

In this model, the GNP arrays were modeled as a collection of regular hemispheres, schematically represented in Fig. 2.7. In these conditions one can defined $A(\nu)$, the local E-field enhancement factor at a frequency ν and at a distance r from the NP center as [12]

$$A(\nu) = (\varepsilon - \varepsilon_0/\varepsilon + 2\varepsilon_0)(a/r)^3, \tag{2.1}$$

where ε represents the dielectric function of the metal, ε_0 the dielectric function of the surrounding medium and a the radius of the hemispheric NPs.

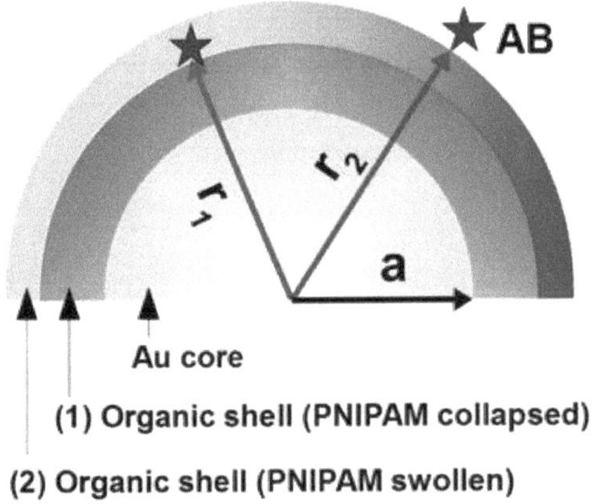

Figure 2.7 Schematic representation of a hemisphere GNP surrounded by an organic shell with an AB photochromic unit. Where **a** is the radius of the hemispheric NP; $\mathbf{r_1}$ and $\mathbf{r_2}$ the distance from the AB unit to the GNP center, in the collapsed (1) and swollen (2) state of PNIPAM.

The corresponding SERS enhancement, G^{EM}, can thus be approximated as [12]

$$G_{SERS}^{EM} \approx |A(\nu_0)|^4 = \left|\frac{\varepsilon(\nu_0) - \varepsilon_0}{\varepsilon(\nu_0) + 2\varepsilon_0}\right|^4 \left(\frac{a}{r}\right)^{12}, \tag{2.2}$$

where $A(\nu_0)$, is the local E-field enhancement factor at the incident laser frequency, ν_0.

This formula, known as the E^4 rule for SERS enhancement factor [12], shows that the Raman gain falls rapidly with the distance r between the probe molecule and the GNP center. If we now consider the two situations experienced by the **GNPA-PNIPAM-AB** samples in water, i.e., (1) when the PNIPAM brushes are collapsed at 50°C with r_1 = 65 nm and (2) when the PNIPAM brushes are swollen at 20°C with r_2 = 70 nm, the expected increase for SERS intensities can be expressed as $\mathbf{I_1/I_2 = (r_2/r_1)^{12} = 2.4}$; where $\mathbf{I_1}$ and $\mathbf{I_2}$ are the SERS intensities at 50 and 20°C, respectively. It should be noted that this simple EM SERS calculation predicts that on increasing the temperature from 20 to 50°C it is expected a twofold increase in SERS intensity, in agreement with the experimental results from Fig. 2.6a. Moreover, these results clearly indicate that AB units are effectively located at PNIPAM end groups.

Since the organic coating layer thickness was reduced from 20 nm (at room temperature) to 15 nm after heating above the LCST, SERS intensities easily probe this distance modulation induced by the thermoresponsive polymer. The full reversibility of the system was demonstrated by several heating/cooling cycles, as shown below in Fig. 2.8 and this definitely confirms that AB molecules are tightly attached at the end of PNIPAM chains.

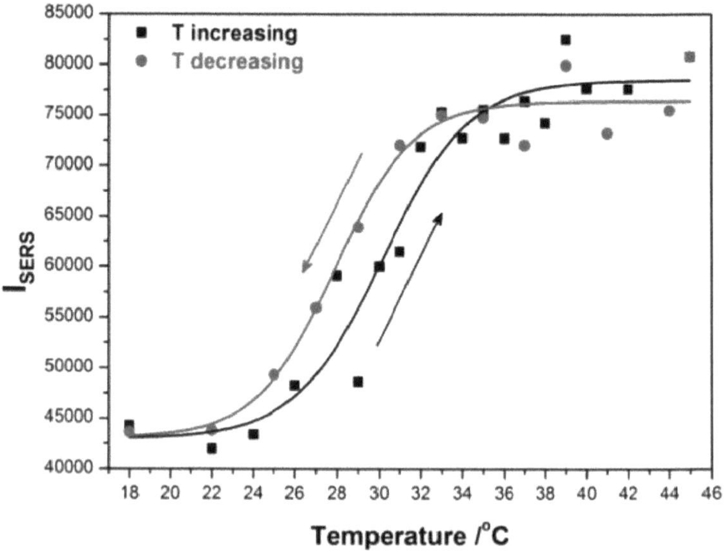

Figure 2.8 Integrated SERS intensity, of the 1142 cm^{-1} Raman line of AB, versus temperature for one selected **GNPA-PNIPAM-AB** sample (D 140).

2.4.2 SERS Intensity Changes upon *cis/trans* Isomerization of Azobenzene

Several 365/540 nm light irradiation cycles were then performed on various **GNPA-PNIPAM-AB** samples and probed by SERS.

First, these experiments were carried out in air at room temperature to verify that SERS intensity changes were able to monitor the reversible *cis/trans* photoisomerization of azobenzene units [11]. Those experiments were performed in water at 20°C and at 54°C, below and above the LCST of PNIPAM, respectively, to detect how the SERS spectra could be changed or modulated by both 365/540 nm photoirradiation and heating.

Figure 2.9 shows SERS spectra of a selected sample after UV (365 nm) and green (540 nm) light irradiations in air at room temperature. At room temperature in air, PNIPAM adopts a collapsed conformation with a polymer thickness of 5±2 nm (*vide supra*). Because, (i) the total distance between the AB chromophore and the GNPs is *ca.* 15±2 nm (see Fig. 2.4, AFM height profiles) and is very close to the near-field area (*ca.* 10 nm) and (ii) the 633 nm excitation was within the LSPR band of this sample (*ca.* 670 nm in air, Fig. 2.5a), these conditions should allow for high SERS enhancement for the **GNPA-PNIPAM-AB** sample under study. When this sample was UV irradiated for 1 h, the SERS spectrum (*cis* form) immediately collected showed a higher intensity, by nearly twofold (Fig. 2.9 left), than the SERS spectrum obtained before UV irradiation (*trans* form). The reasons why the SERS intensity of the *cis* isomer is higher than the *trans* is not clear but enhanced charge transfer of the *cis* state (high dipole moment) along with its bent conformation becoming closer to the GNP core could be at the origin of the higher SERS intensity of the *cis* form. Moreover, these SERS intensity changes recorded upon cycling UV/Vis light irradiations can be likely related to the blue/red shift of the LSPR band. Upon 365 nm irradiation the formation of *cis* isomer led to a blue shift of the LSPR band and thus to a better matching with the 633 nm excitation, leading to a SERS intensity increase. Green light irradiation induced a reversible switching with a bathochromic (red) shift of the LSPR band, which returns to the initial *trans* position value with the concomitant decrease, as observed, of the SERS intensity. Anyway, these observations show that the analysis of SERS intensities could be able to probe the reversible *cis/trans* photoswitching of a very small amount of AB attached to the end of PNIPAM polymer chains.

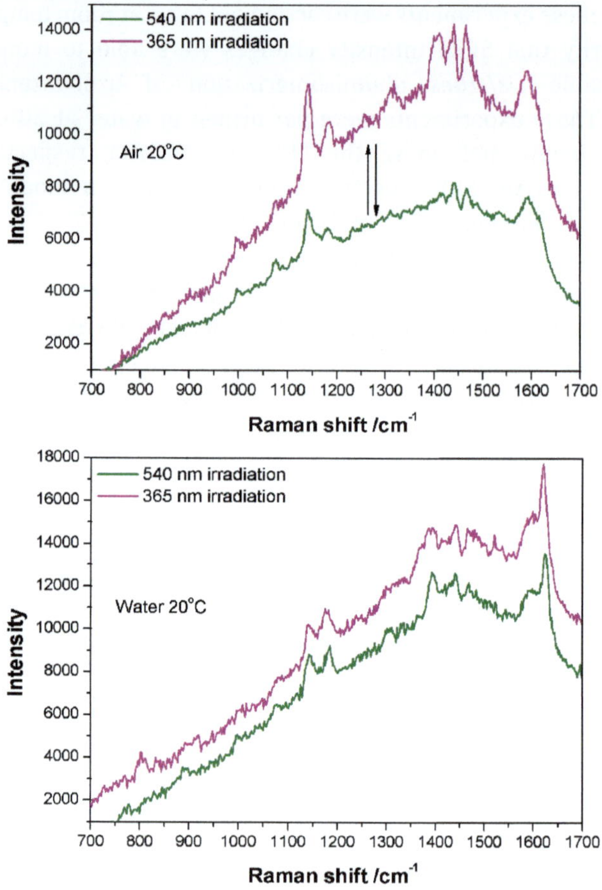

Figure 2.9 SERS spectra excited at 633 nm, of a selected **GNPA-PNIPAM-AB** sample (array D140), upon repetitive cycling irradiations to UV (365 nm, pink line) and visible (540 nm, green line) light, each of 1 h duration. Spectra were recorded in air (up) and in water (down), at room temperature (20°C).

Similar SERS experiments were carried out upon 365/540 nm irradiation in water at 20 and 54°C. SERS spectra of the same selected sample (array D140), recorded after UV (365 nm) and green (540 nm) light irradiation in water at room temperature are shown in Fig. 2.9 (right). It should be noted that these SERS spectra, recorded at 20°C in water upon 365/540 nm cycling irradiations, are almost identical and showed no significant SERS intensity difference.

On going from air to water, the LSPR band makes a large move toward the red for more than 30–40 nm, depending on the arrays (see Fig. 2.5b). In these conditions there is a total mismatch between LSPR (*ca.* 700–710 nm) and laser excitation (633 nm) and a weak EM and molecular coupling takes place, thus leading to poor SERS spectra with no significant difference in intensity between *trans* or *cis* isomer. Moreover, in water at room temperature (below the LCST) the PNIPAM adopts a swollen (extended) conformation with a thickness estimated to *ca.* 10±2 nm and, in these conditions, the total length of the organic linker can be estimated to 20±2 nm, further contributing to the decrease of the SERS intensities with respect to those obtained in air at room temperature. Increasing the temperature up to 50°C in water induces a conformation change of PNIPAM (collapsed state) and the distance between the photoactive AB moiety and the GNP unit is reduced to *ca.* 15 nm (*vide supra*) and it is expected that the concomitant increase in the coupling strength should lead to more important changes in the SERS intensities, upon 365/540 nm light irradiation, than those observed at room temperature. Unfortunately, no significant difference with the SERS spectra at 20°C in water was observed upon reversible cycling photoirradiations at 50°C for all **GNPA-PNIPAM-AB** samples.

In these SERS studies of covalently attached AB moieties in **GNPA-PNIPAM-AB** samples after photoirradiation (UV and visible) in air at room temperature and in water upon heating, promising results were obtained and deserve to be summarized here:

(i) Reproducible SERS intensities, of the AB chromophore, easily probe the distance modulation induced by the thermoresponsive polymer after several reversible 54/20°C heating/cooling cycles in water.

(ii) The analysis of SERS intensity changes of GNPA-PNIPAM-AB upon 365/540 nm irradiation in air at room temperature could be a way to follow the reversible *cis/trans* photoswitching of a very small amount of AB attached to the end of PNIPAM polymer chains. Further experiments are necessary to check the reproducibility of those results.

(iii) As with LSPR detection, we failed to observe any SERS spectral modification at 50°C upon 365/540 nm irradiation in water. Even with the SERS detection, the GNPA-PNIPAM-AB system is not enough sensitive to detect the influence of thermal switching on AB photoisomerization in water.

2.5 Conclusion

All the experiments reported here are in correct agreement with previous recent results either on related photochromic-PNIPAM-Au [13] or in similar Au-SAM-AB systems [11, 14] even if these latter studies highlight some contradictory results compared with ours. Thus, as reported by others, SAMs linker as well as PNIPAM in the present work both ensure reversible *trans/cis* photoswitching of AB upon UV/Vis light irradiation. Interestingly, since the polymer thickness varies with temperature around the LCST, a plasmonic system including PNIPAM is dual and able to respond to both photoirradiation and thermal stimuli. Moreover, although this point is never pointed out in the abundant literature, investigation of AB photoswitching in SAMs is hampered by the fact that exposure to UV light in air causes irreversible damage to the SAMs by oxidation of the thiol bond. This point could perhaps explain some discrepancies and contradiction as compared to previously reported results concerning AB photoswitching [11, 14]. Furthermore, our study highlights an important point in the context of applications to the development of nano-photoswitching or molecular electronic devices: the PNIPAM based plasmonic hybrid system does not suffer from extensive UV degradation and the covalent grafting of this linker onto the GNP substrate ensures longer time stability than the common thiol-based SAMs linker. Finally, the reason for some reverse trends in our results is unclear and need further studies with more suitable EBL arrays and other photochromic systems such as diarylethenes which show very efficient photoisomerization process and large optical contrast between isomers [15, 16]. Therefore, to follow reversible *trans/cis* photoisomerization of AB in our hybrid system we turned toward SERS detection, which was expected to probe the reversible photoswitching of a very small amount of AB attached to the PNIPAM linker with enhanced sensitivity. Unfortunately, even with SERS detection, we were unable to detect the influence of thermal switching on AB photoisomerization in the GNPA-PNIPAM-AB system in water.

Some important points remain and need to be overcome to improve this thermo- and photosensitive hybride plasmonic system. In particular, for the development of applications (e.g., optical switches and electronic devices at the nano scale) one needs to detect the photochromic state nondestructively and this implies to use NIR-SERS excitation. Indeed, as already mentioned above in the main text, SERS excitation at 633 nm is able to induce some *cis* to *trans* photoisomerization, depending on the laser power and on the acquisition time. Unfortunately, it was very difficult to obtain good-quality SERS spectra of ***GNPA-PNIPAM-AB*** at 785 nm because of the rapid thermal degradation of polymer layers preventing to make any quantitative analysis. Therefore, all SERS experiments with ***GNPA-PNIPAM-AB*** samples were conducted at 633 nm laser excitation with short acquisition time (1–4 s) and low laser power (*ca.* 65 μW at the sample), thus allowing SERS spectra to be recorded with moderate fluorescence (SEF) background and probably very limited *cis* to *trans* photoisomerization. In these recording conditions, 633 nm SERS spectra are of medium (low) quality and reliable SERS intensity analysis are "difficult" and it is hopeless in those conditions to detect the weak specific Raman features of *cis* isomer [11].

Finally, the improvement of the thermo- and photosensitive hybride plasmonic system using NIR-SERS detection needs further studies with more suitable EBL arrays and different photochromic systems, such as diarylethenes, which show very efficient photoisomerization process and large SERS contrast between isomers [17]. Furthermore, a higher PNIPAM thickness difference between the collapsed and swollen state might also be taken into account: Rather than the actual 5/10 nm, it would be better to try with 10/20 nm [13]; in this latter case, the theoretical EM SERS calculation (*vide supra*) predicts that on increasing the temperature from 20 to 50°C, a fivefold increase in SERS intensity could be observed. Such experiments are currently under way.

Abbreviations

AB: Azobenzene
AFM: Atomic force microscopy
ATRP: Atomic transfer radical polymerization
CE: Chemical enhancement
DS: Diazonium salt
EBL: Electron-beam Lithography
EE: Electromagnetic enhancement
EF: Enhancement factor
GNPs: Gold nanoparticles
GNPAs: Gold nanoparticle arrays
LCST: Lower critical solution temperature
LSP: Localized surface plasmon
LSPR: Localized surface plasmon resonance
NPs: Nanoparticles
PNIPAM: Poly-N-isopropylacrylamide
SAMs: Self-assembled molecules
SI-ATRP: Surface-initiated atomic transfer radical-polymerization
SEF: Surface-enhanced fluorescence
SERS: Surface-enhanced Raman scattering
XPS: X-ray photoelectron spectroscopy

References

1. Nguyen, M., Aubard, J., Félidj, N., and Mangeney, C. Reversible adsorption of biomolecules on thermosensitive polymer-coated plasmonic nanostructures, chapter 3, this book.
2. Ueno, K., and Misawa, H. (2013). Surface plasmon-enhanced photochemical reactions, *J. Photochem. Photobiol. C*, 15, pp. 31–52.
3. Feringa, B. L. (2001) *Molecular Switches* (Wiley-VCH Verlag GmbH, Darmstadt).
4. Crano, J. C., and Guglielmetti, R. J. (1999) *Organic Photochromic and Thermochromic Compounds*, vol. 1 and 2 (Plenum Press, New York).
5. Irie, M. (2000). Diarylethenes for memories and switches, *Chem. Rev.*, 100, pp. 1685–1716.
6. Bandara, H. M. D., and Burdette, S. C. (2012) Photoisomerization in different classes of azobenzene, *Chem. Soc. Rev.*, 41, pp. 1809–1825.

7. Gehan, H., Fillaud, L., Chehimi, M., Aubard, J., Hohenau, A., Félidj, N., and Mangeney, C. (2010). Thermo-induced electromagnetic coupling in gold/polymer hybrid plasmonic structures probed by surface-enhanced Raman scattering, *ACS Nano*, 4, pp. 6491–6500.
8. Gehan, H., Mangeney, C., Aubard, J., Lévi, G., Hohenau, A., Krenn, J. R., Lacaze, E., and Félidj, N. (2011) Design and optical properties of active polymer-coated plasmonic nanostructures, *J. Phys. Chem. Lett.*, 2, pp. 926–931.
9. Nguyen, M., Kanaev, A., Sun, X., Lacaze, E., Lau-Truong, S., Lamouri, A., Aubard, J., Félidj, N., and Mangeney, C. (2015) Tunable electromagnetic coupling in plasmonic nanostructures mediated by thermoresponsive polymer brushes, *Langmuir*, 31, pp. 12830–12837.
10. Le Ru, E. C., Etchegoin, P. G., Grand, J., Félidj, N., and Lévi, G. (2007) Mechanisms of spectral profile modification in surface-enhanced fluorescence, *J. Phys. Chem. C*, 111, pp. 16077–16079.
11. Joshi, G. K., Blodgett, K. N., Muhoberac, B. B., Johnson, M. A., Smith, K. A., and Sardar, R. (2014) Ultrasensitive photoreversible molecular sensors of azobenzene-functionalized plasmonic nanoantennas, *Nano Lett.*, 14, pp. 532–540.
12. Le Ru, E. C., Grand, J., Félidj, N., Aubard, J., Lévi, G., Hohenau, A., Krenn, J. R., Blackie, E., and Etchegoin, P. G. (2009) Experimental verification of the SERS electromagnetic model beyond the $|E|^4$ approximation: Polarisation effect, *J. Phys. Chem. C*, 112, pp. 8117–8121.
13. Imao, S., Nishi, H., and Kobatake, S. (2013) Thermo- and photo-responsive reversible changes in localized surface plasmon resonance of gold nanoparticles covered by poly(*N*-isopropylacrylamide) with photochromic diarylethene end group, *J. Photochem. Photobiol. A*, 252, pp. 37–45.
14. Zheng, Y. B., Payton, J. L., Chung, C. H., Liu, R., Cheunkar, S., Pathem, B. K., Yang, Y., Jensen, L., and Weiss, P. S. (2011) Surface-enhanced Raman spectroscopy to probe reversibly photoswitchable azobenzene in controlled nanoscale environments, *Nano Lett.*, 11, pp. 3447–3452.
15. Boubekri, R., Yasukuni, R., Lau Truong, S., Grand, J., Perrier, A., Maurel, F., and Aubard, Raman, J. (2013) Study of a photochromic diarylethene molecule: A combined theoretical and experimental study, *J. Raman Spectrosc.*, 44, pp. 1777–1785.
16. Yasukuni, R., Boubekri, R., Grand, J., Felidj, N., Maurel, F., Perrier, A., Métivier, R., Nakatani, K., Pei, Yu., and Aubard, J. (2012) Specific

and nondestructive detection of different diarylethene isomers by NIR-SERS, *J. Phys. Chem. C*, 116, pp. 16063–16069.

17. Yasukuni, R., Ouhenia-Ouadahi, K., Boubekeur-Lecaque, L., Félidj, N., Maurel, F., Métivier, R., Nakatani, K., Grand, J., and Aubard, J. (2013) Silica-coated gold nanorod arrays for nanoplasmonics devices, *Langmuir*, 29, pp. 12633–12637.

Chapter 3

Reversible Adsorption of Biomolecules on Thermosensitive Polymer-Coated Plasmonic Nanostructures

Nguyen Thi Tuyet Mai,[a,b] Jean Aubard,[a] Claire Mangeney,[a,c] and Nordin Felidj[a]

[a]*Laboratoire ITODYS, Université Paris Diderot, Sorbonne Paris Cité, CNRS UMR 7086, 15 rue Jean-Antoine de Baïf, 75205 Paris Cedex 13, France*
[b]*Present address: University of Science and Technology of Hanoi, Vietnam Academy of Science and Technology, 18 Hoang Quoc Viet, Cau Giay, Ha Noi, Viet Nam*
[c]*Univ Paris 05, Lab Chim & Biochim Pharmacolog & Toxicol, UMR 8601, F-75006 Paris, France*

claire.mangeney@parisdescartes.fr

3.1 Introduction

Biomedical applications of stimuli-responsive surfaces have attracted much interest in the mid-1980s, particularly from the pioneer work developed by the Hoffman group [1]. In this context, bio-interfaces with stimuli-responsive/switchable bio-affinity have received considerable attention during the past 15 years

Plasmonics in Chemistry and Biology
Edited by Marc Lamy de la Chapelle and Nordin Felidj

ISBN 978-981-4800-03-7 (Hardcover), 978-0-429-45875-0 (eBook)
www.jennystanford.com

and their use has been demonstrated in cell culture, drug delivery, and tissue engineering [2–5]. These surfaces are typically prepared by grafting environmentally sensitive macromolecules onto a substrate, where the type of macromolecule and substrate are chosen depending on the intended use of the surface. A typical macromolecule used for this purpose is the poly(*N*-isopropylacrylamide) (PNIPAM), which is hydrated and hydrophilic below the lower critical solution temperature (LCST) at 32°C, within the physiologically relevant temperature range, whereas above the LCST, the polymer chains are thought to become more hydrophobic. This temperature-dependent hydrophobicity/hydrophilicity transition has been exploited to selectively switch the interfacial properties of PNIPAM and its consequent interactions with cells and biomolecules [6, 7]. The change in the hydration property of PNIPAM renders its surface adhesive or non-adhesive to bio-molecules [8–12].

Adhesion between particles, the sizes of which are in the micron to nanometer range, and solid surfaces is a key factor in many industrial applications. Unfortunately, real surfaces are no longer smooth at the submicroscopic level and often present significant roughness or they are nanostructured for specific applications such as tissue engineering and cell culture [13]. Surface roughness or structuration plays an important role in adhesion since it reduces the contact area between the species, leading to significantly reduced interaction. Surfaces may possess features in several length scales, but due to the short range of the van der Waals interaction, structuration in nanoscale ultimately determines the strength of adhesion. Progress and improvement of colloidal probe techniques have boosted the studies of adhesion and the effects of nanoscale structuration or roughness. Recent studies have quantified the relative importance of the geometry, surface roughness and medium properties on the adhesion of the approaching bodies as their sizes are scaled from micro- to nano-scale.

In this work, we describe the design of novel hybrid plasmonic nanostructures coated by PNIPAM brushes as smart platforms for the reversible confinement of biomolecules. The combination of lithographic gold nanostructures and thermoresponsive polymer brushes leads to a new generation of hybrid gold/polymer system able to efficiently switch from bio-adherent to bio-repellent surface, by convenient temperature stimuli.

3.2 Experimental

3.2.1 Materials

Reagent grade solvents were purchased from VWR, Prolabo and Alfa Aesar. 2-Bromopropionyl bromide (BPB) (97%, Aldrich), triethylamine (TEA) (99%, Merck), CuBr (98%, Sigma-Aldrich), N,N,N′,N″,N″-pentamethyldiethyltriamine (PMDETA) (99%, Acros Organics) were used as received. *N*-Isopropylacrylamide (NIPAM) (99%, Acros Organics) was purified by recrystallization in *n*-hexane solution. Bovine Serum Albumin (BSA) 98% was obtained from Sigma-Aldrich.

3.2.2 Elaboration of Gold Nanostructure Arrays

Gold nanoparticle arrays were fabricated by electron beam lithography (EBL). The starting material is a flat substrate that is conducting to prevent charging. For optical applications, ITO covered glass plates are widely used. The substrate is spin coated with an electron sensitive resist, typically 90 nm thick. Resists such as polymethylmethacrylate (PMMA) consist of macromolecules that are modified upon exposure to high-energy electrons, resulting in a modification of the solubility. A desired sample pattern can thus be transferred to the resist by writing the pattern with an electron beam and subsequent wet chemical development. The resulting patterned resist layer serves as a mask for depositing the sample material by vacuum evaporation. Finally, a lift-off step removes the resist mask and excess material, leaving only the metal nanostructures on the surface.

3.2.3 Functionalization of Gold Nanostructures by PNIPAM Brushes

3.2.3.1 Synthesis of diazonium salt

The 4-hydroxyethylbenzene diazonium tetrafluoroborate salt ($^{+}N_2$-C_6H_4-CH_2-CH_2-OH) was synthesized following the following protocol: a stirred solution of 2-(4-aminophenyl)ethanol (0.69 g) and HBF_4 (1.8 mL) in acetonitrile (3 mL) at −10°C was added dropwise to a solution of tert-butylnitrite (0.63 g) in acetonitrile (3 mL) at −10°C. The resulting mixture was kept overnight at −10°C. The precipitate was washed three times with 50 mL of diethyl ether, then with 50 mL of acetone, and finally evaporated under vacuum. The 4-(2-hydroxyethyl)benzene diazonium tetrafluoroborate salt was stored at −10°C.

3.2.3.2 Initiator-modified gold surfaces

The atom transfer radical polymerization initiator was grafted in two steps: (i) spontaneous grafting of 4–hydroxyethylbenzene diazonium tetrafluoroborate salt was achieved on cleaned gold nanostructure array by incubation in aqueous solutions (at 0.003 M) for 6 h at room temperature. (ii) the terminal hydroxyl groups were then treated with 2-bromoisobutyryl bromide (0.1 M, toluene) in the presence of TEA (0.12 M) for 5 min to produce bromo-terminated ester groups.

3.2.3.3 Atomic Transfer Radical Polymerization (ATRP) of NIPAM

Solutions were prepared and kept at room temperature during degassing by passing a continuous stream of argon through the solution while being stirred. The polymerization solution was prepared by adding a solution of an organometallic catalyst to a solution of NIPAM monomer. The extremely oxygen-sensitive organometallic catalyst was prepared by adding a 5 mL solution of PMDETA in MeOH (200 μL, 1 mmol) to 30 mg of CuBr (0.2 mmol). A 3 mL portion of the resulting green solution (which could possibly turn blue due to the presence of $CuBr_2$ and provide unsuccessful ATRP) was added to a solution of NIPAM monomer (2 g, 18 mmol) in 11 mL of deionized water under a continuous stream of argon. The polymerization solution was allowed to stir during degassing for 15 min and then transferred into a flask containing the initiator-modified gold surface. The resulting solution was allowed to stir at room temperature under argon for 20 min. The substrates were then removed from the flask and rinsed thoroughly with ethanol and water and subsequently dried under a flush of argon.

Instrumentation

AFM in air was performed on Nanoscope III digital instrument microscope in tapping mode to map the morphology. AFM images were processed and analyzed using the application WSxM and FabViewer. Extinction spectra were recorded with a spectrometer (LOT ORIEL, 74050 model, 400–1000 nm spectral range) equipped with a CCD camera (ANDOR, CCD-8855) under an upright optical microscope (Olympus, BX51TF) through, either a 50× objective (Olympus, numerical aperture, NA: 0.35) or a 100× objective (Olympus, NA: 0.8) for experiments in air. Extinction spectra recorded in water at various temperatures were performed using a 100× immersion objective (Olympus, N.A: 1).

3.3 Results and Discussion

3.3.1 Characterization of PNIPAM-Coated Gold Nanodots

Gold nanodot (GND) arrays were fabricated by EBL on an ITO-covered glass substrate; a selected array with a GND diameter of ~220 nm and a height of 50 nm is presented in Fig. 3.1. This array shows a strong LSPR band peaking at *ca.* 600 nm in air (see Fig. 3.3c, black line). The PNIPAM brushes were grafted on the GND array by using our multi-step strategy (*vide infra*, Fig. 3.2). The stepwise strategy for the preparation of the hybrid plasmonic system GND@PNIPAM consists in three major steps: (i) The spontaneous grafting of the aryl group derived from 4-hydroxyethylbenzene diazonium tetra-fluoroborate salt (HEBDT) to produce –OH terminated aryl moieties, covalently anchored to the surface; (ii) Esterification of the anchored –OH groups with 2-bromoisobutyryl bromide leading to bromo-terminated gold surface which allows to initiate the atomic transfer radical polymerization (ATRP) and (iii) The grafting of PNIPAM brushes from the surface via surface-initiated ATRP.

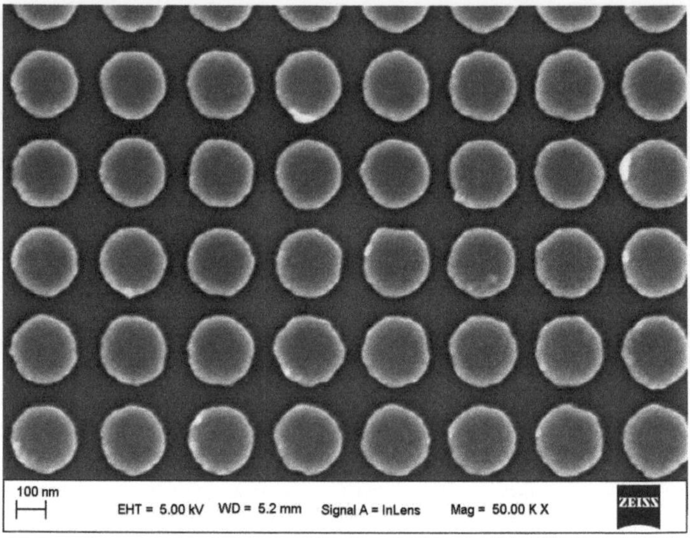

Figure 3.1 SEM image of a gold nanodot array.

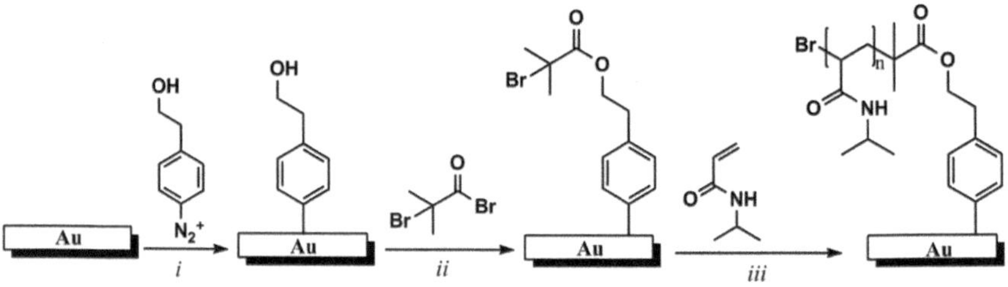

Figure 3.2 General scheme summarizing the stepwise strategy for the functionalization of the GNDs array: (i) spontaneous grafting of hydroxethyl–aryl groups; (ii) esterification with 2-bromoisobutyryl bromide; (iii) SI–ATRP of N–isopropylacrylamide.

The thickness of PNIPAM grafted on the GNDs array was measured by AFM in air at room temperature. The AFM images of GNDs before and after grafting of PNIPAM (Fig. 3.3a,b) allow estimating the dry PNIPAM thickness which is around 10 nm (in air at room temperature, PNIPAM adopts a collapsed conformation). In water at room temperature (i.e., below the LCST), the PNIPAM adopts a swollen conformation. From the AFM measurements in air at room temperature and taking into account that the swelling ratio, α, defined as, $\alpha = h_{\text{swollen}}/h_{\text{dry}}$ (where h_{swollen} and h_{dry} corresponds to the swollen and dry brush thickness) was previously reported to be ~2 [14, 15], the PNIPAM thickness in water at room temperature was estimated to *ca.* 20 ± 2 nm. Increasing the external temperature up to 40°C in water (above the LCST of PNIPAM) induces a dramatic change of the polymer conformation of which the thickness significantly decreases upon heating to, *ca.* 10±2 nm. The extinction spectra of GNDs array, recorded before and after the polymerization process, indicate a red-shift of 35 nm of the LSPR wavelength upon polymerization (Fig. 3.3c). This red-shift is attributed to an increase of the local refractive index around the GNDs in the presence of the polymeric layer.

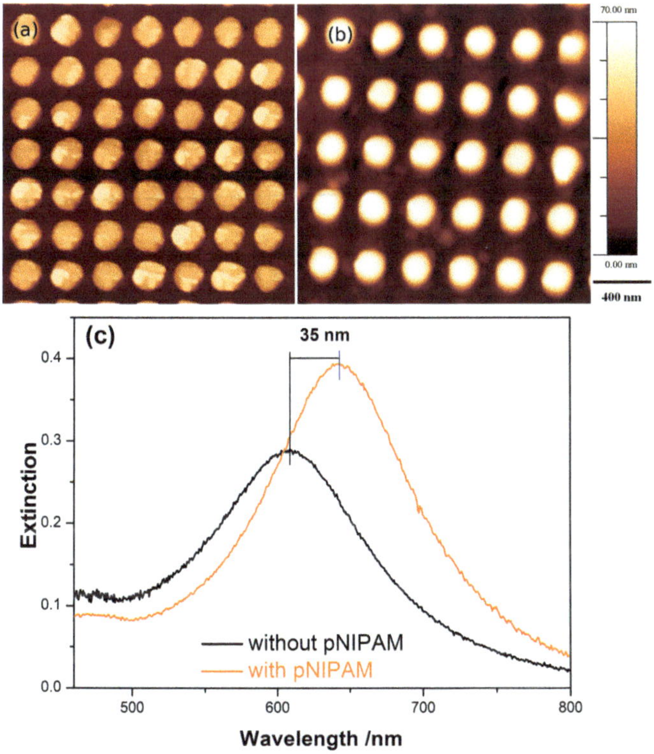

Figure 3.3 AFM images of the bare GND array (a) and the PNIPAM-coated GND array (b). Extinction spectra of GND array in air, before (black line) and after (orange line), grafting of PNIPAM (c).

Interestingly, when the PNIPAM-coated GND array is immersed in water, increasing the temperature from 20°C (below the LCST) to 40°C (above the LCST), leads to a red-shift of 8 nm of the LSPR wavelength, as shown in Fig. 3.4. This red-shift is attributed to the collapse of the polymer brushes above the LCST, leading to an increase of both the polymer density close to the NPs and the refractive index of the surrounding medium. Note that the stability of our system has been checked through six cycles in temperature, indicating that our thermo-sensitive plasmonic device is very stable (inset Fig. 3.4).

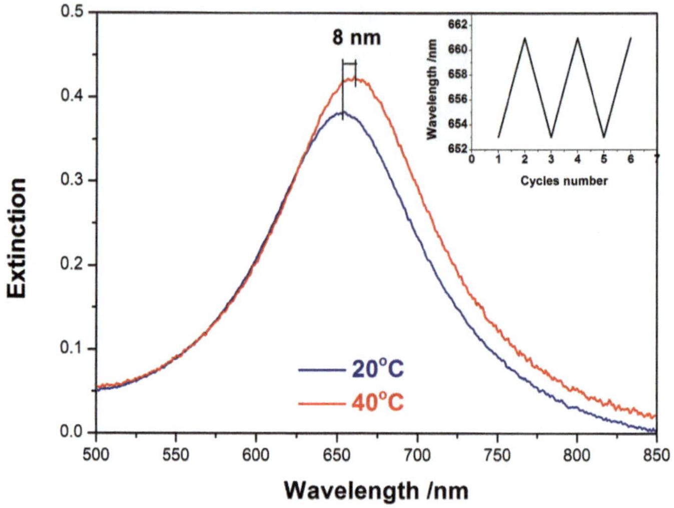

Figure 3.4 Extinction spectra of PNIPAM-coated GND arrays recorded in water at 20°C (blue line) and 40°C (red line). Inset: Variation of the LSP wavelength in water as a function of the external temperature, for repeated temperature cycles between 20 and 40°C.

3.3.2 Adsorption of Proteins on the PNIPAM-Grafted Gold Nanostructured Surface

As already mentioned, in water PNIPAM brushes are hydrated below the LCST (32°C), while above the LCST the polymer segments become more hydrophobic. The change in hydration of the PNIPAM renders the PNIPAM surface adhesive or non-adhesive to bio-molecules. In order to investigate this behavior, we considered the thermo-controlled adsorption of a model protein, the BSA, on GND@PNIPAM. The dry grafted PNIPAM thickness was $h_{dry} \sim 10$ nm. It is known that the hydrophobic interaction has a major role in protein adsorption phenomena. When the temperature increases above the LCST, the surface becomes more hydrophobic due to the collapse of polymer brushes and is susceptible to adhere larger amounts of protein. By using the plasmonic GND array structures, the adsorption of BSA as a function of temperature can be monitored by the modification of UV-vis spectra. Adsorption of BSA on GND@PNIPAM substrates was carried out by incubation of the substrates in aqueous solutions of BSA (10^{-2} M) at 20 and 50°C for 2 h. The substrate was then rinsed in water at the same temperature (20 or 50°C), dried with nitrogen gas and characterized by UV-vis absorption spectroscopy. Figure 3.5 displays the extinction spectra recorded in air at room temperature of GND@PNIPAM before (black curve) and after incubation in a BSA solution at 20°C (blue curve) and 50°C (red curve). These results show that the LSPR shift is negligible when the sample was incubated in BSA solution at 20°C. In contrast, a red shift of the LSPR wavelength of $\Delta\lambda \sim 12$ nm is observed after incubation of the sample in the BSA solution at 50°C. This can be attributed to the adsorption of proteins on the polymer brushes leading to an increase of the surrounding medium refractive index in the vicinity of GNDs.

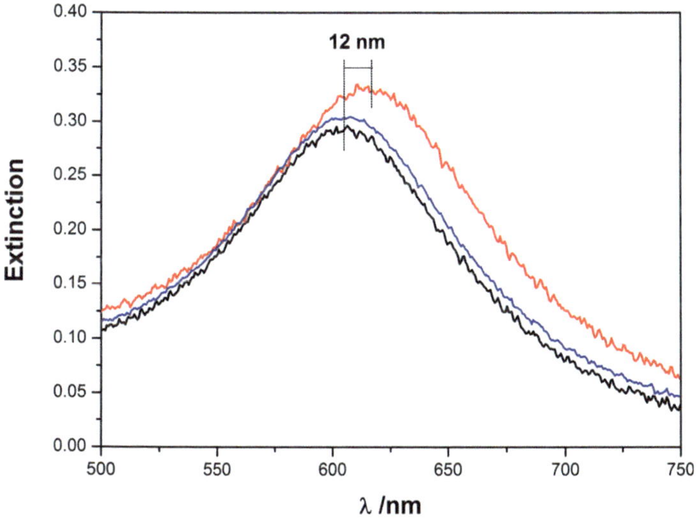

Figure 3.5 Extinction spectra recorded in air at room temperature of GND@PNIPAM nanostructures before (black curve) and after incubation in the BSA solution (10^{-2} M) at 20°C (blue curve) and 50°C (red curve).

The desorption of BSA from PNIPAM brushes was carried out by immersing the sample (after BSA adsorption at 50°C) in water at 20°C. Then, the sample was rinsed in water at 20°C and dried with nitrogen gas. Figure 3.6 shows the evolution of the extinction spectra of GND@PNIPAM as a function of desorption time. A progressive blue-shift of the LSP wavelength is observed after immersion of the sample in water at 20°C, due to the desorption of BSA molecules from the polymer brushes, thereby decreasing the refractive index of the surrounding medium around the GNDs. Nevertheless, we can observe that by incubating the sample in water at 20°C for one day, the extinction spectrum could not be restored to the initial spectrum of the GND nanostructures, recorded before BSA adsorption. This observation indicates that some BSA molecules still remain robustly adsorbed to the gold nanostructures or are embedded into the polymer brushes. In fact, the roughness of the metallic nanostructures can provide several sites for protein adsorption.

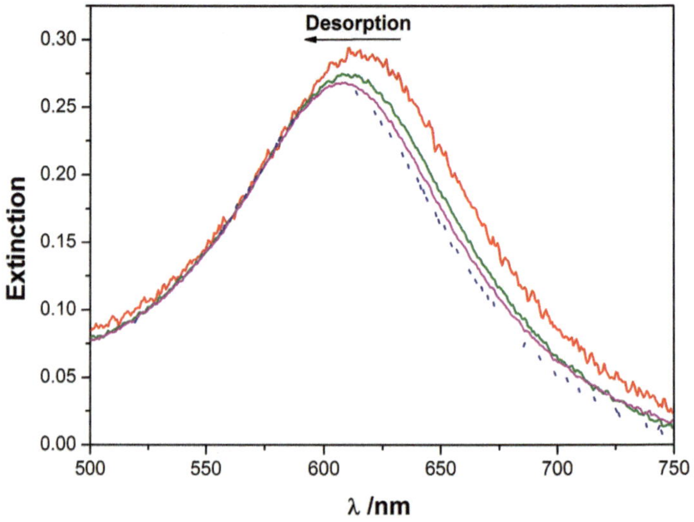

Figure 3.6 Extinction spectra of GND@PNIPAM with adhered BSA protein (red curve); after immersion in water at 20°C, for 10 min (green curve) and 60 min (pink curve); all spectra were recorded in air at room temperature. The pink curve is close to the blue dashed curve that is the extinction spectrum of GND@PNIPAM array after incubation in BSA solution at 20°C.

The reversibility of the thermo-induced adsorption/desorption process of BSA molecules on PNIPAM-coated GNDs was evaluated by performing several heating/cooling cycles. The results are shown in Fig. 3.7. For the first cycle, when the temperature decreased, the LSP wavelength could not be restored to the initial position, before adsorption of BSA, probably because several BSA molecules were trapped as mentioned above. However, from the second cycle, the shift of the LSP wavelength becomes fully reversible and reproducible. Thus, the hybrid PNIPAM-coated GNP structures can be efficiently switched from bio-adherent to bio-repellent by temperature stimuli.

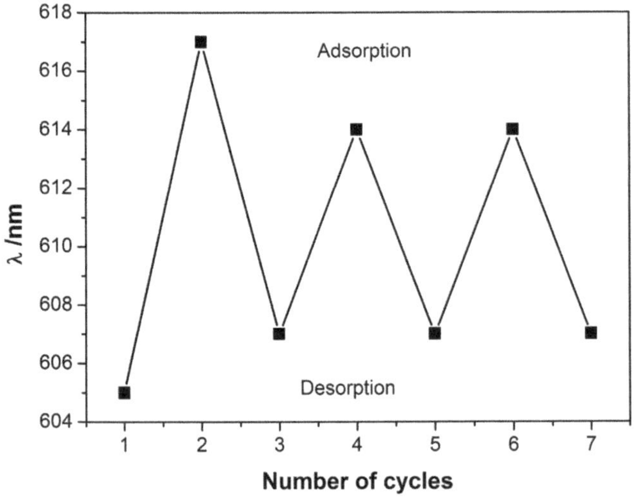

Figure 3.7 Variation of the LSP wavelength of GNP@PNIPAM arrays recorded in air as a function of adsorption/desorption cycles of BSA molecules.

3.4 Conclusion

In conclusion, we described in this chapter the design of a new generation of hybrid plasmonic nanostructures, made of gold nanodots coated by ultra-thin layers of PNIPAM brushes. The temperature-dependent hydrophobic/hybrophilic property of PNIPAM, grafted on the lithographic gold nanostructures, was exploited to switch the adsorption/desorption of BSA proteins.

These thermo-sensitive plasmonic devices are able to efficiently switch from bio-adherent to bio-repellent surface, by convenient temperature stimuli. Moreover, the protein adsorption/desorption process appears to be reversible, opening new perspectives in the field of smart nanomaterials. With the above characteristics, these hybrid nanostructures can be used in biomedical and biological applications such as enzyme immobilization, cell sorting, protein adsorption and purification.

References

1. Hoffman, A. S. (1987). Applications of thermally reversible polymers and hydrogels in therapeutics and diagnostics, *J. Control. Release*, **6,** pp. 297–305.
2. Kikuchi, A., and Okano, T. (2002). Pulsatile drug release control using hydrogels, *Adv. Drug Deliv. Rev.*, **54**, pp. 53–77.
3. Jagur-Grodzinski, J. (2006). Polymers for tissue engineering, medical devices, and regenerative medicine. Concise general review of recent studies, *Polym. Adv. Technol.*, **17**, pp. 395–418.
4. Gupta, P., Vermani, K., and Garg, S. (2002). Hydrogels: From controlled release to pH-responsive drug delivery, *Drug Discov. Today*, **7**, pp. 569–579.
5. Robert Langer, N. A. P. (2003). Advances in biomaterials, drug delivery, and bionanotechnology, *AIChE J.*, **49**, pp. 2990–3006.
6. Xu, F. J. (2004). Surface-active and stimuli-responsive polymer–Si(100) hybrids from surface-initiated atom transfer radical polymerization for control of cell adhesion, *Biomacromolecules*, **5**, pp. 2392–2403.
7. Akiyama, Y. (2004). Ultrathin poly(*n*-isopropylacrylamide) grafted layer on polystyrene surfaces for cell adhesion/detachment control, *Langmuir*, **20**, pp. 5506–5511.
8. Ma, H. H. J., Stiller, P, Chilkoti, A. (2004). "Non-fouling" oligo(ethylene glycol)-functionalized polymer brushes synthesized by surface-initiated atom transfer radical polymerization, *Adv. Mater.*, **16**, pp. 338–341.
9. Gautrot, J. E. (2010). Exploiting the superior protein resistance of polymer brushes to control single cell adhesion and polarisation at the micron scale, *Biomaterials*, **31**, pp. 5030–5041.

10. Azioune, A. (2009). Simple and rapid process for single cell micro-patterning, *Lab on a Chip*, **9**, pp. 1640–1642.

11. Patel, N. G. (2012). Rapid cell sheet detachment using spin-coated pNIPAAm films retained on surfaces by an aminopropyltriethoxysilane network, *Acta Biomater.*, **8**, pp. 2559–2567.

12. Kumashiro, Y. (2013). Modulation of cell adhesion and detachment on thermo-responsive polymeric surfaces through the observation of surface dynamics, *Colloids Surf. B: Biointerfaces*, **106**, pp. 198–207.

13. Singhvi, R., Kumar, A., Lopez, G. P., Stephanopoulos, G. N., Wang, D. I., Whitesides, G. M., and Ingber, D. E. (1994). Engineering cell shape and function, *Science*, **264**, pp. 696–698.

14. Nguyen, M. (2015). Engineering thermoswitchable lithographic hybrid gold nanorods as plasmonic devices for sensing and active plasmonics applications, *ACS Photonics*, **2**, pp. 1199–1208.

15. Nguyen, M. (2015). Tunable electromagnetic coupling in plasmonic nanostructures mediated by thermoresponsive polymer brushes, *Langmuir*, **31**, pp. 12830–12837.

Chapter 4

Reactivity and Bio Samples Probed by Tip-Enhanced Raman Spectroscopy

Zhenglong Zhang,[a] **Robert Meyer,**[b,c] **and Volker Deckert**[b,c]

[a]*School of Physics and Information Technology,*
Shaanxi Normal University, Xi'an, 710062,
People's Republic of China
[b]*Leibniz Institute of Photonic Technology (IPHT),*
Albert-Einstein-Strasse 9, 07745 Jena, Germany
[c]*Institute of Physical Chemistry and Abbe Center of Photonics,*
Friedrich Schiller University, Helmholtzweg 4,
07743 Jena, Germany

Volker.Deckert@leibniz-ipht.de

This chapter aims to provide detailed information of the state-of-the-art in tip-enhanced Raman spectroscopy (TERS) and will focus on its application to induce chemical reactions through plasmonics as well as probing biological processes with enhanced spectroscopy methods.

4.1 Introduction—an Explanation of Tip-Enhanced Raman Spectroscopy

After its discovery in 1974 by Fleischmann et al. [1] and a later correct explanation by van Duyne and Albrecht in 1977 [2, 3],

Plasmonics in Chemistry and Biology
Edited by Marc Lamy de la Chapelle and Nordin Felidj

ISBN 978-981-4800-03-7 (Hardcover), 978-0-429-45875-0 (eBook)
www.jennystanford.com

surface-enhanced Raman scattering (SERS) entered scientific discussions. It is out of question that SERS provides the opportunity to enhance the Raman signal and delivers chemical information of the sample. Interestingly, despite the near-field optical origin of the enhancement, spatial resolution is unaffected and still diffraction limited. In order to solve this challenge, numerous researchers further developed this technique. Wessel first demonstrated the idea of a tip-enhanced Raman scattering (TERS) in 1985 [4]. This approach combines the benefits of surface-enhanced Raman spectroscopy and scanning probe microscopy (SPM) techniques, solving the spatial resolution issue. Utilizing a sharp metallic tip increased drastically the spatial resolution while still maintaining sufficient signal enhancement. It is well known that in 2000 four research groups independently reported the technical feasibility of TERS [5–8]. Thus, TERS is a powerful tool for nanoscale structural analysis for several branches of organic, inorganic, and biological chemistry, since it combines atomic force microscopy (AFM) with Raman spectroscopy, thus a high spatial and spectral characterization can be achieved simultaneously.

4.2 Plasmon-Driven Chemical Reactions

Surface plasmons excited on metal nanostructures have been used to initiate chemical reactions, which are so-called plasmon driven/catalyzed chemical reactions, where the plasmonic nanostructure acts as the catalytic active site [9–11]. The hot electrons, generated from plasmon decay, work as a main catalyst in plasmon driven/catalyzed chemical reactions. Hot electrons can scatter into an excited state of the absorbed molecules triggering a chemical reaction by reducing the activation energy. This is also called hot electron–induced chemical reaction. Hot electron chemistry has drawn great attention from materials, energy, sensing, and catalysis applications, which has a great potential for overcoming many intrinsic limitations, and gains significant attention due to its high throughput and low energy requirements as reported in studies on molecular dimerization and dissociation reactions, e.g., dissociation of hydrogen [12, 13], water [14, 15], and hydrocarbon conversion, etc. [16, 17].

A hot electron–induced chemical reaction can be in situ monitored label-free by Raman spectroscopy, either allowing a model-based band assignment or simply providing a spectroscopic fingerprint of the sample studied. However, Raman spectroscopy generally suffers from very low cross sections ($\sim10^{-27}$–10^{-30} cm^2/sr per molecule), which limits the applications of Raman as a sensing tool to monitor the plasmon-driven chemical reactions where often only minute amounts of sample are presented [18]. Localized surface plasmons at nanoscale edges of noble metal surfaces led to the discovery of surface enhanced Raman scattering (SERS) and tip-enhanced Raman scattering (TERS), both strongly enhancing Raman signals of molecules adsorbed on rough nanostructure surfaces. Here, surface plasmons not only generate hot electrons necessary for plasmon-driven reactions but also enable the monitoring of the chemical reaction using SERS simultaneously.

This part mainly highlights recent breakthroughs and tries to identify emerging fields in heterogeneous catalysis based on hot electron chemistry. Important mechanisms and concepts will be discussed and recent advances on the hot electron–induced chemical reactions in SERS and TERS will be summarized.

4.2.1 Hot Electron–Induced Chemical Reactions

Generation of hot electrons and the respective transfer behavior are crucial for understanding hot electron chemistry applications and achieving practically useful efficiency. Surface plasmons generally decay by either emitting a photon (radiative) or generating an electron-hole pair (non-radiative) through Landau damping [19]. The excited hot electrons have a much larger energy (1–4 eV) than the carriers at the Fermi energy, and quickly diffuse while exchanging their energy with other hot electrons or phonons, producing a Fermi–Dirac distribution and inducing an elevated temperature within 100 fs [20]. Then the electrons are scattered by phonons until the electron and lattice temperatures equilibrate.

In one case, when a semiconductor comes in direct contact with the plasmonic surface, the hot electrons can jump across the Schottky barrier into the conduction band of the semiconductor [21]. On the other hand, when a molecule is adsorbed on the plasmonic surface, hot electrons can be transferred to the molecules lowest unoccupied molecular orbital (LUMO), triggering a chemical reaction by reducing the activation energy [22]. A scheme of the proposed mechanism of plasmon-driven reaction is shown in Fig. 4.1. Hot electrons initially generated from the plasmon decay on the silver or gold nanostructures' surface, soon lose coherence and form a non-equilibrium Fermi–Dirac type distribution. The hot electrons in the high energy level have sufficient energy to transfer to the excited state of molecule A, creating a transient negative ion. This anion of molecule A converts to an excited state of molecule B and after transferring the electron back to the silver or gold nanostructures' surface, returns to the ground state of molecule B, such finishing the overall hot electron–induced reaction of molecule A to B.

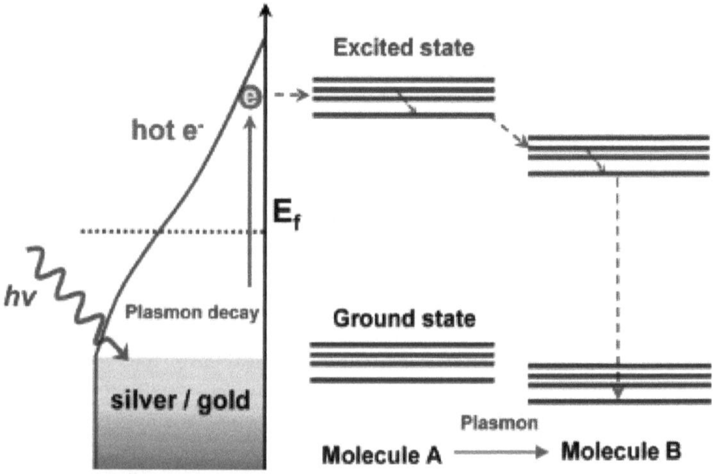

Figure 4.1 Sketch of the hot electron–induced reaction mechanism. Laser excitation of silver or gold nanostructures generates surface plasmons. Due to plasmon decay, the generated hot electrons can be transferred to an excited state of molecule A; from there a driven reaction can be initiated to result in the final product, molecule B [22]. Adapted from [23], published by the Royal Society of Chemistry.

Based on the surface plasmon and induced hot electron principles, plasmon-driven chemical reaction can be tracked at the nanoscale by SERS and TERS. Here we will focus on recent advances in the field of hot electron–induced reactions at the nanoscale, including the early results of plasmon-driven reactions observed by SERS and TERS results. Particularly the latter can be utilized for investigating plasmon-driven reaction with high spatial resolution beyond the diffraction limit.

4.2.2 Plasmon-Driven Chemical Reactions in SERS

Hot electrons arising from surface plasmon decay have a kinetic energy high enough to overcome the reaction activation barrier and can be directly transferred to an unoccupied energy level in the adjacent electron acceptor. As a typical plasmon-driven chemical reaction, p-aminothiophenol (PATP) molecules adsorbed on plasmonic nanostructures react to di-mercaptoazobenzene (DMAB), first reported by Fang, Huang and Capean in 2010 [16, 17, 24]. The plasmon-driven dimerization of PATP to DMAB is characterized by the appearance of new Raman bands around 1140 (β_{C-H}), 1387 ($\nu_{NN} + \nu_{CC} + \nu_{C-N}$) and 1432 ($\nu_{NN} + \nu_{CC} + \beta_{C-H}$) cm^{-1}. This observation can be explained by the formation of the –N=N– unit in DMAB. As shown in Fig. 4.2, the reaction of PATP to DMAB was investigated using different substrate dependent experiments, namely SERS, surface mass spectra, and potential dependent SERS. The results agree well with theoretical calculations utilizing density functional theory. These experiments investigate several aspects to fully understand previous (mis-)interpretations of PATP SERS spectra [25–28].

Similar to the PATP system, another plasmon-driven reaction of p-nitrothiophenol (PNTP) to DMAB was reported by Dong et al. in 2011 [29] The plasmon-driven dimerization of PNTP to DMAB is characterized by the disappearance of the Raman band at 1332 cm^{-1} (ν_{NO2}), and the appearance of new Raman bands of the –N=N– unit in DMAB. This observation explained by a reduction of the NO_2 group and the formation of the –N=N– unit in DMAB.

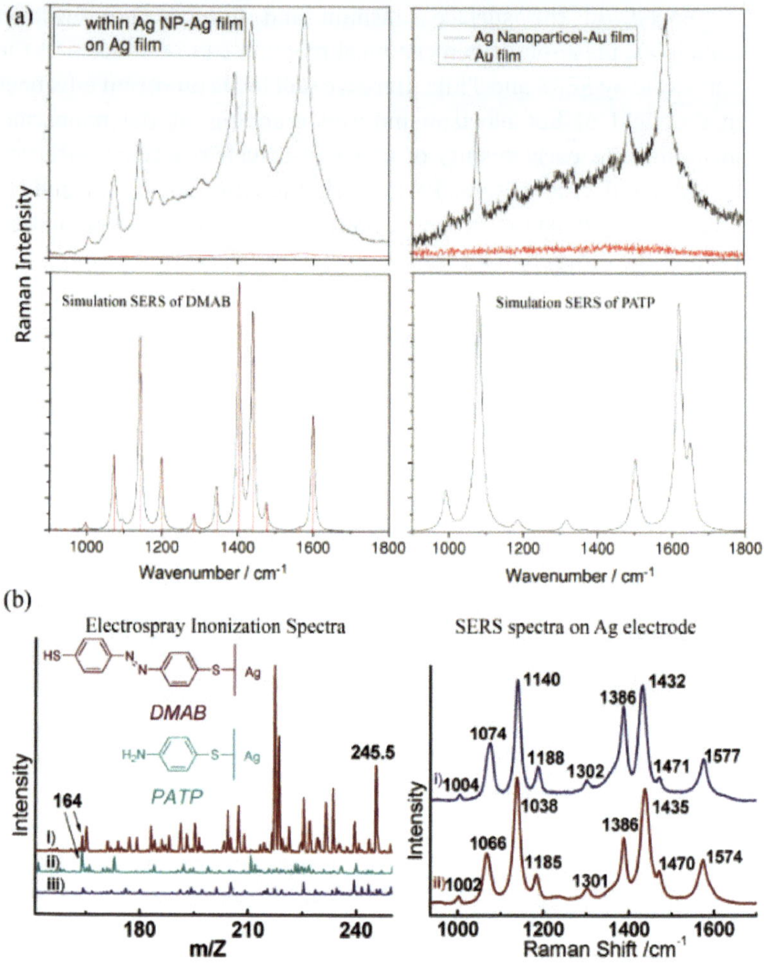

Figure 4.2 Reaction of PATP to DMAB confirmed by SERS, surface mass spectrometry and theoretical calculations. (a) SERS spectra of PATP from nanoparticles on Ag and Au films, which agrees well with theoretical calculations [16]. (b) Surface mass spectra of PATP absorbed on Ag electrodes, illuminated with laser light (i), without any irradiation (ii), and a roughened Ag-free spectrum of PATP (iii), and the SERS spectra of PATP and DMAB on the Ag electrode [17]. Adapted from [16] and [17], published by the Royal Society of Chemistry.

Although the possibility of azobenzene production from aromatic nitro and amine compounds has been reported, the reaction mechanism is unclear. Zhao et al. explained the selective formation of azobenzene derivatives from *p*-substituted nitrobenzenes and anilines using the photoinduced charge transfer model based on time-dependent density functional theory (TD-DFT) methods [30]. The calculations show that the initial reaction occurs via electron transfer from Ag to the LUMO of the adsorbed $-NO_2$ group, and from the HOMO of the adsorbed $-NH_2$ group to the unoccupied energy levels of Ag under visible light irradiation. Subsequently, the *p*-substituted $-NO_2^{-\bullet}$ and $-NO_2^{+\bullet}$ surface species react to form azobenzene derivatives. PATP was converted into DMAB due to the energy transfer from hot electrons to the surface-adsorbed PATP. Under these conditions, oxygen, which acts as an electron acceptor, was essential for the conversion reaction. On the other hand, for a plasmon-driven reaction of PNTP to DMAB, four hot electrons are directly injected from plasmonic nanostructures into the nearby two PNTP molecules, participating in the conversion reactions.

In addition, Xu et al. reported that plasmon-driven reactions can be observed on a single nanoparticle by SERS [31, 32]. Sun and Moskovits reported a remote plasmon-driven reaction of PATP and PNTP to DMAB by Remote-SERS using plasmonic nanowire waveguides [33, 34]. Xie et al. reported another hot electron–induced reaction of PNTP, which reacts to PATP in acidic condition by using plasmonic core-satellite nanoparticles [32]. Zhang, et al. demonstrated that plasmon-driven reactions can even be occur at the single-molecule level, and a new plasmon-driven reaction channel for PNTP was found, which strongly differs from previous reports, as single PNTP exclusively reacts to thiophenol (TP) [23].

4.2.3 Plasmon-Driven Chemical Reaction at the Tip of a Probe

Tip-enhanced Raman scattering is a technique that provides molecular information at the nanoscale [5, 6, 8]. In contrast to SERS, TERS can spatially resolve beyond the diffraction limit and is used for single-molecule sensing [35–40] and bio-molecule detection [41–43], plasmonic catalysis [44–46], and plasmonic gradient effects [47–49]. In TERS, Raman microscopy is combined with SPM, normally including scanning tunneling microscopes (STM) and atomic force microscopes (AFM). In both cases, an ultra-sharp metal SPM tip creates a hot spot to strongly enhance the Raman scattering intensity. The highly localized enhancement provides a high spatial-resolution for optical analysis (<5 nm), which ultimately depends on the nanoparticle material size and shape at the apex of the SPM tip. If the tip and substrate were placed in a high vacuum chamber, so called high vacuum TERS (HV-TERS), a clean environment for TERS study is provided which is particularly suitable for studying model plasmon-driven reactions [50].

Plasmon-driven chemical reactions revealed by TERS were first reported by Lantman and Sun in 2012 [44, 45]. As shown in Fig. 4.3a, a monolayer of PNTP was assembled on a smooth gold nanoplate on glass, and probed with an AFM-TERS system. Excitation and collection of the Raman signals were performed through sample support and sample, and a silver-coated AFM tip was used to generate a strong plasmon intensity to drive PNTP to DMAB conversion. The incident laser was switched between green (532 nm) and red (632.8 nm) to respectively initiate the reaction (green) and monitor (red). Figure 4.3b shows the plasmon-driven reaction of PNTP to DMAB under high vacuum conditions based on a STM-TERS system. The hot electrons generated from the plasmon decay around the gap between gold tip and silver substrate served as the actual agent. Two PNTP molecules localized in the gap between tip and substrate react to a DMAB molecule with the help of hot electrons.

In contrast to the plasmon-driven reaction by SERS, an effective control of the reaction process can be achieved with TERS. Figure 4.4a–e gives incident laser power dependent TERS spectra of the reaction of PNTP to DMAB from a STM-TERS experiment. Initially, the spectrum shows the three main peaks of PNTP at 10 μW. Increasing the laser intensity to 60 μW, no spectral changes were observed. When the laser intensity was raised to 200 μW the intensity of the characteristic band at 1332 cm^{-1} for the $-NO_2$ stretching mode decreased while DMAB bands at 1140, 1387 and 1432 cm^{-1} appeared indicating a gradual reaction. Almost all PNTP molecules dimerized immediately when the laser intensity was increased to 2 mW. Lastly, the spectra did not change when the laser intensity was reduced to the initial 10 μW. Evidently, a series of in situ spectra in Fig. 4.4a were performed in a long-time TERS measurement.

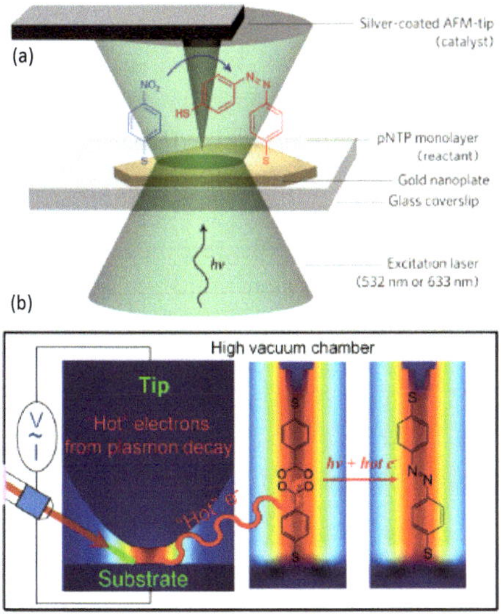

Figure 4.3 Schematic of the plasmon-driven reaction of PNTP in TERS. (a) a monolayer PNTP adsorbed on gold nanoplate on glass was driven to DMAB by TERS based on an AFM system [44]. (b) same plasmon-driven reaction in HV-TERS based on a STM system [45]. Adapted from [44] and Sun, published by the Royal Society of Chemistry.

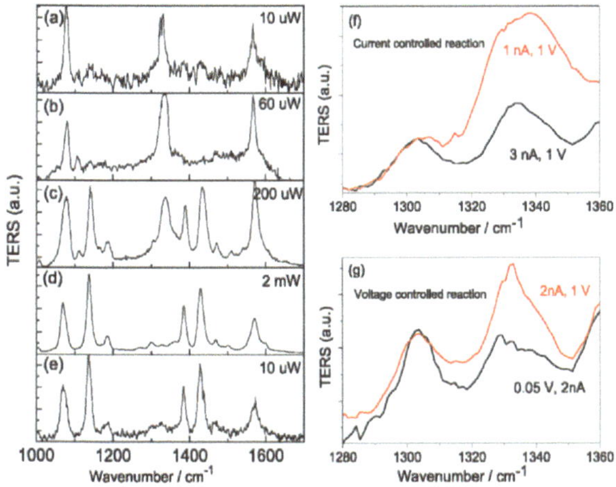

Figure 4.4 Laser intensity and gap distance dependent TERS spectra. Laser intensity controlled plasmon-driven reaction of PNTP to DMAB at a power of (a) 10, (b) 60, (c) 200, (d) 2000 μW, and back to (e) 10 μW; peak of PNTP at 1332 cm^{-1} can decrease when increasing the tunneling current (f) or decreasing the bias voltage (g) [45]. Adapted from 45, published by the Royal Society of Chemistry.

In TERS, a plasmon-driven reaction can be controlled not only by the laser intensity but also by the tip-substrate gap distance. Normally, the decrease of the gap distance can increase the local EM field, and excites a higher surface plasmon intensity at the apex of metallic tip [51]. Similar to the increase of the incident laser intensity, a faster dimerization of PNTP to DMAB can be obtained by decreasing the gap distance. For example in STM-TERS, the gap distance can be decreased when increasing the tunneling current or decreasing the bias voltage. As shown in Fig. 4.4f,g, an increase of the tunneling current of 1 to 3 nA at a constant bias voltage of 1 V, results in a decrease of the gap distance, the marker band at 1332 cm^{-1} decreases, which indicates that more PNTP molecules react to DMAB. Similarly, with a decrease of the bias voltage of 1 to 0.05 V at a constant tunneling current of 2 nA, the peak of PNTP at 1332 cm^{-1} also decreases. It should be noted that normally chemical bonds could be selectively broken or formed by the tunneling electron current. Since decreasing the bias voltage at a constant tunneling current also promotes the plasmon-driven reaction, the contribution of the tunneling current in the experiments can be disregarded.

Due to the high-sensitivity and high spatial-resolution of TERS, the localized temperature can be calculated from the Stokes and anti-Stokes spectra [52, 53]. As shown in Fig. 4.5a, the Stokes and anti-Stokes signals from a STM-TERS experiment were observed at a tunneling current of 2.5 nA and a bias voltage of 1 V. The marker bands of DMAB at 1140, 1387, and 1432 cm^{-1} can also be clearly observed experimentally in Stokes and anti-Stokes TERS spectra, which further confirmed that almost all PNTP reacted to DMAB.

The local reaction temperature can be estimated by fitting the intensities to

$$I_s/I_{as} = a \times e^{(h\nu/k_B T)}, \tag{4.1}$$

where I_s and I_{as} are the intensities of Stokes and anti-Stokes Raman, and h, ν, k_B and T are the Planck constant, frequency, Boltzmann constant, experimental temperature, and a is an experimental constant, which can be written as

$$a = \left(\frac{\sigma(\alpha_\nu)_{\text{Stokes}}}{\sigma(\alpha_\nu)_{\text{anti-Stokes}}} \right) \left(\frac{\nu_0 - \nu}{\nu_0 + \nu} \right)^{-4} \sum_{i=1}^{N} \frac{A^2_{\text{Stokes}}}{A^2_{\text{anti-Stokes}}} \tag{4.2}$$

where $\sigma(\alpha_v)$ are frequency dependent Stokes and anti-Stokes Raman scattering cross section, respectively, ν_0 is the frequency of laser, N is the number of active molecules, and A^2 are the enhancement of Stokes and anti-Stokes Raman scattering in TERS. As the TERS intensity is strongly wavelength dependent and since the temperature is fitted by four peaks at different frequencies (wavelengths), the ratio of the enhancement of Stokes and anti-Stokes at different wavelengths is used to correct the fitting equation. The simulated wavelength dependence of the TERS enhancement is shown in Fig. 4.5b. A strong plasmon resonance with a broad maximum around 633 nm (identical to the incident laser wavelength), and almost no position changes can be found when changing the gap distance from 1 to 1.4 nm. Due to the symmetrical distribution of the plasmon enhancement spectra around 633 nm, the method to estimate the localized experimental temperature is valid for 632.8 nm laser excitation wavelength. The fitted experimental temperature is 330 (±10) K, and the experimental parameter a in equation 2 is 2.06 (±0.19).

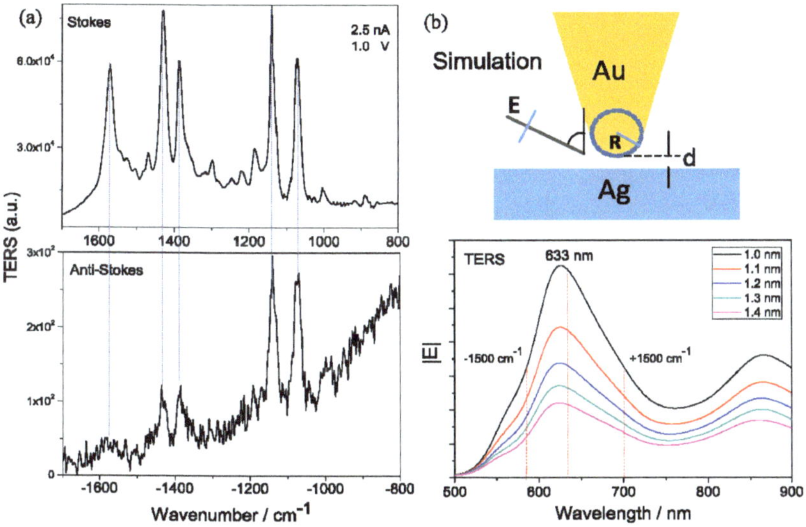

Figure 4.5 (a) Stokes and anti-Stokes TERS spectra at a tunneling current of 2.5 nA and a bias voltage of 1 V, and (b) gap distance dependent simulated spectra of the wavelength scanning plasmon enhancement of TERS, where R is 25 nm and θ = 60° [53]. Adapted from [53], published by the Royal Society of Chemistry.

4.3 Probing Biological Samples

Tip-enhanced Raman spectroscopy on biological samples is still hotly debated. Due to its power to combine a high spatial resolution, as well as the chemical information of the sample, TERS enables the possibility to access and observe biological mechanisms down to nanometer length scale. As mentioned in the introduction, TERS has been reported independently by four research groups in 2000 [5–8]. Therefore, one might think that TERS research for biological samples might only go back in the year 2000, but in 1994 the Morris group studied surface-enhanced Raman scattering in zebrafish embryos by utilizing a coated microelectrode. As a result, they demonstrated that the microelectrode is able to enter biological membranes and restricted volumes [54]. The most interesting fact is the coated microelectrode with its 1–3 μm active tip diameter. Whether intentional or unintentional, this serves as a first hint of the usage of a tip-enhanced Raman approach. With this in mind, the following subchapters will point out the strengths of TERS and discuss experiments from large and complex biological molecules like cells, going down to viruses and bacteria, as well as amino acids, peptide and fibrils, ending up with the smallest structures, like DNA and RNA (see Fig. 4.6).

Figure 4.6 Schematic of tip-enhanced Raman spectroscopy for biological samples. Covering cells, virus and bacteria, amino acids, peptides and fibrils, as well as DNA/RNA.

4.3.1 Human Cells and Its Components

Cells are very complex biological systems and known as weak Raman scatterers. At this point, TERS can demonstrate its strengths. The high spatial resolution offers the opportunity to clearly discriminate the components of the investigated sample volume. Compared to other microscopic techniques like fluorescence microscopy, where the resolution is defined by its focus diameter or the size of the labeled dye molecules, the spectral information in TERS potentially originates only from a few molecules and this label-free.

This potential is used for a great variety of biological samples, beginning with combined measurements on human cells. In 2009, Böhme et al. applied TERS on human dermal derived keratinocyte (HaCaT) and artificial lipid membranes and investigated the correlation of spectral properties of lipids with respect to the cell [55].

It could be found that by induced dynamic structural changes, as well as for nanoscale effects, a spectral feature of the lipid TERS spectra varies. Additionally, a similar spectral deviation appears for the measured TERS cell spectra. This fast detection of structural changes at the nanometer scale, like viewing inside the cell, can be further used for more complex systems. Since lipids in general are crucial elements of a cell and interfere with its environment like, e.g., drugs, pathogens and neighbored cells, TERS provides a label-free investigation of the molecular distribution on the cell membranes, down to the nanoscale.

This can also be comprehended in TERS experiments of supported lipid bilayers (SLBs) with embedded streptavidin molecules [56]. By measuring a spectral map of the biomembranes with a step size below the diffraction limit, the resulting spectra could be classified for lipids, proteins or a mixture of both and successfully correlated with the AFM topography of the membranes (see Fig. 4.7).

The same approach, the label-free characterization of cellular membrane components including spectral mapping, was also applied for human colon cancer cells (cell line HT29) [57]. Additionally, advanced statistical tools, like the N-FINDR algorithm [58] and a principal component analysis (PCA) were utilized to encrypt the huge amount of spectral information. Thus, a successful discrimination of protein- and lipid domains could be provided. Taking into account that the organization of biomembranes is primarily influenced by so-called lipid-rafts, Opilik et al. investigated nanoscale chemical imaging of phase-separated lipid domains [59]. They measured TERS maps of a mixed monolayer (deuterated and non-deuterated lipid domains), analyzing the Raman bands from C–H stretching vibrations. The results further emphasize that TERS can be used for investigations of cell membranes down to nanoscale, since reproducible Raman spectra from only a small number of lipid molecules were obtained.

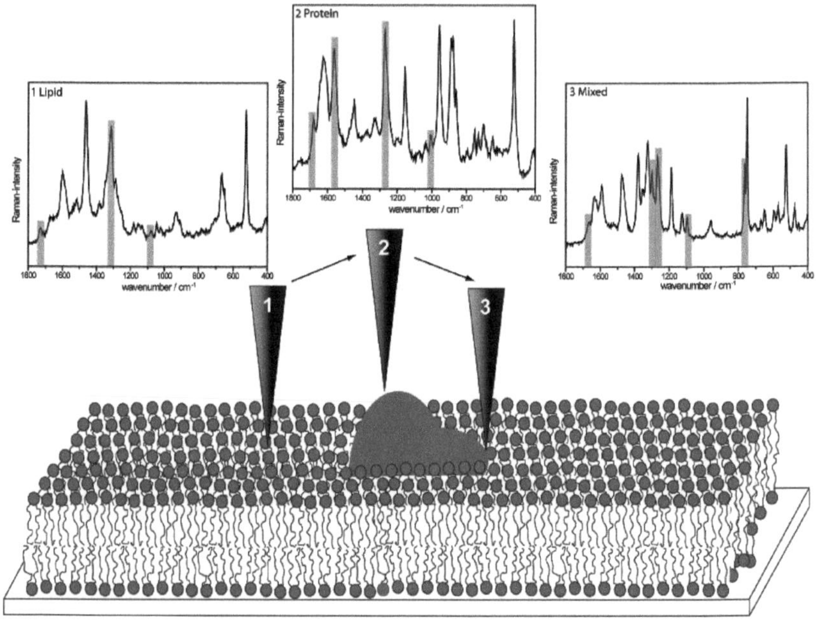

Figure 4.7 Schematic TERS scans. TERS spectra were taken at three positions: (1) lipid domain, (2) protein domain, (3) mixed domain. Adapted from [56]. Copyright Wiley-VCH Verlag GmbH & Co. KGaA. Reproduced with permission.

Another example, highlighting the strengths of TERS to monitor molecular distributions with high spatial resolution and sensitivity in large biomolecules and not being stuck with reporter molecules for fluorescent-based statistical approaches, is provided in investigations of the heme protein cytochrome c. It can be found in the inner membrane of mitochondria. In a first study, Yeo et al. conducted TERS measurements on isolated cytochrome c molecules and were able to detect and distinguish between the heme and the amino acid vibrations [60].

Furthermore, cytochrome c was investigated under increasingly complex conditions. Böhme et al. investigated cytochrome c by TERS in whole yeast cell mitochondria, referring to specific spectral features, which describe the electronic and the actual spin state of the heme moiety [61]. Spectral features are mainly characterized by the central heme moiety. Since the central iron ion can switch between the two different oxidation states Fe(II) and Fe(III), the sensitivity of the signal enables the description of cytochrome c's electronic states. The spectral features show strong lipid contributions and can be classified in three groups. Fe(II) and Fe(III) in the heme center according to their oxidation state markers and at the same time a mixture of both oxidation states. This points out that even very few cytochrome c structures within the mitochondrial membrane system could be detected with highest sensitivity by TERS.

A new different promising method was exploited by the Schultz group in order to image biomolecular interactions on cell and tissue surfaces [62]. Therefore, they used TERS to study antibody conjugated nanoparticles on intact cellular membranes and demonstrated that the observed spectra provide information about the interaction of the specific biomolecule, which is conjugated to the nanoparticle probe. Acting as a rapid method to identify the membrane structure of interest with increased sensitivity and selectivity in intact membranes.

Further investigations with functionalized nanoparticles were also done on intact cell membranes to selectively enhance the amino acid signals of the integrin receptor under TERS conditions [63]. Therefore, a cyclic arginine (R), glycine (G), aspartic acid (D), phenylalanine (f), cysteine (C), abbreviated (c-RGDfC) ligand functionalized nanoparticles were employed, resulting in the successful detection of $\alpha_5\beta_3$ integrins in the cell membranes including the characterization of binding with membrane receptors.

Ligand–receptor interactions are of general interest in biochemistry. Recently, TERS was used for the detection of arginine-glycine-aspartate (RGD) functionalized nanoparticles bound to integrins [64]. Integrins are transmembrane receptors and utilized as integrin-antagonist, like RGD can hinder tumor growth.

TERS results showing a specific Raman signal to the bound protein, in particular the detection and differentiation between the two different integrins ($\alpha_5\beta_1$ and $\alpha_v\beta_3$) bound to the RGD-conjugated gold nanoparticles in intact human colon cancer cell membranes (SW480). This opens a new technology toward the characterization of the membrane receptors chemical structure.

Besides all above-mentioned cell applications, TERS was recently also employed to depict lipid aggregates during stimulated re-epithelialization by applying an in vitro wound healing model [65]. It was the first time that TERS was combined with a sample analysis right after the initial wounding within a timeframe of 24 h for wound healing. Therefore, the human-derived epidermal keratinocyte cell line (HaCaT) was investigated to follow caveolar bulb distributions and fluctuations during the wound healing process. This allows following cell polarization and collaterally a label-free insight to lipid and protein content changes of cells during this complex wound healing process.

Finally, it is worth to mention that TERS has also been applied to blood cells, in particular toward hemozoin crystals in the digestive vacuole of a sectioned malaria parasite-infected cell [66]. Hemozoin is formed by the disposal of hemoglobin by malaria parasite (e.g., *Plasmodium falciparum*). The malaria-infected cell was imaged by AFM, showing that hemozoin crystals protruded from these sections. Measured TERS spectra on crystals could be distinguished from heme spectra and showing a prominent oxidation state marker band, demonstrating that the iron was in high spin state [Fe(III)]. This study demonstrates the detection of hemozoin crystals without any obligated isolation from the affected cells. Thus, representing the potential toward the detection of anti-malaria drugs binding to hemozoin crystals similar to the previous cytochrome c studies. Since this study almost unlocked TERS capability of characterizing malaria parasites and a possible detection for drugs binding on the pathogens, the following chapter will emphasize TERS potency toward a reliable virus and bacteria identification.

4.3.2 Virus and Bacteria

Looking back in history into the 21st century, unfortunately every couple of years the world has to fight against an epidemic.

Starting with the severe acute respiratory syndrome, better known as the SARS coronavirus, led in 2002/2003 to the first big pandemic [67]. Just six years later in 2009, people were confronted with the flu pandemic, commonly known as swine influenza H1N1 [68]. In 2014, which might be still in mind of all of us, the Ebola virus disease in Africa kept the world in suspense, causing more than 10,000 deaths [69]. Unfortunately, it did not take long to the next virus epidemic. Early of 2015, the almost known Zika virus came up again, affecting especially pregnant women and their fetuses. Zika is evident to cause birth defects like microcephaly and other brain anomalies of newborns [70]. The intervals between new and unknown upcoming epidemics are getting even smaller. Therefore, it is essential that quick and reliable analytical methods are employed to prevent the further spread of these diseases.

Conventional Raman spectroscopy could act as a first approach, but the sensitivity is quite poor. Detailed and very sensitive spectral information can only be obtained by using TERS. So, the virus or bacteria identification could be done by means of topography (AFM) plus the spectral information of each sample. The first reported TERS spectra were done in 2006 on bacterial cells, using as sample the Gram positive bacteria *Staphylococcus epidermidis* [71–73]. After AFM imaging of the bacterial cell, several spectra were taken on top and the corner of the cell. The resulting bands could be in assigned according to expected peptides and polysaccharides, fundamental cell components. Additionally, these TERS investigations were pursued by studies of the surface dynamics on the same bacterium [74]. For bacteria, further TERS investigations on *Halobacterium salinarum* cells were conducted, by isolating the purple membrane patches from them [75]. The identification of purple membrane marker bands on bacteriorhodopsin, which originates from the retinal group of bacteriorhodopsin, allowed subsequently the differentiation of retinal-containing, as well as retinal- free membrane areas. Toward Raman visualization, the Sasso group applied TERS mapping for the surface analysis of the *Bacillus subtilis* spore [76]. The observed spectra were evaluated by a principal component analysis and the spores' surface components like proteins and carbohydrates could be mapped, distinguished and finally correlated with AFM phase maps.

Next to an efficient bacteria characterization, fast and reliable single virus particle discriminations down to molecular level with TERS were successfully demonstrated. The very first one published was in 2009, utilizing the tobacco mosaic virus (TMV) as a simple model system and lack of human pathogenicity [77]. The genomic information is codified in single RNA, related with the inner wall of the capsid. After topographical imaging, TERS spectra were recorded. All measured bands could be clearly assigned to the coat proteins as well as to the single RNA, conform to known parts of the TMV.

Apart from the tobacco mosaic virus, TERS investigations were also conducted to detect different virus strains, explicitly the avipoxvirus (APV) and the adeno-associated virus (AAV) [78]. In order to discriminate the two different virus strains, the larger APV (round or elliptical shape, diameter 200–400 nm) and the smaller AAV (icosahedral shape, average diameter 26 nm) were chosen to be identified by means of TERS. For both, APV and AAV spectra were obtained for three different virus particles and prominent bands of amide I and III, as well as CH_2-vibrations were detected. Differences are noticeable in spectral variations. The APV showed clear variations concerning relative peak intensities plus less prominent variations in peak positions. For AAV, this behavior appears less significant compared to APV. An explanation is given due to the different relative size of both virus particles. For smaller ones, only a few spectra are sufficient, whereby for larger particles a TERS spectral map or many averaged spectra are the method of choice for identification.

As interim conclusion from the above-mentioned experiments and the related spectral peak and intensity variations due to highly sensitive TERS measurements, a further data treatment is needed.

A promising method of choice is provided by chemometrics. In particular, the combination of TERS with chemometrics was investigated in order to distinguish two pathogenic viruses, namely the Varicella-zoster virus (VZV) and the Porcinetescho virus (PTV) [79]. After characterizing the virus topography with AFM, TERS measurements on single particle level were performed (see Fig. 4.8).

The obtained spectra were first chemometrically treated to visual discriminate between the two viruses. Followed by an automatic quality rating of the spectra by employing test criteria. If passed, spectra were utilized to calculate a classification model. This method enables a successful discrimination of the two viruses with high accuracy.

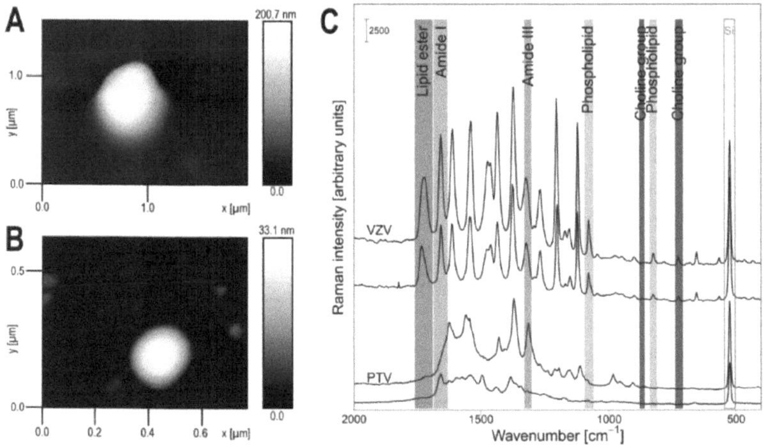

Figure 4.8 Left: AFM topography images of a (A) Varicella-zoster virus (VZV) and of a (B) Porcinetescho virus (PTV) are shown. (C) Two exemplary untreated TERS spectra of each virus. Amide bands (protein markers) are marked in green, lipid bands that only occur in the TERS spectra of VZV are labeled in blue [79]. From [79], published by the Royal Society of Chemistry.

Summarizing this chapter, TERS is a reliable method to identify bacteria and virus particles. For a simple identification and distinction an even lower resolution might be sufficient, but emerging variabilities in high resolved TERS spectra could be further treated with chemometrics, establishing a very precise classification of virus and bacteria strains. Together with a database, current TERS research will proceed as characterization technique in particular for virus particles.

4.3.3 From Amino Acids to Peptides and Fibrils

In the beginning of this chapter, detailed insights in TERS investigations of the building blocks of life—the amino acids—are given. Amino acids contain of an amine group (-NH_2) and a carboxyl group (-COOH) plus a side chain (R-group). At present, approximately 500 amino acids are known. However, there are 23 proteinogenic amino acids, acting as essential building blocks for peptides and proteins. Our human genetic code covers 20 amino acids. For this reason, TERS investigations on selected proteinogenic amino acids are discussed with respect to the single-molecule detection.

As necessary condition for every (TERS) measurement, sample preparation is of vital importance. Since it has been reported that TERS spectra change depending on the orientation of the amino acid with respect to the tip, atomically flat gold nanoplate substrates have been found to achieve a consistent adsorption of amino acids. These plates, showing a roughness of approximately 100–200 pm and a height of 20 nm are essentially atomically flat and thus optically transparent [80]. Due to the so-called gap mode configuration, they further enhance the field between the metal tip and the metal substrate. For TERS experiments of the amino acid histidine (His), silver nanoplates were applied. The experiment revealed that His adsorbed in two preferred states on the nanoplates and the two isomers of His could be identified [81].

In another experiment using gold nanoplates, an essential step toward the distinction of amino acids from proteins in TERS was studied. Thus, the amino acids tyrosine (Tyr), phenylalanine (Phe) and tryptophan (Trp), were utilized. In contrast to Raman and SERS observations, the characteristic ring breathing modes of the aromatic unit of the molecules were only rarely detected in TERS. This behavior could be explained by a parallel orientation of the molecules with respect to the gold layer. Consequently, the main interaction of the molecules between the silver tip and the gold nanoplate takes place via the carboxyl and amino compounds, noticeable in an enhanced TERS spectra of such bands. Furthermore, theoretical calculations confirmed the experimental results in the gap-mode configuration [82].

The molecular orientation in gap-mode configuration could be further investigated by using a small peptide, namely glutathione [83]. Peptides are formed, when two amino acids react. In detail, the acidic carboxyl group of the first amino acid reacts with the alkaline amino group of the second amino acid, consequently forming a peptide and water. Therefore, a long polypeptide chain is forming a protein. To study the orientation of the small peptide, oxidized glutathione (GSSG) adsorbed on gold nanoplates was chosen. GSSG is formed from glutamic acid (Glu), glycine (gly) and cysteine (Cys), which are linked via peptide bonds. The peptide structure could be successfully determined with TERS, indicating a preferred orientation of the molecules on the substrate.

A further experiment aimed to investigate the TERS spectra of peptide bonds. For this reason, dipeptide, tripeptide and a protein were chosen as model systems for Raman, SERS and TERS studies [84]. For the peptides, it could be shown that SERS and TERS spectra are similar to each other, in contrast to normal Raman spectra compared to those from TERS. Differences in the three techniques were experienced in band intensities and number of bands. Additionally, a general guideline for the interpretation of TERS spectra is provided, facilitating the full understanding the spectra in complex biological environment.

Considering proteins, it is noteworthy to mention their structures, because they can give access about mechanisms contributing to several cellular processes. As mentioned above, a long polypeptide chain forms a protein. Its primary structure describes simply the amino acid sequences. The secondary structure gives access to the composition and the spatial arrangement of the protein. A distinction is made in α-helix, β-sheet, β-turn, and β-helix. The first two- and most common structures, revealing a distance of only a few Ångströms between the amino acids in β-sheet and α-helix conformation, subsequently involving a starting point for TERS and its métier to differentiate the chemical composition onto this small scale. Furthermore, the tertiary structure describes the general shape of a single protein molecule, as the quaternary structure is formed by several proteins that have to merge in order to stay functional.

Nevertheless, apart from a correct protein folding, also misfolding plays a crucial role in biochemistry. Misfolding of proteins leads to fiber-like structures and are related to cause Alzheimer's and Parkinson's disease. During a fibrillation process the secondary structure changes in a drastically manner, resulting into a non-active protein, which could even be toxic. Therefore, a detailed understanding of these wrong folding mechanisms, its prevention, as well as the treatment are key points in the development of efficient drugs.

A detailed insight into the proteins secondary structure of amyloid fibrils and the detection of conformational changes demonstrating TERS capability to distinguish selected compounds of fibrils was investigated in several groups [85–89]. Furthermore, a lateral resolution below 2 nm enabled the distinction of the different conformations of the peptide chain, by means of the amid I–III bands and individual contributions of histidine, phenylalanine, proline, tyrosine and cystine could be addressed by their marker bands [75].

A hotly debated story in SERS and TERS experiments is the presence as well as the absence of the amide I band, which is widely used to identify the secondary structure of the proteins.

In 2012, Kurouski et al. [85] detected the amide I band in two regions, either indicating an unordered α-helix structure or - measured on a different location - acting as indicator for a β-sheet conformation.

Nevertheless, apart from this study, the suppression of amide I bands is still prominent in SERS and TERS and was in detail examined by Blum et al. [90] as well as from Kurouski et al. [91]. Blum et al. found an explanation due to the dominance of the aromatic mode and considered next to a too low signal-to-noise ratio, a possible electromagnetic shielding of the protein/peptide backbone by the amino acid side chains. Kurouski et al. experienced that the absence of the band could not be attributed to the experimental setup or an aggregation state of the protein, than more to the spacing of the amino acid side chains between the metal and peptide backbone. This hypothesis could be maintained by SERS spectra of short model peptides, supporting that the bulkiness of the amino acid side chain influences the interaction of enhanced field and the peptide backbone. However, a final answer could not be found yet, but future investigations might enlighten the darkness and uncover the underlying effect.

Apart from this, the further development of TERS setups and its data acquisition could help to give access to such phenomena. One step toward this direction was applied by Paulite et al. with a full spectroscopic imaging of self-assembled peptide nanotapes form amyloid fibril β(16–22) fragments [92]. In detail, the advantage of taking pixel-wise spectra (spectral map) and the simultaneous topography measurement are obvious. It is exposed that even almost invisible structures in the STM topography appear in the mapping image (band reference set on ring breathing mode) and could be clearly discriminated from the surrounding substrate.

The spectral mapping approach was recently used as a powerful tool to identify alternating hydrophobic, hydrophilic, or respective mixtures of nanometer-sized domains on insulin fibrils (see Fig. 4.9) [93].

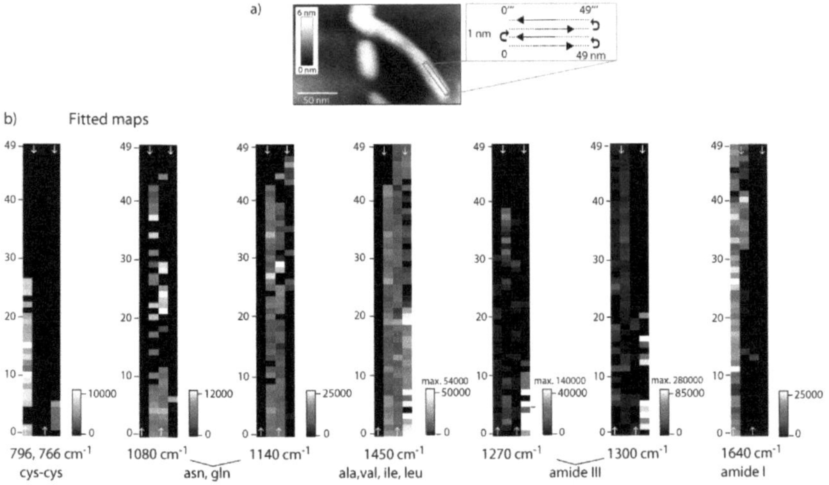

Figure 4.9 AFM topography of a fibril for which 200 spectra on four 1 nm laterally shifted lines were recorded consecutively; each line: 49 nm with 50 equidistant measuring points; (λ = 532 nm, acquisition time = 5 s). (b) Maps generated from the fitted band intensities; pixel brightness corresponds to band intensity [93].

Such properties are crucially connected to their biological activity and can also be associated with the fibrillation mechanisms and consequently the fibrils toxicity. Detecting spectra on the insulin fibrils with a point-to-point distance between 0.5 nm and 1 nm provides access to the secondary structure based on amide III bands. This is unusual, however important, as in TERS and SERS experiments the amide I bands are less reliable markers. Spectral mapping is accompanied with large data set. Therefore, the spectra in this particular study were also processed with the N-FINDR algorithm, resulting in nice agreement maps generated either by direct band fitting or the N-FINDR approach.

4.3.4 DNA and RNA

Discovered in 1953, the molecule that genes are made of—the DNA—was evaluated by Watson and Crick [94]. Deoxyribonucleic acid (DNA) carries all genetic information of all known living organisms and consists of two biopolymer strands, forming the well-known double helix. From a chemical point of view the two strands consist of nucleotides. Each nucleotide involves one of the four nucleobases guanine (G), adenine (A), thymine (T) and cytosine (C), as well as a sugar called deoxyribose and a phosphate group. As mentioned, the two DNA strands are mainly bound together and organized in a helical structure, forming a double-stranded DNA (dsDNA). If the two strands fall apart by denaturation, single-stranded DNA molecules (ssDNA) are created. In comparison to DNA, Ribonucleic acid (RNA) is organized in a single-strand and has an additional OH-group at the sugar (ribose) and thymine (T) is replaced by uracil (U).

For DNA/RNA analysis TERS can benefit from its inherent advantage of ultrahigh resolution, specificity and sensitivity. The first TERS experiment on DNA bases has been published by in 2004 by Watanabe et al. [95]. They reported experiments and quantum chemical calculations based on density-functional theory (DFT) of a single nanocrystal of adenine. In comparison to calculated spectra, several band shifts were observed and related to tip-pressurization effects, explicitly a mechanical deformation of the adenine crystal by contact mode TERS. In ongoing experiments, it could be shown that an adsorption–desorption process in contact-mode TERS might be responsible for the observed band shifts [96].

Furthermore, TERS experiments proceeded in studying the DNA pyrimidine bases cytosine and thymine. All spectra were compared with standard SERS and Raman measurements of the nucleotides and the pure bases to evaluate the influence of near- and far field contributions to the spectra [41]. Additionally, spectra of all DNA bases and nucleotides have been measured, acting as database for TERS experiments on single DNA strands. According single strand sensitivity in TERS, calf thymus strands adsorbed on gold layer were investigated, resulting in a strongly enhanced TERS spectrum [97].

Apart from AFM-TERS, also STM-TERS has to be taken into consideration, providing an excellent measure of the four nucleobases of only a picomole sample amount, which means sub-monolayer coverage. Enabling TERS to access all four nucleobases, set a landmark toward the detection limit for DNA structural analysis via TERS technique.

Additionally, several TERS experiments on pure nucleobases of thymine and cytosine, adenine, and uracil homo-polymers were conducted, demonstrating the reproducibility of this technique [98].

A remarkable step was the detection of hydrogen bonding between the single crystalline DNA base pairs adenine (A) and thymine (T) on a gold surface by Zhang in 2010 [99]. Hydrogen bonds are crucial regarding the base-pairing process in the formation of a DNA double helix. The evidence for hydrogen bonds could be concluded from the presence and absence of vibrational bands. In particular, the ring breathing mode of bound thymine provided evidence for hydrogen bonds because of its change of orientation with respect to pure thymine. Pure thymine only lays flat on the surface and has therefore no ring breathing mode. The strong enhanced in-plane thymine methyl group rocking vibration band proved that thymine was present and bound to adenine via a hydrogen bond. The observations conclude that the base pair was adsorbed in such way that adenine was closer to the gold surface than thymine, since thymine was bound via a hydrogen bond to adenine. Therefore, the base pair must have been oriented in an upright position.

In addition to the analysis of isolated nucleobases, TERS investigations were also performed on RNA and single DNA strands. The initial RNA experiment in 2008 verified TERS' sensitivity again, when measuring along a poly(cytosine) RNA single strand [100]. The obtained spectra along the axis contained the characteristic spectral features of the nucleobase cytosine. In terms of the abovementioned sensitivity and with the knowledge of the tip diameter, a conservative estimation entirely based on the tip diameter concluded that 30–60 molecules were beneath the tip with an estimated SNR of 200. Consequently, the SNR for one molecule was about 3–7, underlining a single base sensitivity in this experiment.

Combed ds-DNA of a λ-phage virus were investigated by Najjar et al. obtaining spectra of DNA nucleobases and backbone, showing phosphodiester and sugar groups [101]. Since the spectra of nucleobases in ds-DNA are matching with those from ss-DNA, additional modes related to the DNA backbone could be experienced in ds-DNA. Hence, single-strand as well as double-strand DNA could be distinguished, allowing the detection of DNA hybridization.

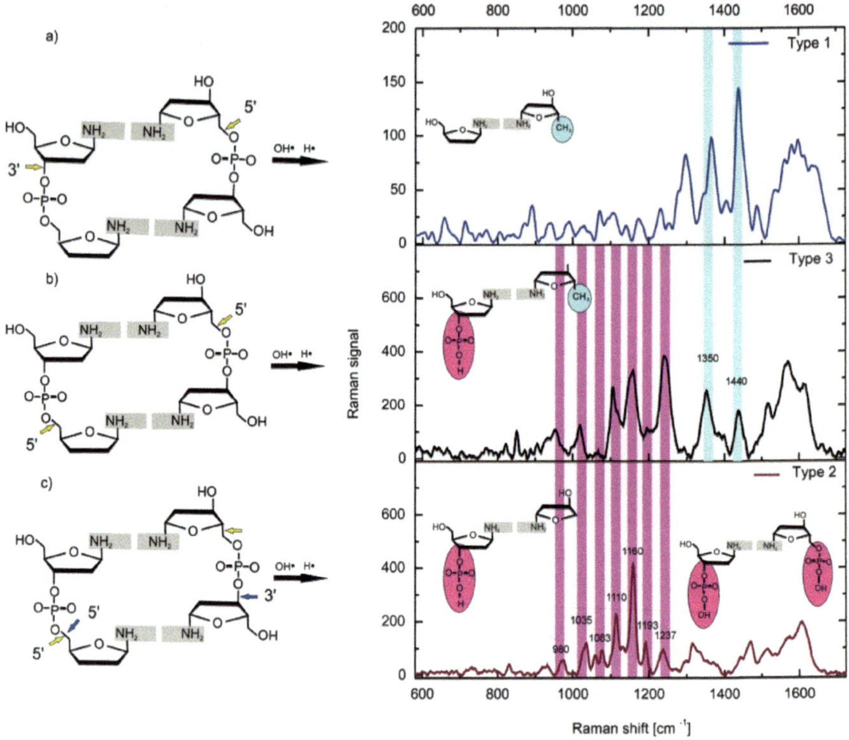

Figure 4.10 Potential cleavage sites and corresponding assigned TERS spectra. (a) Type 1 spectra: showing strong bands corresponding to the CH_3 functional group and do not exhibit the characteristic phosphodiester vibrations. Cleavage occurred at the 5′-bond position in a single DNA strand. (b) Type 3 (mixture of type 1 and type 2) spectra appear to be sum of type 1 and type 2 spectra. Bands from the phosphodiester vibrations and the CH_3 group are observed. (c) Type 2 showing phosphodiester vibrations and strong signal from P-O-H in-plane bending mode confirming that the phosphodiester fragment is hydrogen-terminated at the lesion site. Indicating that 3′-bond cleavage occurred in one strand and 5′-bond cleavage on the second strand (blue arrows). Bands highlighted blue refer to type 1 and type 3 spectra, whereas the pink highlighted bands refer to type 2 and type 3 spectra [102]. Copyright Wiley-VCH Verlag GmbH & Co. KGaA. Reproduced with permission.

Furthermore, the molecular structure of double strand breaks (DSBs) was studied by Lipiec et al. by means of TERS [102]. DSBs are lesions that can lead to genetic defects and cell death. They are caused by the cleavage at the 3′- and 5′-bonds of deoxyribose upon UV exposure. The sample exposed to UV-irradiation was imaged with AFM, showing anomalies in their structure. The control sample kept its supercoiled structure, while the irradiated sample showed a linear shape. Potential bond cleavage sites could be localized in the TERS spectra. Figure 4.10 shows three general types of representative TERS spectra. The irradiated sample revealed intense spectral features, which could be associated to CH_2/CH_3 and phosphodiester vibrations evidencing the cleavage on the 5′-bond position in a single DNA position. A 3′-bond cleavage could be confirmed by phosphodiester vibrations, which are a clear indicator of the phosphate removal.

Recently, TERS ultrahigh resolution capability was examined using ss-DNA. In detail, direct base-to-base transitions in ss-DNA were exposed by TERS [103]. Specially designed ss-DNA consisting of adenine and cytosine sections were utilized as "nanometric rulers" being probed at 0.3 nm equidistant steps under ambient conditions, demonstrating even direct base-to-base variations in the spectra. This high resolution has been supported independently by quantum mechanical modeling of an adenine molecule-silver tip model system and considering the atomistic structure of the actual TERS probe [104–106].

4.4 Conclusion

Almost two decades after the first experiments, the development and applicability of tip-enhanced Raman spectroscopy has been proven to work for physical, chemical, and biological samples. It is a unique technique for monitoring plasmon-driven chemical reactions. In situ plasmon-driven processes can be monitored by time lapse or intensity dependent measurements with extremely high spatial resolution. The reaction speed can be controlled by wavelength, intensity and, a unique feature, by the gap distance between tip and substrate. Intrinsically, the TERS spectra and the topographical imaging are obtained simultaneously, which provides a powerful tool to investigate driven dynamics down to the single-molecule level.

For biochemical aspects, TERS offers due to its high spatial resolution in combination with maximum sensitivity, access to nanoscale phenomena on length scales from complex system like cells, viruses and bacteria, down to amino acids and proteins and on top to our carrier of genetic code—the DNA. Not only for characterization issues, but also to push the development of new therapeutic products to solve and improve medical treatments, TERS became a appreciated method of choice due to its label-free and non-destructive approach. Future investigations will focus on nanoscale structural dynamics, taking advantage of TERS in combination with highly time-resolved spectroscopic methods.

References

1. Fleischmann, M., Hendra, P. J., McQuillan, A. J. (1974). Raman-spectra of pyridine adsorbed at a silver electrode, *Chem. Phys. Lett.*, **26**(2), 163–166.
2. Jeanmaire, D. L., Vanduyne, R. P. (1977). Surface Raman spectroelectrochemistry. 1. Heterocyclic, aromatic, and aliphatic-amines adsorbed on anodized silver electrode, *J. Electroanal. Chem.*, 84(1), 1–20.
3. Albrecht, M. G., Creighton, J. A. (1977). Anomalously intense Raman-spectra of pyridine at a silver electrode, *J. Am. Chem. Soc.*, **99**(15), 5215–5217.
4. Wessel, J. (1985). Surface-enhanced optical microscopy, *J. Opt. Soc. Am. B*, **2**(9), 1538–1541.
5. Anderson, M. S. (2000). Locally enhanced Raman spectroscopy with an atomic force microscope, *Appl. Phys. Lett.*, 76(21), 3130–3132.
6. Hayazawa, N., Inouye, Y., Sekkat, Z., Kawata, S. (2000). Metallized tip amplification of near-field Raman scattering, *Opt. Commun.*, 183(1–4), 333–336.
7. Pettinger, B., Picardi, G., Schuster, R., Ertl, G. (2000). Surface enhanced Raman spectroscopy: Towards single moleculer spectroscopy (E), *Electrochem. Tokyo*, 68(12), 942–949.
8. Stockle, R. M., Suh, Y. D., Deckert, V., Zenobi, R. (2000). Nanoscale chemical analysis by tip-enhanced Raman spectroscopy, *Chem. Phys. Lett.*, 318(1–3), 131–136.

9. Sun, M., Xu, H. (2012). A novel application of plasmonics: Plasmon-driven surface-catalyzed reactions, *Small*, 8(18), 2777–2786.

10. Ueno, K., Misawa, H. (2013). Surface plasmon-enhanced photochemical reactions, *J. Photoch. Photobio. C*, 15, 31–52.

11. Xiao, M. D., Jiang, R. B., Wang, F., Fang, C. H., Wang, J. F., Yu, J. C. (2013). Plasmon-enhanced chemical reactions, *J. Mater. Chem. A*, 1(19), 5790–5805.

12. Mukherjee, S., Libisch, F., Large, N., et al. (2013). Hot electrons do the impossible: Plasmon-induced dissociation of H_2 on Au, *Nano Lett.*, 13(1), 240–247.

13. Mukherjee, S., Zhou, L., Goodman, A. M., et al. (2014). Hot-electron-induced dissociation of H_2 on gold nanoparticles supported on SiO_2, *J. Am. Chem. Soc.*, 136(1), 64–67.

14. Lee, J., Mubeen, S., Ji, X., Stucky, G. D., Moskovits, M. (2012). Plasmonic photoanodes for solar water splitting with visible light, *Nano Lett.*, 12(9), 5014–5019.

15. Chen, H. M., Chen, C. K., Chen, C. J., et al. (2012). Plasmon inducing effects for enhanced photoelectrochemical water splitting: X-ray absorption approach to electronic structures, *ACS Nano*, 6(8), 7362–7372.

16. Fang, Y., Li, Y., Xu, H., Sun, M. (2010). Ascertaining p,p′-dimercaptoazobenzene produced from p-aminothiophenol by selective catalytic coupling reaction on silver nanoparticles, *Langmuir ACS J. Surf. Colloids*, 26(11), 7737–7746.

17. Huang, Y. F., Zhu, H. P., Liu, G. K., Wu, D. Y., Ren, B., Tian, Z. Q. (2010). When the signal is not from the original molecule to be detected: Chemical transformation of para-aminothiophenol on Ag during the SERS measurement, *J. Am. Chem. Soc.*, 132(27), 9244–9246.

18. Moskovits, M. (2005). Surface-enhanced Raman spectroscopy: A brief retrospective, *J. Raman Spectrosc.*, 36(6–7), 485–496.

19. Narang, P., Sundararaman, R., Atwater, H. A. (2016). Plasmonic hot carrier dynamics in solid-state and chemical systems for energy conversion, *Nanophotonics Berlin.*, 5(1), 96–111.

20. Brongersma, M. L., Halas, N. J., Nordlander, P. (2015). Plasmon-induced hot carrier science and technology, *Nat. Nanotechnol.*, 10(1), 25–34.

21. Knight, M. W., Sobhani, H., Nordlander, P., Halas, N. J. (2011). Photodetection with active optical antennas, *Science*, 332(6030), 702–704.

22. Zhang, Z., Deckert-Gaudig, T., Deckert, V. (2015). Label-free monitoring of plasmonic catalysis on the nanoscale, *Analyst*, 140(13), 4325–4335.

23. Zhang, Z., Deckert-Gaudig, T., Singh, P., Deckert, V. (2015). Single molecule level plasmonic catalysis—a dilution study of p-nitrothiophenol on gold dimers, *Chem. Commun.*, 51(15), 3069–3072.

24. Canpean, V., Iosin, M., Astilean, S. (2010). Disentangling SERS signals from two molecular species: A new evidence for the production of p,p′-dimercaptoazobenzene by catalytic coupling reaction of p-aminothiophenol on metallic nanostructures, *Chem. Phys. Lett.*, 500(4–6), 277–282.

25. Osawa, M., Matsuda, N., Yoshii, K., Uchida, I. (1994). Charge-transfer resonance raman process in surface-enhanced Raman-scattering from P-aminothiophenol adsorbed on silver-Herzberg-Teller contribution, *J. Phys. Chem. US*, 98(48), 12702–12707.

26. Zhou, Q., Li, X., Fan, Q., Zhang, X., Zheng, J. (2006). Charge transfer between metal nanoparticles interconnected with a functionalized molecule probed by surface-enhanced Raman spectroscopy, *Angew. Chem. Int. Edit.*, 45(24), 3970–3973.

27. Maniu, D., Chis, V., Baia, M., Toderas, F., Astilean, S. (2007). Density functional theory investigation of p-aminothiophenol molecules adsorbed on gold nanoparticles, *J. Optoelectron Adv. M.*, 9(3), 733–736.

28. Richter, A. P., Lombardi, J. R., Zhao, B. (2010). Size and wavelength dependence of the charge-transfer contributions to surface-enhanced Raman spectroscopy in Ag/PATP/ZnO junctions, *J. Phys. Chem. C*, 114(3), 1610–1614.

29. Dong, B., Fang, Y., Chen, X., Xu, H., Sun, M. (2011). Substrate-, wavelength-, and time-dependent plasmon-assisted surface catalysis reaction of 4-nitrobenzenethiol dimerizing to p,p′-dimercaptoazobenzene on Au, Ag, and Cu films, *Langmuir ACS J. Surf. Colloids*, 27(17), 10677–10682.

30. Zhao, L. B., Huang, Y. F., Liu, X. M., et al. (2012). A DFT study on photoinduced surface catalytic coupling reactions on nanostructured silver: Selective formation of azobenzene derivatives from para-substituted nitrobenzene and aniline, *Phys. Chem. Chem. Phys.*, 14(37), 12919–12929.

31. Kang, L., Xu, P., Zhang, B., Tsai, H., Han, X., Wang, H. L. (2013). Laser wavelength- and power-dependent plasmon-driven chemical

reactions monitored using single particle surface enhanced Raman spectroscopy, *Chem. Commun.*, 49(33), 3389–3391.

32. Xie, W., Walkenfort, B., Schlucker, S. (2013). Label-free SERS monitoring of chemical reactions catalyzed by small gold nanoparticles using 3D plasmonic superstructures, *J. Am. Chem. Soc.*, 135(5), 1657–1660.
33. Sun, M. T., Hou, Y. X., Li, Z. P., Liu, L. W., Xu, H. X. (2011). Remote excitation polarization-dependent surface photochemical reaction by plasmonic Waveguide, *Plasmonics*, 6(4), 681–687.
34. Lee, S. J., Moskovits, M. (2012). Remote sensing by plasmonic transport, *J. Am. Chem. Soc.*, 134(28), 11384–11387.
35. Liu, Z., Ding, S. Y., Chen, Z. B., et al. (2011). Revealing the molecular structure of single-molecule junctions in different conductance states by fishing-mode tip-enhanced Raman spectroscopy, *Nat. Commun.*, 2, 305.
36. Sonntag, M. D., Chulhai, D., Seideman, T., Jensen, L., Van Duyne, R. P. (2013). The origin of relative intensity fluctuations in single-molecule tip-enhanced Raman spectroscopy, *J. Am. Chem. Soc.*, 135(45), 17187–17192.
37. Deckert-Gaudig, T., Kammer, E., Deckert, V. (2012). Tracking of nanoscale structural variations on a single amyloid fibril with tip-enhanced Raman scattering, *J. Biophotonics.*, 5(3), 215–219.
38. Zhang, R., Zhang, Y., Dong, Z. C., et al. (2013). Chemical mapping of a single molecule by plasmon-enhanced Raman scattering, *Nature*, 498(7452), 82–86.
39. Zhang, W. H., Yeo, B. S., Schmid, T., Zenobi, R. (2007). Single molecule tip-enhanced Raman spectroscopy with silver tips, *J. Phys. Chem. C*, 111(4), 1733–1738.
40. Pettinger, B. (2010). Single-molecule surface- and tip-enhanced raman spectroscopy, *Mol. Phys.*, 108(16), 2039–2059.
41. Rasmussen, A., Deckert, V. (2006). Surface- and tip-enhanced Raman scattering of DNA components, *J. Raman Spectrosc.*, 37(1–3), 311–317.
42. Deckert-Gaudig, T., Deckert, V. (2010). Tip-enhanced Raman scattering (TERS) and high-resolution bio nano-analysis—a comparison, *Phys. Chem. Chem. Phys.*, 12(38), 12040–12049.
43. Treffer, R., Bohme, R., Deckert-Gaudig, T., et al. (2012). Advances in TERS (tip-enhanced Raman scattering) for biochemical applications, *Biochem. Soc. T*, 40(4), 609–614.

44. van Schrojenstein Lantman, E. M., Deckert-Gaudig, T., Mank, A. J., Deckert, V., Weckhuysen, B. M. (2012). Catalytic processes monitored at the nanoscale with tip-enhanced Raman spectroscopy, *Nat. Nanotechnol.*, 7(9), 583–586.

45. Sun, M., Zhang, Z., Zheng, H., Xu, H. (2012). In-situ plasmon-driven chemical reactions revealed by high vacuum tip-enhanced Raman spectroscopy, *Sci. Rep. UK*, 2, 647.

46. Zhang, Z., Sheng, S., Zheng, H., Xu, H., Sun, M. (2014). Molecular resonant dissociation of surface-adsorbed molecules by plasmonic nanoscissors, *Nanoscale*, 6(9), 4903–4908.

47. Sun, M. T., Zhang, Z. L., Wang, P. J., Li, Q., Ma, F. C., Xu, H. X. (2013). Remotely excited Raman optical activity using chiral plasmon propagation in Ag nanowires, *Light Sci. Appl.*, 2.

48. Sun, M. T., Zhang, Z. L., Chen, L., Sheng, S. X., Xu, H. X. (2014). Plasmonic gradient effects on high vacuum tip-enhanced Raman spectroscopy, *Adv. Opt. Mater.*, 2(1), 74–80.

49. Sun, M., Fang, Y., Zhang, Z., Xu, H. (2013). Activated vibrational modes and Fermi resonance in tip-enhanced Raman spectroscopy, *Phys. Rev. E*, 87(2), 020401.

50. Zhang, Z.-L., Chen, L., Sheng, S.-X., et al. (2013). High-vacuum tip enhanced Raman spectroscopy, *Front Phys. Beijing.*, 9(1), 17–24.

51. Sun, M., Fang, Y., Yang, Z., Xu, H. (2009). Chemical and electromagnetic mechanisms of tip-enhanced Raman scattering, *Phys. Chem. Chem. Phys.*, 11(41), 9412–9419.

52. Zhang, Z. L., Tian, X. R., Zheng, H. R., Xu, H. X., Sun, M. T. (2013). Tip-enhanced ultrasensitive stokes and anti-stokes Raman spectroscopy in high vacuum, *Plasmonics*, 8(2), 523–527.

53. Zhang, Z., Chen, L., Sun, M., Ruan, P., Zheng, H., Xu, H. (2013). Insights into the nature of plasmon-driven catalytic reactions revealed by HV-TERS, *Nanoscale*, 5(8), 3249–3252.

54. Todd, E. A., Morris, M. D. (1994). Micron surface-enhanced Raman-spectroscopy of intact biological organisms and model systems, *Appl. Spectrosc.*, 48(5), 545–548.

55. Bohme, R., Richter, M., Cialla, D., Rosch, P., Deckert, V., Popp, J. (2009). Towards a specific characterisation of components on a cell surface-combined TERS-investigations of lipids and human cells, *J. Raman Spectrosc.*, 40(10), 1452–1457.

56. Bohme, R., Cialla, D., Richter, M., Rosch, P., Popp, J., Deckert, V. (2010). Biochemical imaging below the diffraction limit–probing cellular

membrane related structures by tip-enhanced Raman spectroscopy (TERS), *J. Biophotonics.*, 3(7), 455–461.

57. Richter, M., Hedegaard, M., Deckert-Gaudig, T., Lampen, P., Deckert, V. (2011). Laterally resolved and direct spectroscopic evidence of nanometer-sized lipid and protein domains on a single cell, *Small*, 7(2), 209–214.
58. Winter, M. E. (1999). N-FINDR: An algorithm for fast autonomous spectral end-member determination in hyperspectral data.
59. Opilik, L., Bauer, T., Schmid, T., Stadler, J., Zenobi, R. (2011). Nanoscale chemical imaging of segregated lipid domains using tip-enhanced Raman spectroscopy, *Phys. Chem. Chem. Phys.*, 13(21), 9978–9981.
60. Yeo, B.-S., Mädler, S., Schmid, T., Zhang, W., Zenobi, R. (2008). Tip-enhanced Raman spectroscopy can see more: The case of cytochrome c, *J. Phys. Chem. C*, 112(13), 4867–4873.
61. Bohme, R., Mkandawire, M., Krause-Buchholz, U., et al. (2011). Characterizing cytochrome c states--TERS studies of whole mitochondria, *Chem. Commun.*, 47(41), 11453–11455.
62. Alexander, K. D., Schultz, Z. D. (2012). Tip-enhanced Raman detection of antibody conjugated nanoparticles on cellular membranes, *Anal. Chem.*, 84(17), 7408–7414.
63. Wang, H., Schultz, Z. D. (2014). TERS detection of alphaVbeta3 integrins in intact cell membranes, *Chemphyschem*, 15(18), 3944–3949.
64. Xiao, L., Wang, H., Schultz, Z. D. (2016). Selective detection of RGD-integrin binding in cancer cells using tip enhanced Raman scattering microscopy, *Anal. Chem.*, 88(12), 6547–6553.
65. Watkins-Mariani, M., Deckert-Gaudig, T., Deckert, V. (2014). Label-free in vitro visualization and characterization of caveolar bulbs during stimulated re-epithelialization, *Anal. Bioanal. Chem.*, 406(27), 6993–7002.
66. Wood, B. R., Bailo, E., Khiavi, M. A., et al. (2011). Tip-enhanced Raman scattering (TERS) from hemozoin crystals within a sectioned erythrocyte, *Nano Lett.*, 11(5), 1868–1873.
67. Fleck, F. (2003). How SARS changed the world in less than six months, *B World Health Organ*, 81(8), 625–626.
68. World Health Organization. Pandemic (H1N1) 2009. 2016; http://www.who.int/csr/disease/swineflu/en/.

69. World Health Organization. Ebola virus disease.

70. World Health Organization. Zika virus.

71. Neugebauer, U., Rosch, P., Schmitt, M., et al. (2006). On the way to nanometer-sized information of the bacterial surface by tip-enhanced Raman spectroscopy, *Chemphyschem*, 7(7), 1428–1430.

72. Neugebauer, U., Schmid, U., Baumann, K., et al. (2007). Towards a detailed understanding of bacterial metabolism–spectroscopic characterization of Staphylococcus epidermidis, *Chemphyschem*, 8(1), 124–137.

73. Budich, C., Neugebauer, U., Popp, J., Deckert, V. (2008). Cell wall investigations utilizing tip-enhanced Raman scattering, *J. Microsc.*, 229(Pt 3), 533–539.

74. Neugebauer, U., Schmid, U., Baumann, K., et al. (2007). The influence of fluoroquinolone drugs on the bacterial growth of S. epidermidis utilizing the unique potential of vibrational spectroscopy, *J. Phys. Chem. A*, 111(15), 2898–2906.

75. Deckert-Gaudig, T., Bohme, R., Freier, E., et al. (2012). Nanoscale distinction of membrane patches—a TERS study of Halobacterium salinarum, *J. Biophotonics*, 5(7), 582–591.

76. Rusciano, G., Zito, G., Isticato, R., et al. (2014). Nanoscale chemical imaging of Bacillus subtilis spores by combining tip-enhanced Raman scattering and advanced statistical tools, *ACS Nano*, 8(12), 12300–12309.

77. Cialla, D., Deckert-Gaudig, T., Budich, C., et al. (2009). Raman to the limit: Tip-enhanced Raman spectroscopic investigations of a single tobacco mosaic virus, *J. Raman Spectrosc.*, 40(3), 240–243.

78. Hermann, P., Hermelink, A., Lausch, V., et al. (2011). Evaluation of tip-enhanced Raman spectroscopy for characterizing different virus strains, *Analyst*, 136(6), 1148–1152.

79. Olschewski, K., Kammer, E., Stockel, S., et al. (2015). A manual and an automatic TERS based virus discrimination, *Nanoscale*, 7(10), 4545–4552.

80. Deckert-Gaudig, T., Deckert, V. (2009). Ultraflat transparent gold nanoplates—ideal substrates for tip-enhanced Raman scattering experiments, *Small*, 5(4), 432–436.

81. Deckert-Gaudig, T., Deckert, V. (2009). Tip-enhanced Raman scattering studies of histidine on novel silver substrates, *J. Raman Spectrosc.*, 40(10), 1446–1451.

82. Deckert-Gaudig, T., Rauls, E., Deckert, V. (2010). Aromatic amino acid monolayers sandwiched between gold and silver: A combined tip-enhanced Raman and theoretical approach, *J. Phys. Chem. C*, 114(16), 7412–7420.

83. Deckert-Gaudig, T., Bailo, E., Deckert, V. (2009). Tip-enhanced Raman scattering (TERS) of oxidised glutathione on an ultraflat gold nanoplate, *Phys. Chem. Chem. Phys.*, 11(34), 7360–7362.

84. Blum, C., Schmid, T., Opilik, L., Weidmann, S., Fagerer, S. R., Zenobi, R. (2012). Understanding tip-enhanced Raman spectra of biological molecules: A combined Raman, SERS and TERS study, *J. Raman Spectrosc.*, 43(12), 1895–1904.

85. Kurouski, D., Deckert-Gaudig, T., Deckert, V., Lednev, I. K. (2012). Structure and composition of insulin fibril surfaces probed by TERS, *J. Am. Chem. Soc.*, 134(32), 13323–13329.

86. Kurouski, D., Deckert-Gaudig, T., Deckert, V., Lednev, I. K. (2014). Surface characterization of insulin protofilaments and fibril polymorphs using tip-enhanced Raman spectroscopy (TERS), *Biophys. J.*, 106(1), 263–271.

87. Gullekson, C., Lucas, L., Hewitt, K., Kreplak, L. (2011). Surface-sensitive Raman spectroscopy of collagen I fibrils, *Biophys. J.*, 100(7), 1837–1845.

88. Krasnoslobodtsev, A. V., Deckert-Gaudig, T., Zhang, Y., Deckert, V., Lyubchenko, Y. L. (2016). Polymorphism of amyloid fibrils formed by a peptide from the yeast prion protein Sup35: AFM and tip-enhanced Raman scattering studies, *Ultramicroscopy*, 165, 26–33.

89. VandenAkker, C. C., Schleeger, M., Bruinen, A. L., et al. (2016). Multimodal spectroscopic study of amyloid fibril polymorphism, *J. Phys. Chem. B*, 120(34), 8809–8817.

90. Blum, C., Schmid, T., Opilik, L., Metanis, N., Weidmann, S., Zenobi, R. (2012). Missing amide i mode in gap-mode tip-enhanced Raman spectra of proteins, *J. Phys. Chem. C*, 116(43), 23061–23066.

91. Kurouski, D., Postiglione, T., Deckert-Gaudig, T., Deckert, V., Lednev, I. K. (2013). Amide I vibrational mode suppression in surface (SERS) and tip (TERS) enhanced Raman spectra of protein specimens, *Analyst*, 138(6), 1665–1673.

92. Paulite, M., Blum, C., Schmid, T., et al. (2013). Full spectroscopic tip-enhanced Raman imaging of single nanotapes formed from beta-amyloid(1–40) peptide fragments, *ACS Nano*, 7(2), 911–920.

93. Deckert-Gaudig, T., Kurouski, D., Hedegaard, M. A., Singh, P., Lednev, I. K., Deckert, V. (2016). Spatially resolved spectroscopic differentiation of hydrophilic and hydrophobic domains on individual insulin amyloid fibrils, *Sci. Rep. UK*, 6, 33575.
94. Watson, J. D., Crick, F. H. (1953). Molecular structure of nucleic acids; a structure for deoxyribose nucleic acid, *Nature*, 171(4356), 737–738.
95. Watanabe, H., Ishida, Y., Hayazawa, N., Inouye, Y., Kawata, S. (2004). Tip-enhanced near-field Raman analysis of tip-pressurized adenine molecule, *Phys. Rev. B*, 69(15), 155418.
96. Ichimura, T., Watanabe, H., Morita, Y., Verma, P., Kawata, S., Inouye, Y. (2007). Temporal fluctuation of tip-enhanced Raman spectra of adenine molecules, *J. Phys. Chem. C*, 111(26), 9460–9464.
97. Hennemann, L. E., Meixner, A. J., Zhang, D. (2010). Surface- and tip-enhanced Raman spectroscopy of DNA, *Spectrosc. Int. J.*, 24(1–2), 119–124.
98. Treffer, R., Lin, X., Bailo, E., Deckert-Gaudig, T., Deckert, V. (2011). Distinction of nucleobases—a tip-enhanced Raman approach, *Beilstein J. Nanotechnol.*, 2, 628–637.
99. Zhang, D., Domke, K. F., Pettinger, B. (2010). Tip-enhanced Raman spectroscopic studies of the hydrogen bonding between adenine and thymine adsorbed on Au (111), *Chemphyschem*, 11(8), 1662–1665.
100. Bailo, E., Deckert, V. (2008). Tip-enhanced Raman spectroscopy of single RNA strands: Towards a novel direct-sequencing method, *Angew. Chem. Int. Edit.*, 47(9), 1658–1661.
101. Najjar, S., Talaga, D., Schue, L., et al. (2014). Tip-enhanced Raman spectroscopy of combed double-stranded DNA bundles, *J. Phys. Chem. C*, 118(2), 1174–1181.
102. Lipiec, E., Sekine, R., Bielecki, J., Kwiatek, W. M., Wood, B. R. (2014). Molecular characterization of DNA double strand breaks with tip-enhanced Raman scattering, *Angew. Chem. Int. Edit.*, 53(1), 169–172.
103. Lin, X.-M., Deckert-Gaudig, T., Singh, P., et al. (2016). Direct base-to-base transitions in ssDNA revealed by tip-enhanced Raman scattering, *ArXiv e-Prints*, 1604. http://adsabs.harvard.edu/abs/2016arXiv160406598L. Accessed April 1, 2016.

104. Latorre, F., Kupfer, S., Bocklitz, T., et al. (2016). Spatial resolution of tip-enhanced Raman spectroscopy—DFT assessment of the chemical effect, *Nanoscale*, 8(19), 10229–10239.

105. Barbry, M., Koval, P., Marchesin, F., et al. (2015). Atomistic near-field nanoplasmonics: Reaching atomic-scale resolution in nanooptics, *Nano Lett.*, 15(5), 3410–3419.

106. Trautmann, S., Aizpurua, J., Gotz, I., et al. (2017). A classical description of subnanometer resolution by atomic features in metallic structures, *Nanoscale*, 9(1), 391–401.

Chapter 5

Surface-Enhanced Spectro-Electrochemistry of Biological and Molecular Catalysts on Plasmonic Electrodes

Patrycja Kielb and Inez M. Weidinger

Faculty of Science, Department of Chemistry/Electrochemistry, Technische Universität Dresden, 01062 Dresden, Germany

inez.weidinger@tu-dresden.de

This chapter gives an overview on how plasmonic noble metal supports can be used to study the structure of biological and chemical catalysts at a given electrochemical potential. By integrating a plasmonic support as a working electrode into an electrochemical cell, attached molecules can be studied via surface-enhanced vibrational spectro-electrochemistry. Here, both plasmonic-enhanced surface-sensitive vibrational methods, SERS and SEIRA, are introduced and different examples using enzymes and bioinspired molecular complexes as heterogenized catalysts are given.

Plasmonics in Chemistry and Biology
Edited by Marc Lamy de la Chapelle and Nordin Felidj

ISBN 978-981-4800-03-7 (Hardcover), 978-0-429-45875-0 (eBook)
www.jennystanford.com

5.1 Principles of Electrocatalysis

5.1.1 Why Do We Need to Understand Electrocatalytic Reactions?

Electrochemical sensors and fuel cells are key devices in modern energy or heath care technologies. The functional principle of these devices is shown in based on electrocatalytic half reactions shown in Fig. 5.1. To obtain the required efficiency and selectivity, good catalysts are required that not only decrease the activation energy for a reaction as much as possible but also promote very specifically a certain target reaction. In electrocatalysis, the catalytic process takes place at the surface of a conductive support and is coupled to an electron transfer process. These catalyst/support systems, when incorporated as a working electrode in an electrochemical cell, can directly transform electrical energy in chemical energy and vice versa.

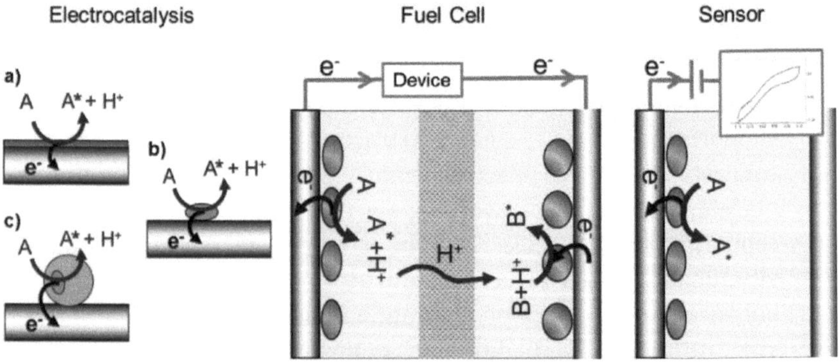

Figure 5.1 On the left-hand side, different electrocatalytical systems are shown. One distinguishes between heterogeneous (a), molecular (b) und enzymatic (c) electrocatalysis. On the right-hand side, the systematic setups of a fuel cell and an electrochemical sensor are drawn.

One distinguishes generally between heterogeneous, molecular, and enzymatic catalysts. In principle, enzymes are the best catalysts regarding efficiency and selectivity. Nevertheless, heterogeneous catalysts show a better performance in functional systems [1]. Molecular catalysts often do not reach the activity of the heterogeneous and enzymatic systems; however, they play a key role in elucidating mechanistic details of the catalytic reaction.

Electrocatalytic reactions involve many different steps as shown in Fig. 5.2. These steps have to occur in a defined sequence to guarantee efficiency and, even more important, selectivity of the process. The reaction starts in most cases with one or multiple electron transfer (ET) steps followed by substrate binding and conversion. Electron transfer is often accompanied by proton transfer (PT) as many important reactions require electrons and protons to proceed. Good catalysts consider coupling between these two reaction steps. The kinetics of the reaction is in most cases controlled by the heterogeneous ET between electrode and catalyst. Therefore, it is important to optimize this reaction. Often simple adsorption of the catalysts does not yield the fastest rates and therefore a suitable functionalization of the electrode has to be performed prior to catalyst immobilization.

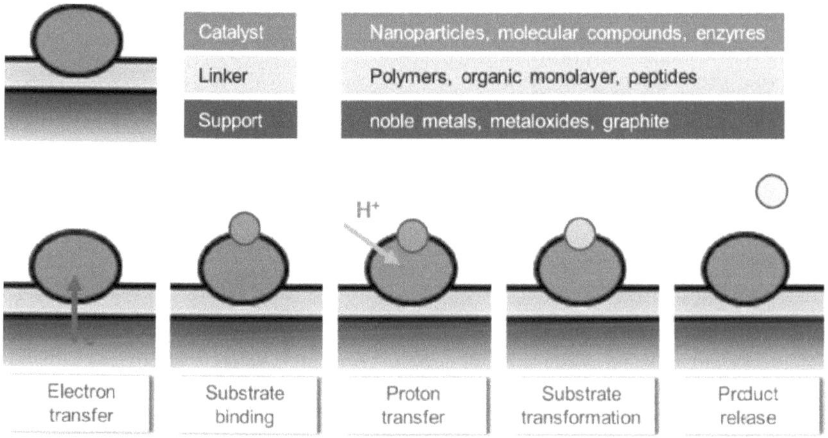

Figure 5.2 Top: Constituents of an electrocatalytic system. Bottom: reaction steps of an electrocatalytic reaction. The sequence of the steps 1–3 can vary from system to system.

A typical example where all these things have to be taken into account is the oxygen reduction reaction (ORR), which is the general cathodic part reaction in fuel cells. To obtain the highest electric energy from the reaction oxygen has to be reduced to water in a 4-electron-4-proton reaction:

$$O_2 + 4e^- + 4H^+ \rightarrow H_2O$$

This reaction is highly sensitive to the kinetics of electron and proton supply. It could be shown for example that a typical molecular catalyst was able to perform this reaction with good selectivity if the electron transfer from the electrode to the molecule was fast. If it became slower due to a different type of binding the catalysts instead performed the 2-electron-2-proton reduction of oxygen to hydrogen peroxide [2].

This example shows how important the type of binding to the electrode is for such systems.

To design a good electrocatalytic system, analysis of the catalyst in solution is not sufficient as the system may change upon surface adsorption. Depending on the communication between electrode and catalyst, electron transport can be severely altered, which might inactivate the system although the catalytic site is still intact. Thus, effective electrocatalytic systems have to consider not only the nature of the active site but also the electrode material and if necessary a suitable coating that is able to establish a good electronic communication between catalytic center and electrode while the accessibility to substrate molecules and/or protons is preserved. As a consequence, many different surface compositions have to be considered for the same type of catalyst.

Enzymes are extremely good catalysts both regarding efficiency and selectivity. This would make them very suitable as electrochemical sensors or catalysts in biofuel cells. However, they react very sensitive to changes in the environment and not seldom an enzymatic system becomes inactive or less active when the biomolecule is attached to a surface. The reasons for the observed inactivity are different for each enzyme and they are difficult to understand without having structural information about the molecule in its immobilized state.

In contrast to enzymes, molecular catalysts can be constructed from scratch. A good strategy is hereby to design them in a way such that they mimic certain enzymatic active sites. By doing so, one is able to copy the very efficient reaction mechanism of the biological analogue but combine it with a longer stability, higher surface coverage and better electrical accessibility; problems that are intrinsic in enzyme/electrode systems. Such biomimetic systems are especially promising as the active site of an enzyme often consists of earth abundant metals like iron or nickel.

Elucidation of mechanistic patterns of individual systems on electrodes is therefore of immense importance to develop novel types of electrocatalysts. These methods have to be surface sensitive and give information about structure and function of the system at the same time. Plasmonic support systems selectively enhance light at the electrode surface and thus enable spectroscopic analysis of surface bound molecules. Combined with electrochemistry surface-enhanced vibrational spectroscopy is able to study the structure of specific catalysts in their resting state and under catalytic turnover conditions. Explaining these plasmon-based methods together with specific examples from biology and chemistry will be the central aim of this chapter. A special focus will hereby lie on metal porphyrin systems, which are very common active sites in catalysis and which, due to their strong color, are ideally suited for spectroscopy.

5.1.2 Metal–Porphyrin Complexes in Biology and Chemistry

A variety of enzymatic catalysts are based on metal-porphyrins and their abundance has a great impact in a variety of life-sustaining reactions. The most known protein cofactors with incorporated porphyrins are hemes. Biosynthesis of all hemes in a cell begins from the formation of protoporphyrin IX, which structure is depicted in Fig. 5.3 [3]. The porphyrin macrocycle consists of four pyrrole rings connected to each other by methine bridges. Such a structure allows the incorporation of different metal atoms, which are coordinated by four pyrrolic nitrogens [4].

In naturally existing biological molecules, the protoheme incorporates an iron ion in case of cytochromes, a magnesium ion in chlorophylls or a cobalt ion in vitamin B_{12}. In addition, the peripheral positions of the macrocycle can be substituted with various side-chains that allow, i.e., to vary the redox properties of the metal-porphyrin. In biology, the presence of such a structure in a variety of different heme proteins with diverse functions has been known for many years [5, 6]. For instance, in the group of cytochromes a, cytochrome c oxidase (CcO) is the most famous protein containing heme a and a_3 being directly involved in process of cellular respiration taken place in the membrane of mitochondria. Cytochromes b are widely spread in both plants and animals and they function greatly differs. Some of them are electron carriers like cytochrome b_5, another play a significant role in steroid metabolism or drug detoxication like cytochrome P450 or oxygen transport like myoglobin and hemoglobin. Cytochromes c are biologically significant in many metabolic pathways, like cellular respiration, photosynthetic processes, nitrate or sulfate reduction. They broad existence in so many biological aspects of life implies their presence from millions of years. Indeed, evolutionary, cytochromes were present in aerobic and anaerobic photosynthetic organisms, which represented early forms of life [7].

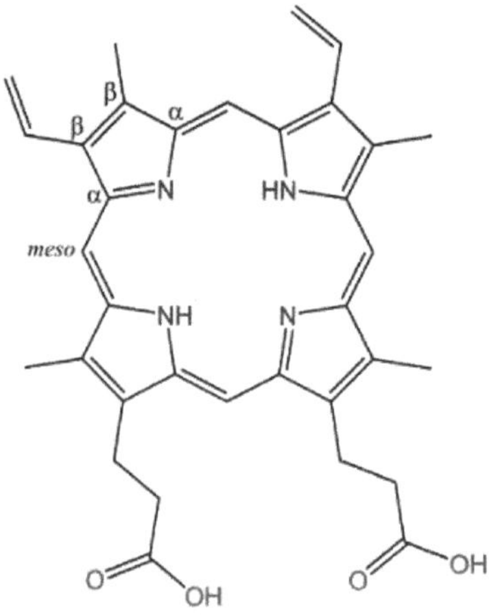

Figure 5.3 Structure of protoporphyrin IX.

Nowadays, extensive research on porphyrins resulted not only on expanded knowledge of naturally existing cytochromes which can have a great variety of function but also implied their use in biomedical and technological applications. The presence of neighboring amino acids from the protein matrix provides an additional proton relay. Knowledge about how these reactions proceed and which parameters are influencing the overall performance allows to tune their properties for specific applications in industry and medicine.

In synthetic chemistry, porphyrin complexes offer an impressive array of applications as models for biological electron transport or many catalytic reactions. The metal core serves as reactive active center responsible for performing a desired reaction such as oxygen or proton reduction [8–10]. Additional substituents at C_{meso} positions of pyrrole rings can further tune the redox properties of a compound, whereas substituents at C_{β} positions can modify their molecular flexibility [11].

5.2 Methods to Probe Structure and Function of Catalysts on Electrodes

5.2.1 Electrochemistry

Electrochemistry is an efficient method to study redox reactions and catalytic activity of a surface bound system. Hereby cyclovoltammetry (CV) is one of the most commonly used techniques. In CV measurements the potential is swept from an initial value with a certain scan rate v (which is a change of potential as a function of time $v = dE/dt$) to a final potential and then reversed [12]. During such a cycle, the flowing current is recorded. A typical CV curve is depicted in Fig. 5.4A. If a redox active molecule is electrically connected to the electrode the resulting current voltage curve consists of a faradaic current as a result of the redox transition on the electrode and a capacitive current arising as a response of the changed charged density on the metal/solution interface. If no additional catalytic reaction takes place cathodic and anodic peaks in the voltammogram are observed when the sweeping potential passes the potential of the redox transition. Detecting these peaks with different scan rates allows determining the formal potential of the redox reaction, the number of transferred electrons per transition and the electron transfer rate. For one-electron redox center, the formal redox potential E^0 can be determined by the arithmetic mean between the potential of the two peaks.

Under conditions, in which substrate molecules are present the voltage current curves change. Electrons are now not only exchanged between electrode and redox center but are created or consumed by the catalytic reaction. As a result, the current follows an S-shape as observed in Fig. 5.4B. Often the catalytic reaction requires initial oxidation or reduction of the active site prior to substrate transformation. Therefore, catalytic turnover is only observed above or below the specific redox potential of the catalyst, which is denoted as the onset potential for catalysis. The value of this potential can be directly determined from the voltammogram by defining the potential at which the catalytic current starts to rise. The onset potential is often obtained by extrapolating the catalytic response of the reaction [13]. One has to note that, if the scan rate is sufficiently low, no redox peaks are observed in the voltammogram and the obtained maximum current is dependent on species kinetics and substrate mass transfer.

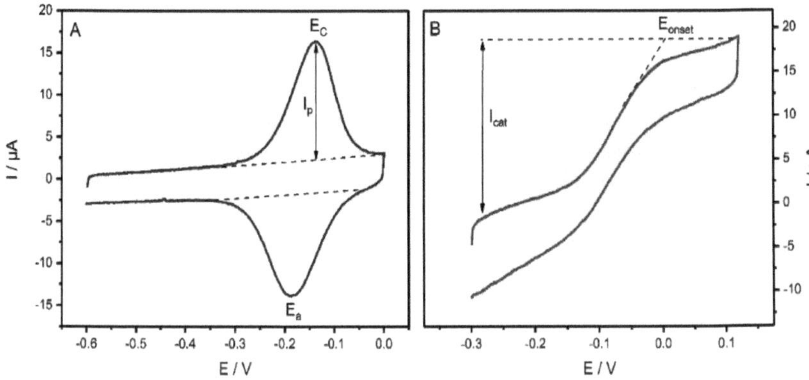

Figure 5.4 (A) Cyclovoltammogramm of an immobilized heme protein (B) CV under turnover conditions. In this case, the signal of cellobiose dehydrogenase (CDH) in the presence of lactose is shown.

Electrochemical measurements are very common to evaluate the catalytic activity of a given electrode/catalyst system. They, however, only yield information about the redox active species on the surface and is only applicable if good electrical communication between the active site and the electrode is present. To understand the reactions that take place on a structural basis, spectroscopic methods are needed. Hereby vibrational spectroscopies such as IR and Raman are very helpful to analyze the structure of molecules, but they do not exhibit the surface sensitivity that is needed to selectively monitor surface reactions. This problem can be overcome by using plasmonic support materials that are able to enhance the light at the electrode/catalyst interface.

5.2.2 Infrared and Resonance Raman Spectroscopy of Porphyrins

Infrared (IR) spectroscopy displays the absorbance of infrared light if the energy of the light matches a vibrational transition of the molecule. In IR spectroscopy the light directly interacts with the vibrating nuclei of the molecule. IR activity is given if the dipole moment during the vibration is changing thus it is sensitive to polar groups such as C=O or C–H. Raman spectroscopy, on the other hand, yields information on molecular vibrations due to the inelastic scattering of visible light by the electron cloud of the molecule. A dipole moment is hereby induced in the electron cloud and thus Raman intensity is a function of electron polarizability. Therefore, it is sensitive to conjugated systems such as phenols or porphyrins. For porphyrins also, the resonance Raman effect can be utilized. Here a strong enhancement of the Raman signal is observed if the laser used for Raman scattering coincidences with a molecular absorbance of the target probe. That means by choosing the right laser line for excitation not only the sensitivity can be improved by several orders of magnitude, but special sensitivity to a certain part of the molecule is achieved. The latter point is especially important for complex systems such as proteins as the contribution of the protein matrix to the Raman signal can be blend out and solely the spectroscopic fingerprint of the chromophore is visible.

RR spectra of metal porphyrins have been investigated in detail and especially for heme proteins a complete assignment could be done [14]. The spectroscopic fingerprint in the region between 1300 and 1700 cm^{-1} is dominated by the breathing vibration of the porphyrin macrocycle. Typical vibrational modes named ν_4, ν_3, ν_2 are shown in Fig. 5.5. These modes are sensitive to expansion of the porphyrin pocket size, which results in weakening of the vibration force constant and thus in downshifting of the vibrational frequency. Increased core size could be a result of changed oxidation or coordination state of the central metal. For example, the iron in the oxidized (ferric, Fe^{III}) state occupies smaller space than in the reduced (ferrous, Fe^{II}) state and thus the porphyrin modes are downshifted for the latter as shown in Fig. 5.5. Also the type of axial ligation defines the size of the heme pocket. In general, iron can be found either in 5-coordinated (5c) or 6-coordinated (6c) state, but depending on the character of axial ligand, it can form a high-spin (HS) or low-spin (LS) state. Which coordination and spin state is formed is determined by the energy gap between d-orbitals of iron. Thus, large energy gaps result in a LS state, whereas small energy gaps result in a HS state. As in the HS state, the radius of the iron ion increases in respect to the LS state, the heme pocket enlarges and heme vibrations become downshifted.

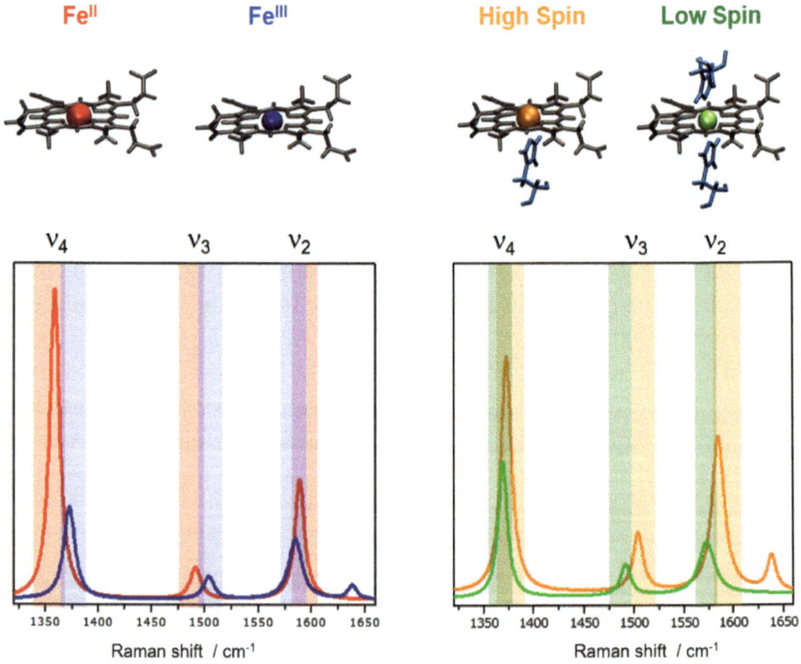

Figure 5.5 Simulated RR intensities of a heme for different oxidation states (reduced Fe^{II}, oxidized Fe^{III}) and axial coordination (high spin, HS; low spin, LS).

Following the porphyrin breathing modes makes it possible to study redox properties but also changes in axial ligation can become visible. Hemes that catalyze the reaction with oxygen (cytochrome c oxidase (CcO), cytochrome P450) or hydrogen peroxide (catalases, peroxidases) are generally present in a 5-coordinated high-spin (5cHS) state that allows binding of oxygen to the iron at the sixth coordination site. Other proteins contain hemes not for catalysis but for effective transport of electrons to or from the catalytic center. These are mostly 6-coordinated low-spin (6cLS) hemes (i.e., cytochrome c, heme a in CcO, cytochrome b_5 in sulfite oxidase or cellobiose dehydrogenase).

In contrast to RR spectroscopy, IR spectra will feature vibrations of the whole system without a specific enhancement of the chromophore. In proteins thus contributions from the protein matrix will dominate the spectrum. Here analysis of vibrational bands in the region between 1300 and 1800 cm^{-1} allows distinguishing between α-helices, β-sheets and unstructured amino acids. Also in metal porphyrin complexes the IR spectra will mostly show vibrations of polar functional groups attached to the porphyrin. Thus, RR and IR give complementary information about the system.

5.2.3 Surface-Enhanced Vibrational Spectroscopy

The combination of RR and IR spectroscopy with the plasmon-induced surface enhancement of nanostructured noble metal electrodes can be used to study the structure of surface bound metal–porphyrins. Surface-enhanced Raman (SER) scattering was the first method that made use of the plasmon-induced field enhancement [15]. Fewer studies followed that applied surface-enhanced infrared absorbance (SEIRA) [16, 17], which relays on the same principle as SER spectroscopy but uses support materials that exhibit surface plasmon resonances in the infrared region. Surface-enhanced resonance Raman (SERR) spectroscopy is a combination of the surface enhancement with the resonance Raman effect and allows selective analysis of surface bound chromophores.

Monodisperse nanoparticles are known to exhibit a defined plasmon resonance. While these systems can display a huge surface enhancement for the right laser line they are not as suitable for spectro-electrochemical SERR investigations where the choice of laser line is given by the absorbance profile of the molecule. Electrochemically roughened electrodes exhibit a randomly distributed nanostructure leading to broad electric field enhancement over the whole range of visible to infrared light [18]. The roughening procedure depends on the material and is well established these days [19]. A typical morphology of an electrochemically roughened Ag electrode is shown in Fig. 5.6A [20]. Here the electrode was functionalized with a very thin silica layer for biocompatible protein attachment. SEIRA electrodes require a more connected nanostructure. A convenient way to create a SEIRA-active gold surface on a Silica surface is by electroless deposition of an Au plating solution composed of 2% HF, $NaAuCl_4 \cdot 2H_2O$, Na_2SO_3, $Na_2S_2O_3 \cdot 5H_2O$ and NH_4Cl, at 65°C [21, 22]. The resulting Au island film is shown in Fig. 5.6B.

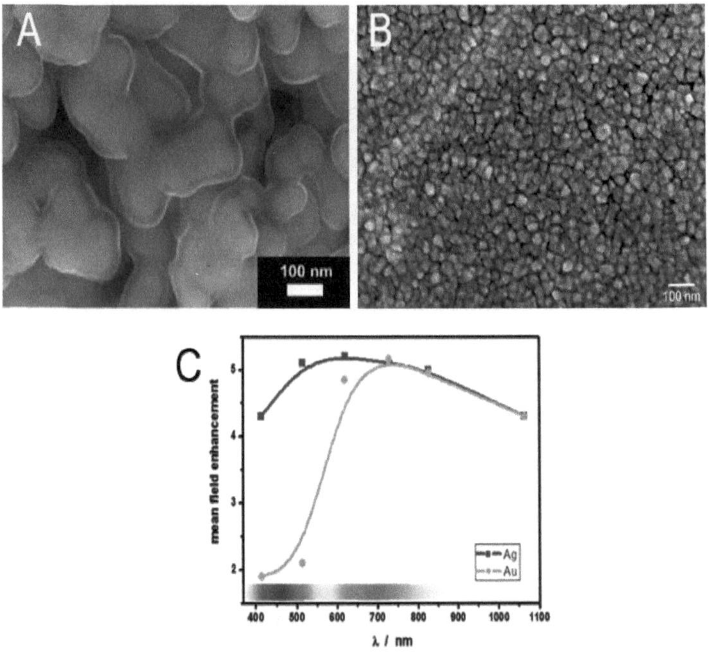

Figure 5.6 Nanostructured SER and SEIRA electrodes. Calculated plasmonic field enhancement of randomly nanostructured noble metal supports (Redrawn from [18]).

In order to combine surface-enhanced vibrational spectroscopy with electrochemistry the nanostructured electrodes have to be inserted as working electrodes into a spectro-electrochemical cell. Different setups of such a systems are depicted in Fig. 5.7. For SER spectroscopy a roughened cylindrical Ag electrode is placed as a working electrode in a spectro-electrochemical cell that allows illumination of the electrode surface by laser light and Raman measurements in a backscattering setup [23]. The electrode is placed on a shaft that can be constantly rotated to minimize laser-induced degradation of the sample.

For SEIRA measurements, the nanostructured Au film is placed on a silica ATR crystal that is functioning as a working electrode in an electrochemical cell [24, 17]. The ATR setup is needed for SEIRA experiments to minimize the contribution of the aqueous buffer to the IR spectrum.

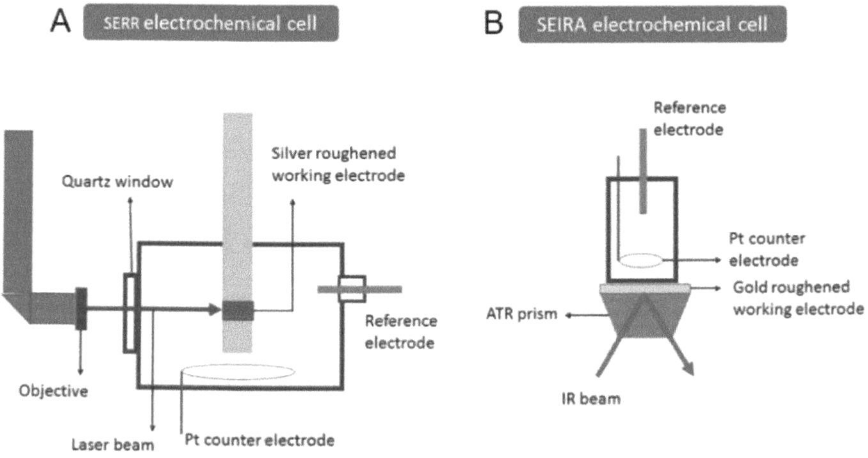

Figure 5.7 Schematic representation of a spectro-electrochemical cell for (A) SER(R) or (B) SEIRA spectro-electrochemistry.

5.2.4 Surface-Enhanced Spectro-Electrochemistry on Porphyrin Systems

Potential-dependent SERR spectroscopy with laser excitation in the violet region makes it possible to follow the redox transition of a surface bound metal–porphyrin. A typical example is shown in Fig. 5.8 for the heme protein cytochrome b_5 (Cyt b_5) attached to a nanostructured Ag electrode that has been functionalized with 8-aminothiophenol prior to protein adsorption.

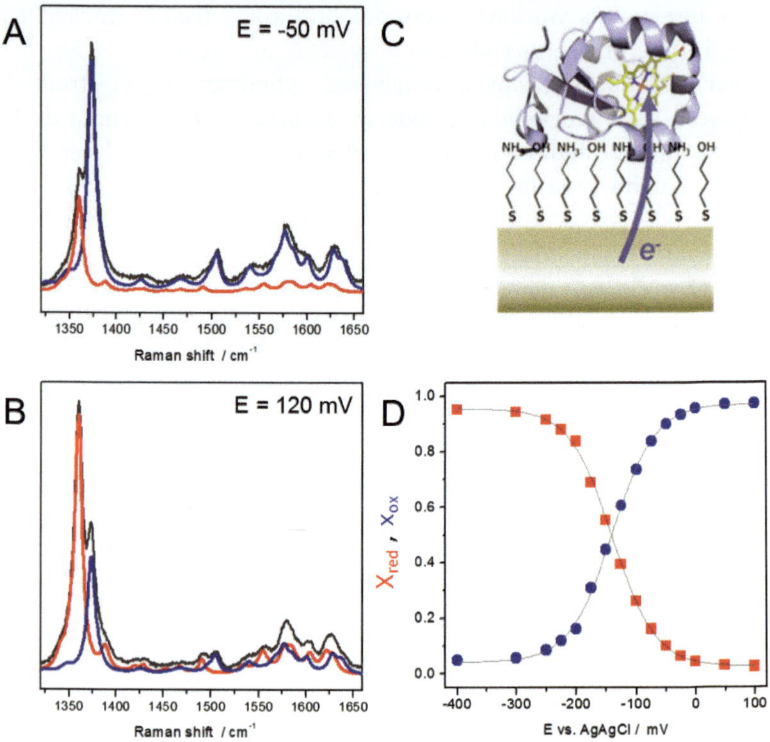

Figure 5.8 Measured SERR spectra of Cyt b_5 at (A) −50 and (B) −120 mV vs. Ag/AgCl. The red and blue lines indicate the contribution of the Fe^{II} and Fe^{III} species respectively. (C) Schematic illustration of Cyt b_5 attached to 8-aminothiophenol functionalized electrodes. (D) Plotting the contributions of each component (*the intensities have to be multiplied by a factor that accounts for the different visibility of each component in the Raman spectrum) as a function of potential creates a titration curve that can be fitted via the Nernst equation.

In this case, the SERR spectrum of the Cyt b_5 heme chromophore can be fitted with two components, one belonging to the reduced Fe^{II} and the other to the oxidized Fe^{III} LS state. Plotting the contributions of each component as a function of potential leads to a redox titration curve from which the redox potential E^0 can be derived via the Nernst equation:

$$E = E^0 - \frac{RT}{nF}\ln\left(\frac{x_{\mathrm{ox}}}{x_{\mathrm{red}}}\right), \tag{5.1}$$

where F is the Faraday constant and n denotes the number of transferred electrons. In the case of Cyt b_5 a redox potential of –30 mV vs. Ag/AgCl was obtained. The spectroscopic data could confirm additionally that no inactive species was present at the metal surface.

5.2.5 Time Resolved SER Spectroscopy

SER spectra give information on the steady state structures of a given surface bound system. The time-dependent response of a given system, i.e., to a change in electrode potential can be investigated with the setup depicted schematically in Fig. 5.9 [23]. Here the change in potential is connected to the laser illumination time of the sample by a delay generator that can control the delay time δ between these two events. With this setup, electron transfer rates between electrode and redox center can be derived.

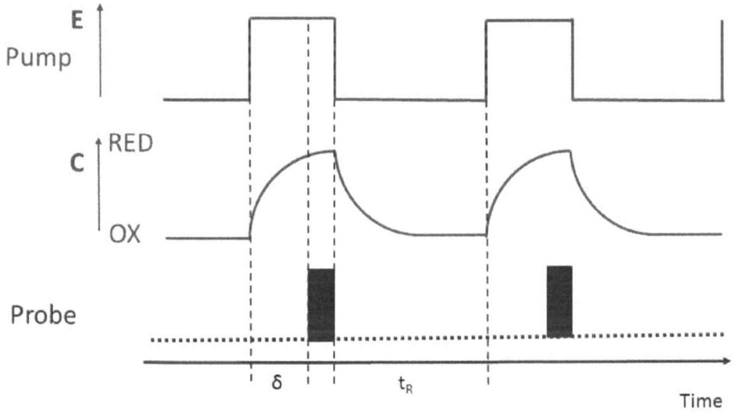

Figure 5.9 Design of a pump-probe potential-dependent SER experiment, with a change of potential serving as the pump. The delay time between potential jump and a probe is depicted as δ, t_R refers to recovery time.

5.3 Examples

In this part, three heme-based systems will be analyzed using the methods described before. Two of them are heme proteins and the third one is a heme-based molecular complex.

5.3.1 Cellobiose Dehydrogenase

Cellobiose dehydrogenase (CDH) catalyzes the oxidation of cellulose to cellobionolactone [25, 26]. As CDH also catalyzes the same oxidation reaction with other hydrocarbons it is considered a promising enzyme for biomass fuel cells [27–29]. The enzyme is depicted in Fig. 5.10. It consists of a flavin domain, which serves as the active site for oxidation of substrate, and a heme-containing domain. The heme domain is flexibly attached to the flavin domain and is responsible for taking up the electrons, generated by the oxidation process, and transferring them to an external redox partner. Such a flexible group acts as a mediator and therefore it might help to generate a good electrical contact with an electrode via this electron transfer pathway.

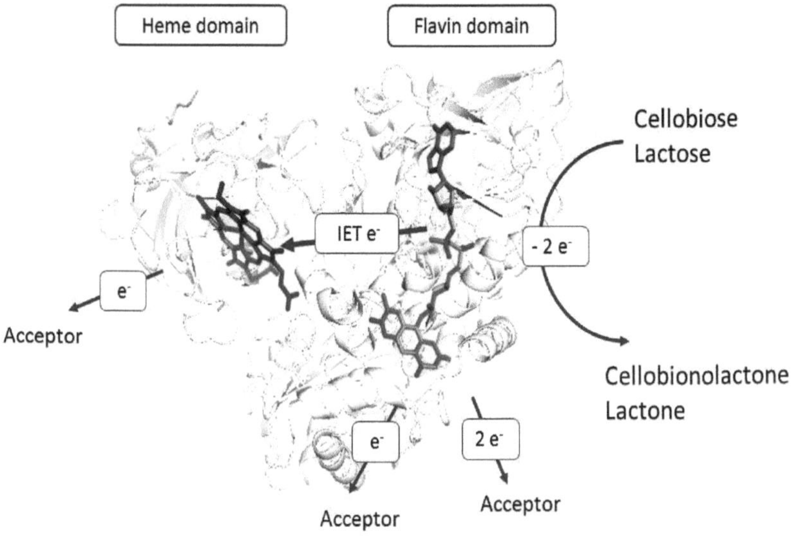

Figure 5.10 Simplified catalytic reaction of cellobiose dehydrogenase oxidizing cellobiose to cellobionic acid with marked possible pathways for electron/s donation.

CDH works quite well on electrodes if the latter are functionalized with a layer containing positively charged groups. When CDH is attached to electrodes that are coated with the polymer PDADMAC, a voltage of −0.072 V is built up in the presence of lactose, which functions in this case as a model substrate.

In additional presence of Ca^{2+}, the voltage is shifted to −0.1 V, which would be preferential for its use as a catalyst in biofuel cells. While this phenomenon was observed before its origin remained unknown [30]. Spectroscopic data can give more insight into the electron transfer pathways and interaction with Ca^{2+} ions. Typical SERR and SEIRA spectra of the protein are shown in Fig. 5.11. The SERR spectra of the enzyme show selectively the heme, while SEIRA spectra display the amide bands of the protein matrix. SERR spectra recorded at different applied potentials show reduction and oxidation of the heme. This confirms that the electron transfer between the heme and the electrode is working. Addition of lactose shows reduction of the heme in the absence of an electrode potential. This feature confirms the presence of a preserved internal electron transfer (IET) between the flavin cofactor and the heme in the immobilized state. In the presence of Ca^{2+} ions the spectrum remained reduced but its intensity decreased. A signal decrease can be interpreted either by desorption of the protein or by protein reorientation that results in a greater distance or altered orientation of the heme in respect to the electrode. The SEIRA spectra show the protein backbone and here no calcium-induced signal decrease is observed. Therefore, protein desorption can be excluded and it can be confirmed that the Ca^{2+} directly interacts with the heme.

Electron transfer pathways can be analyzed very well by comparing electrochemical data with SERR spectroscopy. This is shown for CDH in Fig. 5.12. Cyclovoltammetry in absence of Ca^{2+} shows catalytic turnover current the moment the heme gets reduced by the electrode. This can be seen nicely in Fig. 5.12A, where the molar fraction of the oxidized heme, as determined by SERR spectroscopy, is overlaid with the CV signal. Putting these observations together a complete catalytic cycle that includes electron transfer from the active site to the electrode via the heme as shown schematically in Fig. 5.12C can be postulated in absence of Ca^{2+}.

The situation is different in the presence of Ca^{2+}, as can be seen in the comparison of electrochemical and spectroscopic data in Fig. 5.12B. The SERR spectroscopic titration of the heme yields a shifted redox potential at −0.015 V. However, catalytic turnover currents are observed at more negative potentials where the heme is still reduced and thus cannot be involved as electron acceptor in the electrocatalytic cycle. Therefore, an alternative pathway must have been established by the addition of Ca^{2+}. A possible explanation is demonstrated in Fig. 5.12D. Ca^{2+} is known to attach to the heme propionates once the heme iron gets reduced. But because Ca^{2+} has two positive charges this process generates an excess charge. Due to this extra charge, the heme domain might attach to a negatively charged amino acid of the flavin domain. This could cause a reorientation of the protein on the surface that changes the distance and orientation of the heme to the electrode but decreases the distance between the flavin center and the electrode. Under these circumstances electrons might be transferred directly from the flavin to the electrode. As the flavin has a more negative redox potential than the heme, a larger voltage can be created.

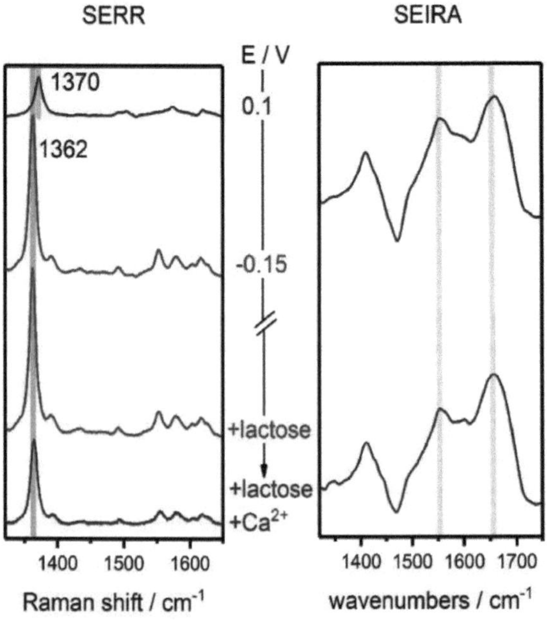

Figure 5.11 SERR and SEIRA spectra of CDH on Ag (SERR) and Au (SEIRA) electrodes coated with PDADMAC measured at an applied potential of 0.1 V, −0.15 V, at open circuit in the presence of lactose and in addition of Ca^{2+} ions. Blue and pink lines indicate oxidized and reduced ν_4 porphyrin marker band, respectively. Green lines indicate amide I and II bands of the protein backbone.

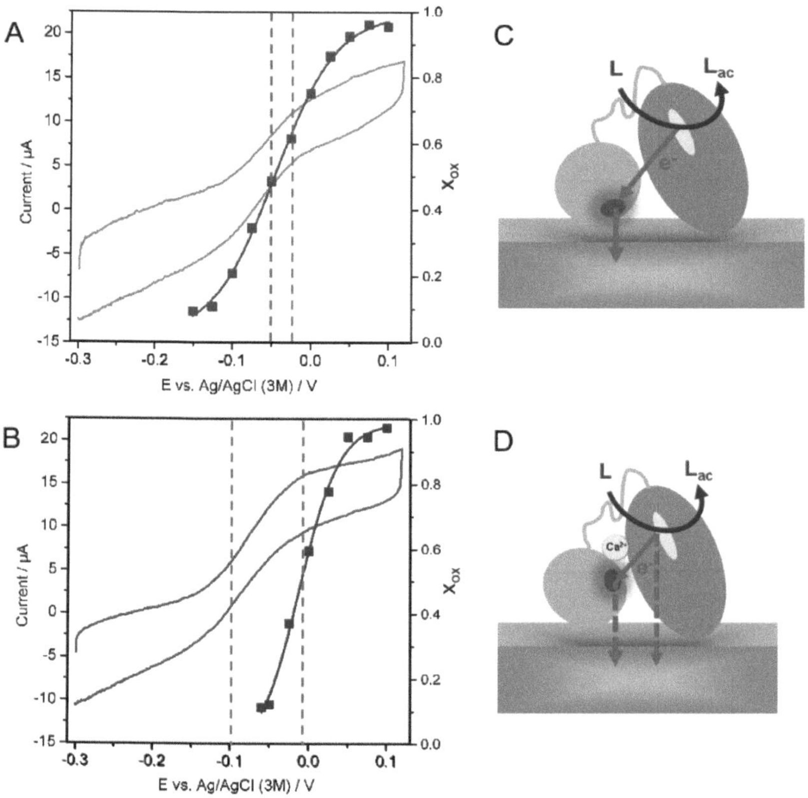

Figure 5.12 Cyclovoltammetric turnover currents and SERR spectroscopic titration of the heme in absence (A) and presence (B) of Ca^{2+}. The data can be interpreted as in absence of Ca^{2+} ET exclusively occurs via the heme cofactor (C) but in the presence of Ca^{2+} (D) can also go directly via the flavin domain.

5.3.2 Cytochrome c Oxidase

Cytochrome c oxidase (CcO) is a membrane protein that catalyzes the reduction of oxygen to water in the respiratory chain of living organisms. The catalytic reaction is, however, not the main function of this enzyme. Instead the ORR is used as energy source to pump protons through the membrane making the interplay between electrons and protons in this system highly complicated. Enzymes that catalyze the ORR are important as cathodic catalysts in biofuel cells; therefore, it is surprising that CcO does not play a role in this scientific field. The reason for this lies in the low stability of this enzyme on electrodes and its bad accessibility for electrons in the immobilized state. As a result high overpotentials have to be applied to create sufficient electrocatalytic currents [31].

For spectroscopists on the other hand CcO has always been a much investigated protein due to its central role in respiration and its complex coupling mechanism between ORR and proton translocation. Especially Resonance Raman spectroscopy has been applied extensively in the past to analyze the redox chemistry and oxygen reaction of the active site.

The analysis of CcO is, nevertheless, challenging as here two hemes contribute to the RR spectra. Heme a functions as electron relay, whereas heme a_3 is part of the catalytic binuclear center. A spectroscopic distinction can be made because heme a is a 6-coordinated low-spin heme and heme a_3 a 5-coordinated high-spin heme. The respective spectral contributions of both hemes have been identified such that a profound assignment can be made [32, 33]. Given the large amount of vibrational spectroscopic data of CcO, it is surprising that there are only a few publications that do show surface-enhanced vibrational spectroscopic analysis [34, 35].

In Fig. 5.13C, SERR spectra of CcO attached to an amino functionalized Ag electrode are presented for different applied potentials. The spectra have very low intensity as the heme cofactors are much buried within the protein and therefore a large distance between the plasmonic electrode and the SERR active hemes is present. Furthermore the spectra show contributions of the reduced and oxidized species of both hemes leading to minimum four species present in the spectra that change with potential. A decent assignment of SERR intensities to certain bands is therefore almost impossible. There is, however, the formyl vibration of heme a_3, which is relatively isolated at 1663 cm^{-1} for the reduced and at 1672 cm^{-1} for the oxidized species. If the protein is illuminated with 442 nm laser light, only the reduced species will be visible with high intensity. Looking at the changes of the 1663 cm^{-1} band allows then following the individual redox titration of heme a_3. This is shown in Fig. 5.14A,B. Under highly anaerobic conditions a nice redox transition is observed with a redox potential at +20 mV, which is in line with the value that has been reported in literature.

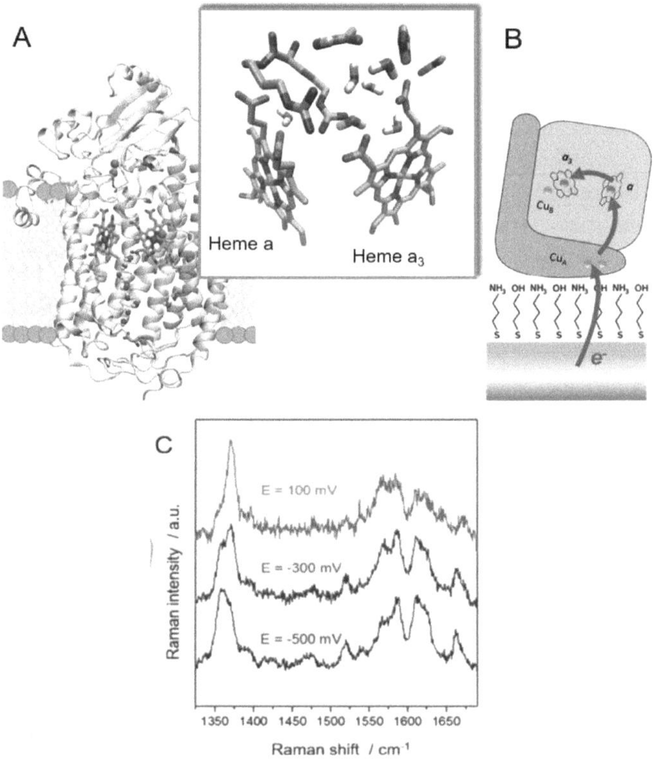

Figure 5.13 (A) Picture of cytochrome c oxidase and its heme cofactors (PDB 1M56). Heme a acts as an electron relay while heme a_3 is part of the catalytic heme-Cu_B center where oxygen is reduced to water. (B) Schematic of CcO attached to amino functionalized electrodes with proposed electron transfer pathway. (C) SERR spectrum of CcO at different applied potentials vs. Ag/AgCl.

Careful inspection of the spectra, however, reveals the bands that do not belong to the native hemes. The most prominent one is at 1490 cm^{-1}, which can be seen very well in Fig. 5.14D. Plotting this band as a function of potential reveals a redox potential of −300 mV, which is much more negative than the natural redox potentials of both hemes. This band cannot belong to heme a_3 as no changes occur for the 1663 cm^{-1} band in this potential range. Therefore, it must belong to a non-natural variation of heme a. The position of this band suggests that it belongs to a 5-coordinated heme a species that might be formed in the presence of the electrode's electric field. Formation of such a non-native heme would very efficiently cut of electron transfer from the electrode to the catalytic center via heme a, which is the natural and optimized electron transfer pathway. Only at potentials that are lower than –300 mV the ET pathway becomes open, which would be an explanation for the high overpotentials that are observed in electrochemical measurements. Such information can only be derived from SERR spectroscopic measurements as the protein in solution does not show this denaturation effect.

5.3.3 Hangman Complexes

The principles of enzymatic catalysis can be adapted to design artificial molecular catalysts. Here the structure of the active site is rebuilt in the hope to achieve the same efficiency and selectivity as in the case of the enzyme. To succeed with this ambitious goal the mechanistic patterns of the catalytic cycle in the enzyme have to be understood and transferred to the molecular system. In this context spectroscopic investigations become evident, which has been used extensively, i.e., to investigate heme complexes that are modeled after cytochrome P450 or cytochrome c oxidase [36–38]. If these patterns can be transferred, these molecular catalysts have very high potential in electrocatalysis as they can be arranged on electrode surfaces with much higher active site density and much smaller distance between catalyst and electrode. The latter point is mandatory to create faster electron transfer.

Simple bioinspired systems are hangman complexes, shown in Fig. 5.15A, that exhibit a metal–porphyrin as catalytic center and a carboxylic acid hanging group that shields the metal from undesired binding of solution molecules and that might act as a relay for protons to the catalytic reaction [39, 40]. Choosing iron as the central metal these complexes have been tested for the electrocatalytic reduction of hydrogen peroxide [40]. For this reaction the coupling between electron transfer and proton supply is essential. In situ information on both processes can be achieved via SERR and SEIRA spectro-electrochemistry, which will be demonstrated in the following paragraph.

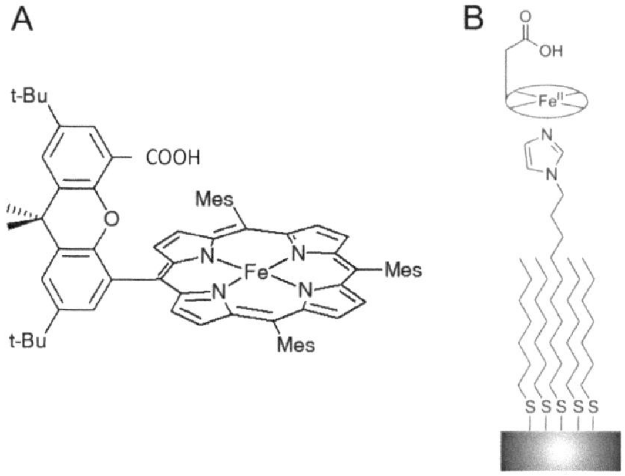

Figure 5.15 (A) Schematic of a heme-based hangman complex and (B) its immobilization on electrodes via an imidazole linker.

In these measurements the porphyrin was bound to a functionalized electrode via an imidazole linker (See Fig. 5.15B). Such a binding mimics the histidine coordination present in cytochrome c oxidase and might be beneficial for the catalytic activity of the hangman complex. The SERR spectrum of immobilized hangman complexes is much broader than usually observed for heme cofactors inside proteins. Component analysis reveals that at least one 5-coordinated high-spin (HS) and one 6-coordinated low-spin species is present. The nature of the 6th ligand is not clear nor its influence on catalytic activity. Potential-dependent SERR analysis reveals two independent redox transitions of the two different species with a similar redox potential. The redox potential itself is pH dependent which points to an involvement of protons in the redox chemistry of hangman complexes.

With SEIRA no specific sensitivity for the metal porphyrin is given. However, since IR intensities scale with the dipole moment changes during the initiated vibration, a high sensitivity for the carboxylic acid group of the hangman motif is given. The C=O stretching vibration of the protonated carboxylic acid was found to be at 1737 cm^{-1}, whereas the delocalized vibration of the COO- carboxylate is at 1565 cm^{-1} (Fig. 5.17A).

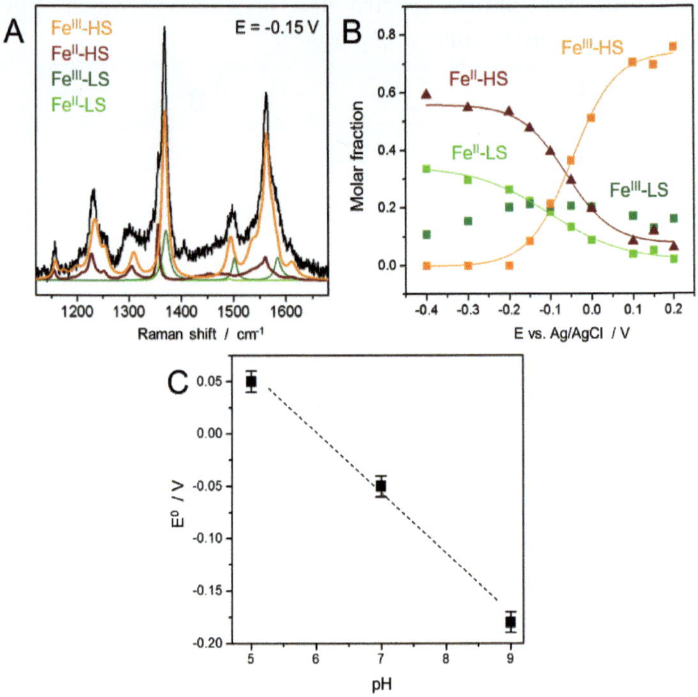

Figure 5.16 (A) SERR spectrum of a hangman complex on the electrode surface with fitted component spectra. (B) Potential-dependent contribution of the redox components. From a fit of the data with two independent Nernst functions the redox potential of both components can be determined. (C) pH dependence of the redox potential of the HS species.

The intensities of these vibrations were plotted for different pHs and are shown in Fig. 5.17B. Fitting these data points with the Henderson–Hasselbach equation

$$pH = pK_a + \log \frac{(1 - x_{COOH})}{x_{COOH}} \tag{5.2}$$

yields the pK_a of the hanging group. It should be pointed out that it is in general very difficult to obtain these pK_a values in aqueous solution as the metal complexes are not water soluble.

The measurements yielded a pK_a of the hangman carboxylic acid group of 3.4. Interestingly a much higher pK_a value of 5.2 was determined in D_2O.

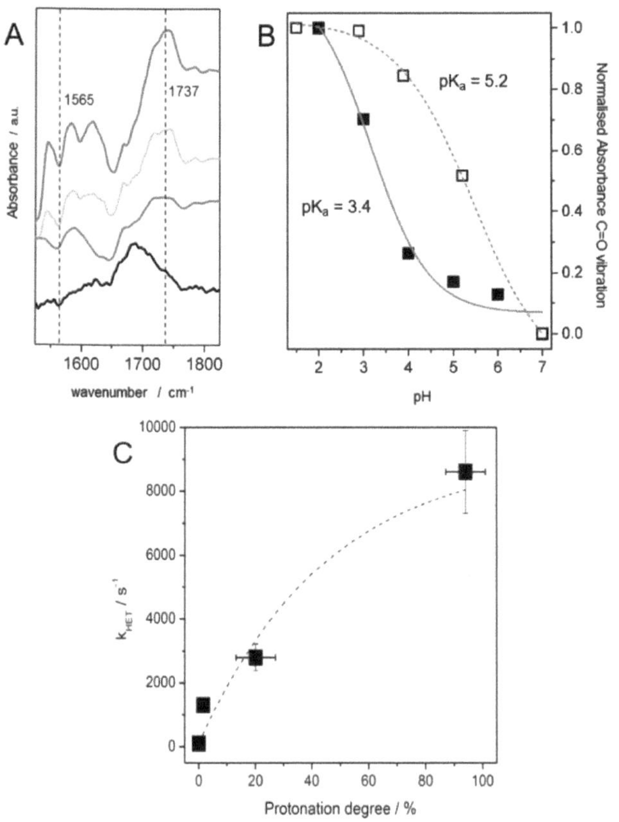

Figure 5.17 (A) SEIRA spectrum of the hangman compound as a function of pH. (B) Intensity of the normalized C=O vibration vs. pH in H_2O and D_2O. A fit with the Henderson-Hasselbach equation yields the pK_a of the hanging group in aqueous solution. (C) Heterogeneous electron transfer rate between hangman and electrode as a function of the protonation grade of the carboxylic hanging group.

This trend is in line with the general higher binding strength of COOD compared to COOH. If the pH and pK_a is known the protonation grade x_{COOH} of the hanging group can be determined via Eq. 5.2. These values were compared with electron transfer rates obtained from time resolved SERR spectroscopy described in Section 2.5. Interestingly this way a significant correlation of ET rates with the protonation grade was observed which confirms the proposed coupling of electron and proton transfer in these systems.

5.4 Conclusions

In this chapter, the insight gained from surface-enhanced vibrational spectroscopy in the field of biological and molecular electrocatalysis was demonstrated. By choosing a plasmon active nanostructured support as working electrode, electrochemical measurements could be combined with surface-sensitive spectroscopy. The combination of both techniques allows correlating structure and function of catalysts under in situ conditions. Examples were given for two enzymatic and one molecular catalyst. All the investigated systems are of potential use in fuel cells to catalyze important reactions for activation of oxygen, hydrogen peroxide or hydrocarbons. It was demonstrated how the different sensitivity of SERR and SEIRA spectroscopy can be used to get a more complete picture of the processes that control redox chemistry and catalytic substrate conversion on electrodes.

References

1. McKone, J. R., Marinescu, S. C., Brunschwig, B. S., Winkler, J. R. and Gray, H. B. (2014). Earth-abundant hydrogen evolution electrocatalysts, *Chem. Sci.*, **5**, 865–878.
2. Collman, J. P., Devaraj, N. K., Decréau, R. A., Yang, Y., Yan, Y.-L., Ebina, W., Eberspacher, T. A., and Chidsey, C. E. D. (2007). A cytochrome c oxidase model catalyzes oxygen to water reduction under rate-limiting electron flux, *Science,* **315**, pp. 1565–1568.
3. Beale, S. I., (2005). Biosynthesis of hemes, *Knowl. Creat. Diffus. Util.*, **3**, 2406–2408.
4. Barkigia, K. M., Chantranupong, L., Smith, K. M., and Fajer, J. (1988). Structural and theoretical models of photosynthetic chromophores. Implications for redox, light-absorption properties and vectorial electron flow, *J. Am. Chem. Soc.*, **110**, 7566–7567.
5. Poulos, T. L. (2014). Heme enzyme structure and function, *Chem. Rev.*, **114**, 3919–3962.
6. Mense, S. M., and Zhang, L. (2006). Heme: A versatile signaling molecule controlling the activities of diverse regulators ranging from transcription factors to MAP kinases, *Cell Res.*, **16**, 681–692.
7. Lemberg, R., and Barrett, J. (1973). *Cytochromes* (Academic Press).

8. Soper, J. D., Kryatov, S. V., Rybak-Akimova, E. V., and Nocera, D. G. (2007). Proton-directed redox control of O–O bond activation by Heme hydroperoxidase models, *J. Am. Chem. Soc.*, **129**, 5069–5075.
9. Chang, C. J. Chng, L. L., and Nocera, D. G. (2003). Proton-coupled O–O activation on a redox platform bearing a hydrogen-bonding scaffold, *J. Am. Chem. Soc.*, **125**, 1866–1876.
10. Roubelakis, M. M., Bediako, D. K., Dogutan, D. K., and Nocera, D. G. (2012). Proton-coupled electron transfer kinetics for the hydrogen evolution reaction of hangman porphyrins, *Energy Environ. Sci.*, **5**, 7737–7740.
11. Auwärter, W., Écija, D., Klappenberger, F., and Barth, J. V. (2015). Porphyrins at interfaces, *Nat. Chem.*, **7**, 105–120.
12. Kissinger, P. T., and Heineman, W. R. (1983). Cyclic voltammetry, *J. Chem. Educ.*, **60**, 702.
13. Harnisch, F., and Freguia, S. (2012). A basic tutorial on cyclic voltammetry for the investigation of electroactive microbial biofilms, *Chem. An Asian J.*, **7**, 466–475.
14. Hu, S., Morris, I. K., Singh, J. P., Smith, K. M., and Spiro, T. G. (1993). Complete assignment of cytrochrome c resonance Raman spectra via enzymic reconstitution with isotopically labeled hemes, *J. Am. Chem. Soc.*, **115**, 12446–12458.
15. Stiles, P. L., Dieringer, J. A., Shah, N. C., and Van Duyne, R. P. (2008). Surface-enhanced Raman spectroscopy, *Annu. Rev. Anal. Chem.*, **1**, 601–626.
16. Neubrech, F., Pucci, A., Cornelius, T. W., Karim, S., García-Etxarri, A., and Aizpurua, J. (2008). Resonant plasmonic and vibrational coupling in a tailored nanoantenna for infrared detection, *Phys. Rev. Lett.*, **101**, 157403.
17. Wisitruangsakul, N., Zebger, I., Ly, K. H., Murgida, D. H., Ekgasit, S., and Hildebrandt, P. (2008). Redox-linked protein dynamics of cytochrome c probed by time-resolved surface enhanced infrared absorption spectroscopy, *Phys. Chem. Chem. Phys.*, **10**, 5276–5286.
18. García-Vidal, F. J., and Pendry, J. B. (1996). Collective theory for surface enhanced Raman scattering, *Phys. Rev. Lett.*, **77**, 1163–1166.
19. Tian, Z.-Q., and Wu, D.-Y. (2002). Surface-enhanced Raman scattering: From noble to transition metals and from rough surfaces to ordered nano-structures, *J. Phys. Chem. B*, **106**, 9463–9483.
20. Feng, J.-J., Gernert, U., Hildebrandt, P., and Weidinger, I. M. (2010). Induced SER-activity in nanostructured Ag-Silica-Au supports via long-range plasmon coupling, *Adv. Funct. Mater.*, **20**, 1954–1961.

21. Miyake, H., Ye, S., and Osawa, M. (2002). Electroless deposition of gold thin films on silicon for surface-enhanced infrared spectroelectrochemistry, *Electrochem. Commun.*, **4**, 973–977.
22. Kozuch, J. A. (2012). *Dissertation Technische Universität Berlin*, p. 127.
23. Wackerbarth, H., Klar, U., Gunther, W., and Hildebrandt, P. (1999). Novel time-resolved surface-enhanced (resonance) Raman spectroscopic technique for studying the dynamics of interfacial processes: Application to the electron transfer reaction of cytochrome c at a silver electrode, *Appl. Spectrosc.*, **53**, 283–291.
24. Ataka, K., and Heberle, J. (2004). Functional vibrational spectroscopy of a cytochrome c monolayer: SEIDAS probes the interaction with different surface-modified electrodes, *J. Am. Chem. Soc.*, **126**, 9445–9457.
25. Eriksson, K.-E., Pettersson, B., and Westermark, U. (1974). Oxidation: An important enzyme reaction in fungal degradation of cellulose, *FEBS Lett.*, **49**, 282–285.
26. Bao, W. J., Usha, S. N., and Renganathan, V. (1993). Purification and characterization of cellobiose dehydrogenase, a novel extracellular hemoflavoenzyme from the white-rot fungus Phanerochaete chrysosporium, *Arch. Biochem. Biophys.*, **300**, 705–713.
27. Ludwig, R., Harreither, W., Tasca, F., and Gorton, L. (2010). Cellobiose dehydrogenase: A versatile catalyst for electrochemical applications, *Chemphyschem*, **11**, 2674–97.
28. Stoica, L., Ludwig, R., Haltrich, D., and Gorton, L. (2006). Third-generation biosensor for lactose based on newly discovered cellobiose dehydrogenase, *Anal. Chem.*, **78**, 393–398.
29. Tasca, F., Gorton, L., Harreither, W., Haltrich, D., Ludwig, R., and Nöll, G. (2008). Highly efficient and versatile anodes for biofuel cells based on cellobiose dehydrogenase from Myriococcum thermophilum, *J. Phys. Chem. C*, **112**, 13668–13673.
30. Schulz, C., Ludwig, R., Micheelsen, P. O., Silow, M., Toscano, M. D., and Gorton, L. (2012). Enhancement of enzymatic activity and catalytic current of cellobiose dehydrogenase by calcium ions, *Electrochem. Commun.*, **17**, 71–74.
31. Meyer, T., Melin, F., Xie, H., von der Hocht, I., Choi, S. K., Noor, M. R., Michel, H., Gennis, R. B., Soulimane, T., and Hellwig, P. (2014). Evidence for distinct electron transfer processes in terminal oxidases from different origin by means of protein film voltammetry, *J. Am. Chem. Soc.*, **136**, 10854–10857.

32. Babcock, G. T., Callahan, P. M., Ondrias, M. R., and Salmeen, I. (1981). Coordination geometries and vibrational properties of cytochromes a and a_3 in cytochrome oxidase from Soret excitation Raman spectroscopy, *Biochemistry*, **20**, 959–966.

33. Heibel, G. E., Hildebrandt, P., Ludwig, B., Steinrücke, P., Soulimane, T., and Buse, G. (1993). Comparative resonance Raman study of cytochrome c oxidase from beef heart and Paracoccus denitrificans, *Biochemistry*, **32**, 10866–10877.

34. Hrabakova, J., Ataka, K., Heberle, J., Hildebrandt, P., and Murgida, D. H. (2006). Long distance electron transfer in cytochrome c oxidase immobilised on electrodes. A surface enhanced resonance Raman spectroscopic study, *Phys. Chem. Chem. Phys.*, **8**, 759–766.

35. Jiang, X., Zaitseva, E., Schmidt, M., Siebert, F., Engelhard, M., Schlesinger, R., Ataka, K., Vogel, R., and Heberle, J. (2008). Resolving voltage-dependent structural changes of a membrane photoreceptor by surface-enhanced IR difference spectroscopy, *Proc. Nat. Acad. Sci. U. S. A.*, **105**, 12113–12117.

36. Liu, J.-G., Naruta, Y., and Tani, F. (2005). A functional model of the cytochrome c oxidase active site: Unique conversion of a Heme-μ-peroxo-Cu^{II} intermediate into Heme-superoxo/Cu^{I}, *Angew. Chem. Int. Ed.*, **44**, 1836–1840.

37. Collman, J. P., Ghosh, S., Dey, A., Decréau, R. A., and Yang, Y. (2009). Catalytic reduction of O_2 by cytochrome c using a synthetic model of cytochrome c oxidase, *J. Am. Chem. Soc.*, **131**, 5034–5035.

38. Sengupta, K., Chatterjee, S., Samanta, S., Bandyopadhyay, S., and Dey, A. (2013). Resonance Raman and electrocatalytic behavior of thiolate and imidazole bound iron porphyrin complexes on self assembled monolayers: Functional modeling of cytochrome P450, *Inorg. Chem.*, **52**, 2000–2014.

39. Rosenthal, J., and Nocera, D. G. (2007). Role of proton-coupled electron transfer in O–O bond activation, *Acc. Chem. Res.*, **40**, 543–553.

40. Graham, D. J., Dogutan, D. K., Schwalbe, M., and Nocera, D. G. (2012). Hangman effect on hydrogen peroxide dismutation by Fe(III) corroles, *Chem. Commun.*, **48**, 4175–4177.

Chapter 6

Fluorescence Spectroscopy Enhancement on Photonic Nanoantennas

Jérôme Wenger

Aix Marseille University, CNRS, Centrale Marseille, Institut Fresnel, UMR 7249, Marseille, France

jerome.wenger@fresnel.fr

6.1 Introduction and Motivation

Despite the significant progress in single-molecule fluorescence microscopy made over the past two decades, the efficient detection of a single molecule remains a major goal with applications in chemical, biochemical, and biophysical analysis (Holzmeister, 2013; Punj, 2014). The phenomenon of light diffraction appears as a main physical limiting factor. Indeed, there is a huge size mismatch between a single molecule (below 5 nm) and the wavelength of light (around 500 nm). This size mismatch prevents the efficient interaction between an incoming light beam from a conventional optical microscope and a single fluorescent molecule, so the net detected fluorescence signal from a single

Plasmonics in Chemistry and Biology
Edited by Marc Lamy de la Chapelle and Nordin Felidj

ISBN 978-981-4800-03-7 (Hardcover), 978-0-429-45875-0 (eBook)
www.jennystanford.com

molecule (ultimately defining the sensitivity and dynamic temporal resolution achievable) remain limited (Novotny, 2006).

Additionally, conventional optical microscopes are restricted to conditions of low density (or concentration) of fluorescent molecules (Holzmeister, 2013; Levene, 2003). In order to isolate a single molecule in the diffraction-limited volume of a confocal microscope, the concentration range must be typically in the pico to nanomolar range. However, a large majority of enzymes and proteins requires concentrations in the micro to millimolar range to reach relevant reaction kinetics and biochemical stability. Monitoring single molecules at high physiological concentrations thus requires overcoming the diffraction limit to confine light in a nanometer spot of volume in the zepto- (10^{-21}) or atto- (10^{-18}) liter range, more than three orders of magnitude below the femtoliter volumes achieved with confocal microscopes (Moran-Mirabal, 2008; Wenger, 2010).

Confining light to the nanoscale can be achieved thanks to optical antennas (Novotny, 2011). Optical antennas are generally metallic nanostructures with dimensions much below the wavelength of light (Schuller, 2010; Biagioni, 2012). Like their radiofrequency counterparts, optical antennas convert propagating radiation into localized energy and vice versa (Novotny, 2011). This opens new routes to enhance and control the emission from a single fluorescent emitter by improving the light–matter interaction between a single fluorescent molecule and the incoming beam, leading to the phenomenon of antenna-enhanced fluorescence emission (Lakowicz, 2009; Halas, 2011; Koenderink, 2017). Over thousand-fold enhancement of the single-molecule fluorescence signal was reported with lithographically fabricated gold nanogap antennas (Kinkhabwala, 2009; Punj, 2013; Flauraud, 2017), with dimers of gold nanoparticles assembled with DNA origami (Acuna, 2012; Puchkova, 2015) and with a single gold nanorod (Yuan, 2013; Khatua, 2014).

In this chapter, we will briefly introduce the key concepts to understand the phenomenon of fluorescence enhancement with optical nanostructures. The equations will remain simple, with the major goal to illustrate the different effects at play. Next, we will provide an overview of different recent approaches to enhance single-molecule fluorescence with plasmonic and non-plasmonic nanoantennas. Finally, we will describe three main biochemical applications that this technique opens.

6.2 Brief Theoretical Background: The Physics of Fluorescence Enhancement

In order to understand the phenomenon of enhanced fluorescence in the vicinity of a nanostructure, one has to go back to the representation of the quantum emitter as a two-energy-level system. Γ_{rad} and Γ_{nr} are the rate constants for radiative and nonradiative transitions from the excited singlet state to the ground state. The total de-excitation rate from the excited singlet state is then $\Gamma_{\text{tot}} = \Gamma_{\text{rad}} + \Gamma_{\text{nr}}$, and corresponds to the inverse of the excited state fluorescence lifetime $\tau = 1/\Gamma_{\text{tot}}$. With these notations, the ratio of the radiative rate to the total decay rate (probability of de-excitation transition accompanied with the emission of a photon) is defined as the fluorescence quantum yield $\phi = \Gamma_{\text{rad}}/\Gamma_{\text{tot}} = \Gamma_{\text{rad}}/(\Gamma_{\text{rad}} + \Gamma_{\text{nr}})$. Lastly, the excitation rate constant is noted σI_{e}, with σ being the excitation cross section and I_{e} the local excitation intensity.

Under steady-state conditions, the fluorescence brightness (or count rate) per molecule Q can be written as (Zander, 2002)

$$Q = \kappa\phi \frac{\sigma I_e}{1 + I_{\text{e}}/I_{\text{s}}}. \tag{6.1}$$

Here κ denotes the light collection efficiency of the optical microscope apparatus, and $I_{\text{s}} = \Gamma_{\text{tot}}/\sigma$ is the so-called saturation intensity. Generally the experiments are performed in the weak excitation regime well below fluorescence saturation (corresponding to $I_{\text{e}} << I_{\text{s}}$), so as to avoid photobleaching the emitters and introducing other types of photodamage to the photonic nanostructure. In this regime, the brightness Q reduces to a simpler form (Lakowicz, 2006):

$$Q = \kappa\phi\sigma I_{\text{e}}. \tag{6.2}$$

We retrieve here that the fluorescence rate per molecule is proportional to the collection efficiency, the quantum yield, and the excitation intensity.

In the presence of the optical nanoantenna, all the transition rates are potentially modified (here, we note with a^* the quantities modified by the nanoantenna). The local excitation intensity becomes I^*_{exc} and can be significantly greater than the incoming intensity by taking advantage of the subwavelength confinement of light at the nanoantenna (similar to the lightning rod effect) which can be further additionally enhanced by the plasmonic resonances of the nanoantenna. Locally, the excitation intensity I^*_{exc} can thus overcome I_{exc} by more than a hundred fold. The radiative decay rate is also enhanced (as a consequence of the reciprocity theorem) and becomes Γ^*_{rad}.

A major challenge in enhancing fluorescence with metal nanostructures comes from the presence of additional losses (with rate constant noted Γ^*_{loss}) which are introduced by the presence of the metal nearby. These losses account for non-radiative energy transfer to the free electron gas in the metal, whose energies are further dissipated by Joule (ohmic) losses. Consequently, the total decay rate with the nanostructure becomes $\Gamma^*_{\text{tot}} = \Gamma^*_{\text{rad}} + \Gamma_{\text{nr}} + \Gamma^*_{\text{loss}}$.

It is greater than the initial decay rate Γ_{tot} so the net photodynamics decay is accelerated by the presence of the nanoantenna. However, this increase in the total decay rate $\Gamma^*_{\text{tot}}/\Gamma_{\text{tot}}$ must not be confused with the fluorescence enhancement (which is the ratio of the brightness per emitter Q^*/Q). Indeed, a significant fraction of the total decay rate Γ^*_{tot} can be due to non-radiative losses Γ^*_{loss}, which lead to a decrease of the apparent fluorescence brightness also known as quenching. Said differently, a fluorescent emitter can experience an ultrashort lifetime near a nanoantenna, corresponding to an increased total decay rate Γ^*_{tot}, but this can be essentially a consequence of strong non-radiative losses damping all the energy away from the radiative transitions. This situation is typically found for a fluorescent molecule in very close (below 5 nm) proximity to a small gold nanoparticle of diameter below 15 nm: The gold nanoparticle acts here as a fluorescence quencher. Therefore, in any fluorescence enhancement experiment with a nanoantenna, a delicate balance must be found between enhancing the radiative decay rate Γ^*_{rad} without introducing too many losses Γ^*_{loss}. As a side note, we point out that the situation is conceptually different in surface enhanced Raman scattering

(SERS), where the scattering is a nearly instantaneous process and where there is no finite lifetime of the excited state and no non-radiative decay rates involved. Note also that we assume here that the non-radiative decay rate Γ_{nr} (intrinsic to the fluorescent emitter in use, set by its chemical structure) is not modified by the presence of the metal nanostructure ($\Gamma^*_{nr} = \Gamma_{nr}$). This assumption will simplify the coming equations and is still fully general as the additional losses effects are included in the term Γ^*_{loss}.

With the previous expression of the brightness per emitter Q, the fluorescence enhancement factor η_F is then simply defined as the ratio of brightness with and without the photonic nanostructure. So the fluorescence enhancement factor writes

$$\eta_F = \frac{Q^*}{Q} = \frac{\kappa^*}{\kappa}\frac{\phi^*}{\phi}\frac{I^*_{exc}}{I_{exc}}. \quad (6.3)$$

This equation indicates the different relevant quantities to enhance the net fluorescence signal per emitter. The collection efficiency κ^* can be improved by the nanoantenna, directing more optical energy into the collection aperture of the microscope setup. The excitation intensity I^*_{exc} can be locally enhanced and lastly the quantum yield ϕ^* can be modified.

Improving the apparent quantum yield of the fluorescent emitter deserves more comments to discuss the ratio ϕ^*/ϕ. First, one has to keep in mind that the quantum yield as a probability of radiative transition cannot exceed unity. Therefore, if the emitter has a high initial quantum yield ϕ without the nanostructure (close to 1, like Rhodamine6G molecules for instance), then there is not much that the optical nanoantenna can do to further increase this quantity, and the ratio ϕ^*/ϕ will also be close to one (if it is not reduced to quenching losses). On the contrary, if one starts with a low quantum yield emitter with ϕ on the order of a few percents, then the ratio ϕ^*/ϕ can become apparently very large, thereby improving the net fluorescence enhancement factor η_F. Low quantum yield emitters experience thus the maximum benefit from the presence of the plasmonic nanoantenna. To better represent the influence of the quantum yield and the various transition rates, Eq. (6.3) can be rewritten as

$$\eta_F = \frac{\kappa^*}{\kappa}\frac{I_{exc}^*}{I_{exc}}\frac{\Gamma_{rad}^*}{\Gamma_{rad}}\frac{1}{1-\phi+\phi(\Gamma_{rad}^*+\Gamma_{loss}^*)/\Gamma_{rad}}. \quad (6.4)$$

Note that the last ratio in the denominator $(\Gamma_{rad}^* + \Gamma_{loss}^*)/\Gamma_{rad}$ describes the influence of the nanoantenna that is only due to the photonic environment, including ohmic losses and compensating for the intrinsic chemical non-radiative decays Γ_{nr}. This ratio thus corresponds to the net increase in the local density of photonic states (LDOS), sometimes also referred to as the Purcell factor. A high LDOS may thus seem to lower the fluorescence enhancement η_F, but this is counter-balanced by the term $\Gamma_{rad}^*/\Gamma_{rad}$ in the numerator, so a high LDOS generally leads to a higher fluorescence enhancement.

Because of the dependence of the apparent value of the fluorescence enhancement η_F on the initial quantum yield ϕ, comparing between experimental results obtained with different molecular emitters must be made with caution. Extreme fluorescent enhancement factors above a thousand fold have been reported, but all use fluorescent dyes of quantum yield below 10%, and a high fluorescence enhancement factor does not necessarily correlate with a bright photon count rate. To compensate for the dependency on ϕ, the product $\eta_F \times \phi$ has been proposed as a figure of merit for the fluorescence enhancement (Gill, 2012). This solves one of the problems, but extra care still needs to be taken while comparing results to take into account the influence of other parameters such as the alteration of the collection efficiency and the saturation of the fluorescence process due to large excitation powers (Wenger, 2012).

6.3 Experimental Approaches to Enhance Fluorescence

To overcome the diffraction limit, nanoantenna designs take advantage of sharp curvature radii, nanoscale gaps and plasmonic resonances, using metal nanostructures (Novotny, 2011). A large number of approaches have been investigated over the last decade in order to enhance the fluorescence emission of single molecules and quantum dots (Holzmeister, 2013; Punj, 2014; Lakowicz, 2009). While classical top-down lithography like electron-beam lithography or focused ion beam milling remains the workhorse for fabricating the nanoantennas, alternative bottom-up strategies using nanoparticle assembly have recently seen much development driven by the seek for a simpler and higher throughput nanofabrication approach.

6.3.1 Top-Down Milling

Electron beam lithography, focused ion beam milling, or deep UV photolithography have a large flexibility in creating a variety of planar antenna designs at specific locations. A nice example is given by the bowtie antenna made of two facing nanotriangles with a 10 nm gap (Fig. 6.1a) (Kinkhabwala, 2009, 2012). Plasmonic coupling between the nanoparticles leads to a local excitation intensity enhancement of 100-fold inside the nanogap region (so-called hot spot) where single fluorescent molecules are randomly deposited. By analyzing all the different single-molecule fluorescence traces on individual nanoantennas, fluorescence enhancement factors up to 1000-fold could be quantified, together with fluorescence lifetime down to 10 ps (Kinkhabwala, 2009).

In order to detect individual molecules diffusing in solution at high concentration, a special must be taken to ensure reducing the background fluorescence from molecules a few tens of nanometers away from the plasmonic antenna, yet still in the confocal detection volume. At the physiologically relevant concentrations of a few micromolar, several thousands of molecules that are still present within the diffraction-limited confocal volume, and their non-enhanced emission can overwhelm the enhanced single-molecule signal from the nanoscale plasmonic hotspot. Specific strategies have been developed to address this issue, using low quantum yield emitters and/or adding a metal cladding layer. The design termed "antenna-in-box" combines a gap-antenna inside a nanoaperture, and has enabled single-molecule detection at concentrations above 20 μM together with fluorescence enhancement above a thousand-fold (Fig. 6.1b) (Punj, 2013; Flauraud, 2017).

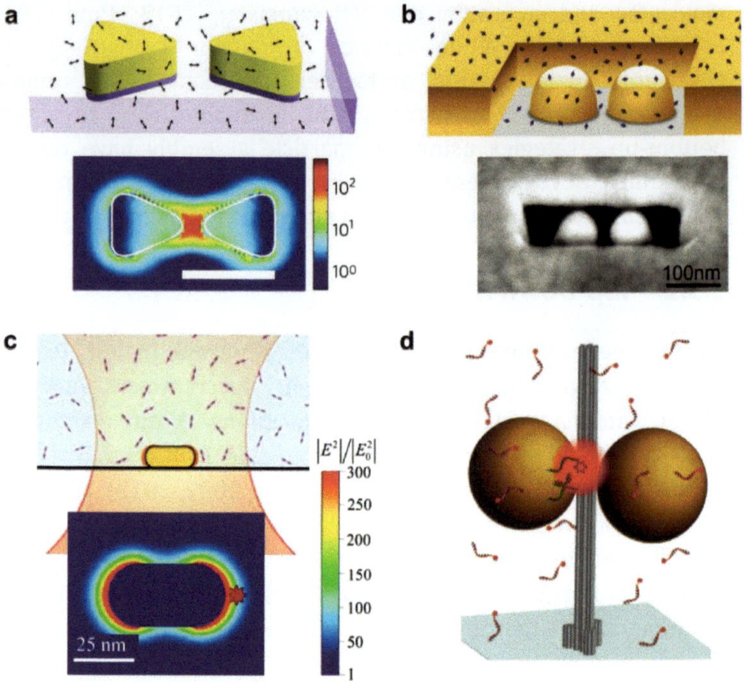

Figure 6.1 Example of nanoantenna designs for single-molecule fluorescence enhancement. (a) Bowtie nanoantenna covered by fluorescent molecules embedded in a PMMA resin (Kinkhabwala, 2009). The simulations represent the local excitation intensity enhancement around the nanoantenna (scale bar 100 nm). (b) Antenna-in-box designed for single-molecule analysis at high concentrations (Punj, 2013). (c) Single gold nanorod for enhanced single-molecule detection in solution (Khatua, 2014). The simulations show the near-field intensity map for a 47 × 25 nm gold nanorod. The excitation wavelength is 633 nm, and the excitation is polarized along the long axis of the nanorod. (d) Dimer nanogap antenna assembled by DNA origami (Acuna, 2012). Figures reproduced with permission: Copyright (a) Nature Publishing Group 2009, (b) Nature Publishing Group 2013 (c) American Chemical Society 2014 (d) American Association for the Advancement of Science 2012.

Enhanced fluorescence spectroscopies applications need nanoantennas featuring a large-scale availability together with narrow gaps accessible to target the fluorescent probes. Recent advances in top-down lithography have been achieved towards this goal, using blurring-free stencil lithography patterning by dry etching through nanostencils (Flauraud, 2015) or a combination of electron-beam lithography followed by planarization, etch back and template stripping (Flauraud, 2017).

6.3.2 Bottom-Up Self-Assembly

Bottom-up assembly of nanoparticles offers an attractive alternative to classical top-down nanofabrication thanks to its low operation cost, narrow gap sizes, use of single crystalline structures and potential large-scale availability (Grzelczak, 2010; Edel, 2013). A simple and direct approach relies on using a single nanoparticle of the shape of a sphere (Estrada, 2008; Wang, 2011; Lu, 2012; Choudury, 2012; Punj, 2013b) or nanorod (Yuan, 2013; Khatua, 2014, 2015). Taking advantage of the local surface plasmon resonance enhancing and confining the electromagnetic field within a few nanometers near the nanorod apex, fluorescence enhancement above a thousand fold could be obtained with an antenna as simple as a single nanorod (Fig. 6.1c) (Yuan, 2013; Khatua, 2014), enabling fluorescence correlation spectroscopy at high concentrations (Khatua, 2015). Drying of nanoparticle colloidal suspension provides another straightforward access to resonant optical antennas with nanometer gap sizes (Fan, 2010). For a dimer of 80 nm gold nanoparticles with 6 nm gap, 600-fold fluorescence enhancement and detection volumes down to 70 zL were achieved (Punj, 2015). It is worth also mentioning that ultrafast emission of quantum dots down to picosecond lifetimes was achieved thanks to a patch antenna realized by a silver nanocube on a gold surface (Akselrod, 2014; Hoang, 2015, 2016).

To provide a better control on the nanogap distance and provide more reproducible antennas, the self-assembly of metal nanoparticles can be templated by DNA double strands (Fan, 2011; Busson, 2011, 2012b; Bidault, 2016) or DNA origami (Acuna, 2012; Puchkova, 2015; Zhang, 2015; Vietz, 2017). This latter approach constitutes a powerful method to assemble nanoparticles into complex 3D designs with gap size tunability (Fig. 6.1d). It also brings the key ability to bind the desired target molecule at the antenna hot spot (Acuna, 2012). Recent results use 100 nm diameter nanoparticles assembled on a DNA origami pillar with 12–17 nm gap featuring a binding site for a single fluorescent molecule. Fluorescence enhancement up to 5000-fold were reported, together with single-molecule detection at concentrations up to 25 μM (Puchkova, 2015).

6.3.3 Dielectric Alternatives to Plasmonic Metals

The metals conventionally used for plasmonics (gold, silver, aluminum) still have significant absorption losses in the visible spectral range used for fluorescence spectroscopy applications. Energy damping to the free electron gas in the metal can severely quench the fluorescence emission, while simultaneously absorption of the laser beam induce Joule heating of the nanostructure (Lakowicz, 2009). To find an alternative featuring lower absorption losses, all-dielectric nanoantennas based on silicon or germanium have been recently introduced (Garcia-Etxarri, 2011; Evlyukhin, 2012; Kuznetsov, 2012; Albella, 2013). Planar concentrators have been used to achieve extremely high collection efficiencies and detection rates of a single fluorescent molecule (Lee, 2011; Chu, 2014). Silicon nanogap antennas (with a design very close to their gold counterpart) have been reported to enhance the fluorescence of a dye layer (Caldarola, 2015; Cambiasso, 2017) and of single molecules (Regmi, 2016), with single-molecule enhancement factors up to 270-fold. Although the gains reported remain lower than for the plasmonic metal counterparts, this approach is a promising alternative compatible with CMOS processing.

To conclude this section, Fig. 6.2 compares the figure of merit for the single-molecule fluorescence enhancement (defined as $\eta_F \times \phi$) for different techniques. As the techniques develop further and further, the figure of merit rises up to values around 300, which represent the current state of the art at the end of 2017.

6.4 Biochemical Applications of Enhanced Fluorescence

In this section, we discuss two driving applications for antenna-enhanced single-molecule fluorescence detection: real-time DNA sequencing and living cell membrane investigations at the nanoscale. We also present the emerging field of single-molecule Förster resonance energy transfer (smFRET) enhanced by nanoantennas.

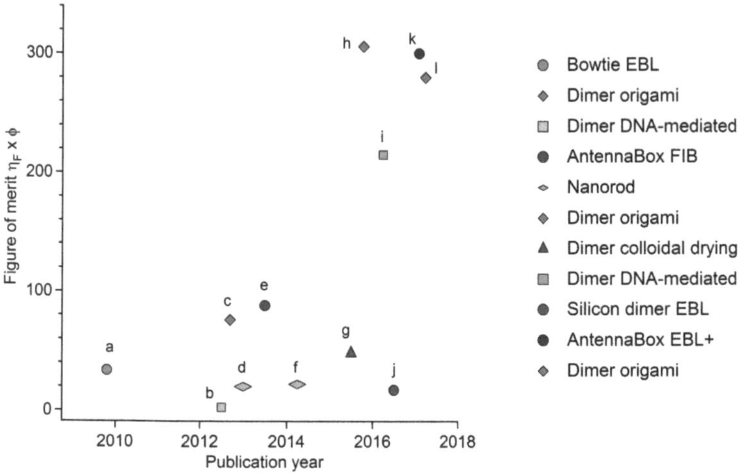

Figure 6.2 Figure of merit for the single-molecule fluorescence enhancement (defined as the product of the measured fluorescence enhancement factor η_F times the initial quantum yield ϕ of the fluorescent emitter used) as a function of the publication year for different nanoantenna designs and fabrication approach. The letters written next to the data points relate to the following references: (a) Kinkhabwala (2009), (b) Busson (2012), (c) Acuna (2012), (d) Yuan (2013), (e) Punj (2013), (f) Khatua (2014), (g) Punj (2015), (h) Puchkova (2015), (i) Bidault (2016), (j) Regmi (2016), (k) Flauraud (2017), (l) Vietz (2017).

6.4.1 Real-Time DNA Sequencing

Achieving personalized quantitative genomics requires the development of novel methods for DNA and RNA sequencing that enables high-throughput, high-accuracy, and low operating costs. Monitoring real-time single-molecule DNA and RNA sequencing by a single polymerase enzyme is a promising approach. So far, it has been achieved using circular nanoholes milled in an aluminum film (also called zero-mode waveguides (Levene, 2003; Eid, 2009; Uemura, 2010)). Each nanohole provides an observation chamber for watching the activity of a single polymerase enzyme at micromolar concentration of fluorescently tagged oligonucleotides, each base (A,T,G,C) is color-coded with a spectrally distinct fluorophore. Every time a fluorescent nucleotide is incorporated into the DNA strand, there is a burst of fluorescence of millisecond duration, which allows retrieving the corresponding sequence as the replication is performed. A few thousand of nanoholes are currently monitored simultaneously, enabling massive parallelization (Eid, 2009). So far, only circular aluminum nanoholes have been used for this application, but the sensitivity and detection rates could greatly benefit from the higher fluorescence enhancement offered by plasmonic nanoantennas.

6.4.2 Nanoscale Organization of Lipid Membranes

Recent progress in cell biology indicates that the cell membrane features transient and fluctuating nanoscale assemblies of sterol and sphingolipids known as lipid rafts (Sezgin, 2017). Investigating the role and formation of these nanodomains is of key interest for cell biology. However, the nanometer and microsecond resolutions that are simultaneously required fall far beyond the reach of standard microscopes (Eggeling, 2009). With their ability to confine light into nanometer dimensions and drastically improve the fluorescence signal from a single molecule, optical nanoantennas offer a promising approach to investigate the nanoscale dynamic organization of living cell membranes (Lohmüller, 2012; Flynn, 2017; Pradhan, 2016; Winkler, 2017; Regmi, 2017). The flat geometry of the lipid bilayer membrane is remarkably well suited for the investigation by planar nanoantenna designs. Demonstrations on model lipid membranes (Winkler, 2017) and CHO cell membranes (Regmi, 2017) highlight the potential of nanoantennas to reveal nanodomains of 10 nm dimensions and sub-millisecond characteristic times. This fully biocompatible approach opens interesting opportunities for living cell biophysics with single-molecule sensitivity at ultrahigh spatial and temporal resolutions.

6.4.3 Förster Resonance Energy Transfer

As we have seen, optical nanoantennas can enhance the fluorescence signal from single quantum emitters. Interestingly, they can also be applied to enhance and modify the Förster resonance energy transfer (FRET) between two nearby fluorescent emitters. FRET is one of the most widely used single-molecule fluorescence techniques, with many applications to investigate the conformation dynamics of large molecules (Weiss, 2000) or perform biosensing (Medintz, 2003). Metal nanoapertures (Ghenuche, 2014; deTorres, 2015) and nanogap antennas (Ghenuche, 2015; Bidault, 2016b; deTorres, 2016) have been shown to significantly enhance the FRET rate while simultaneously enabling single-molecule FRET at micromolar concentrations. Interestingly, the FRET gain is larger for more distant donor and acceptor molecules, which is interesting for the investigation of biochemical structures with donor–acceptor distances much beyond the classical Förster radius (Ghenuche, 2014, 2015). Moreover, the strongly inhomogeneous electromagnetic fields in plasmonic nanoantennas was also demonstrated to enhance FRET between nearly perpendicular donor and acceptor dipoles, enabling FRET detection that would otherwise be forbidden in a homogeneous confocal volume (deTorres, 2016).

6.5 Conclusion

Optical nanoantennas overcome the classical diffraction limits of confocal microscopes and confine light down to dimensions close to the size of a single molecule. This enables to drastically improve the fluorescence emission rate, notably for low quantum yield emitters. Thanks to the recent advances in the nanofabrication, either top-down or bottom-up, the single-molecule fluorescence toolbox is now significantly expanded. New directions involve single-molecule fluorescence dynamics studies at physiologically relevant micromolar concentrations and living cell membrane studies at ultrahigh spatiotemporal resolutions. The potential of nanoantennas is at hand to reveal new information on biological functions and dynamics.

References

Acuna, G. P., Möller, F. M., Holzmeister, P., Beater, S., Lalkens, B., Tinnefeld, P. (2012). Fluorescence enhancement at docking sites of DNA-directed self-assembled nanoantennas. *Science*, 338, 506–510.

Akselrod, G. M., Argyropoulos, C., Hoang, T. B., Ciraci, C., Fang, C., Huang, J., Smith, D. R., Mikkelsen, M. H. (2014). Probing the mechanisms of large Purcell enhancement in plasmonic nanoantennas. *Nat. Photonics*, 8, 835–840.

Albella, P., Poyli, M. A., Schmidt, M. K., Maier, S. A., Moreno, F., Saenz, J. J., Aizpurua, J. (2013). Low-loss electric and magnetic field-enhanced spectroscopy with subwavelength silicon dimers. *J. Phys. Chem. C*, 117, 13573–13584.

Biagioni, P., Huang, J. S., Hecht, B. (2012). Nanoantennas for visible and infrared radiation. *Rep. Prog. Phys.*, 75, 024402.

Bidault, S., Devilez, A., Ghenuche, P., Stout, B., Bonod, N., Wenger, J. (2016). Competition between Förster resonance energy transfer and donor photodynamics in plasmonic dimer nanoantennas. *ACS Photonics*, 3, 895–903.

Bidault, S., Devilez, A., Maillard, V., Lermusiaux, L., Guigner, J.-M., Bonod, N., Wenger, J. (2016). Picosecond lifetimes with high quantum yields from single-photon emitting colloidal nanostructures at room temperature. *ACS Nano*, 10, 4806–4815.

Busson, M. P., Rolly, B., Stout, B., Bonod, N., Bidault, S. (2012). Accelerated single photon emission from dye molecule-driven nanoantennas assembled on DNA. *Nat. Commun.*, 3, 962.

Busson, M. P., Rolly, B., Stout, B., Bonod, N., Larquet, E., Polman, A., Bidault, S. (2011). Optical and topological characterization of gold nanoparticle dimers linked by a single DNA double strand. *Nano Lett.*, 11, 5060–5065.

Busson, M. P., Rolly, B., Stout, B., Bonod, N., Wenger, J., Bidault, S. (2012). Photonic engineering of hybrid metal-organic chromophores. *Angew. Chem. Int. Ed.*, 51, 11083–11087.

Caldarola, M., Albella, P., Corts, E., Rahmani, M., Roschuk, T., Grinblat, G., Oulton, R. F., Bragas, A. V., Maier, S. A. (2015). Non-plasmonic nanoantennas for surface enhanced spectroscopies with ultra-low heat conversion. *Nat. Commun.*, 6, 7915.

Cambiasso, J., Grinblat, G., Li, Y., Rakovich, A., Cortes, E., Maier, S. A. (2017). Bridging the gap between dielectric nanophotonics and the visible regime with effectively lossless gallium phosphide antennas. *Nano Lett.*, 17, 1219–1225.

Choudhury, S. D., Ray, K., Lakowicz, J. R. (2012). Silver nanostructures for fluorescence correlation spectroscopy: Reduced volumes and increased signal intensities. *J. Phys. Chem. Lett.*, 3, 2915–2919.

Chu, X.-L., Brenner, T., Chen, X.-W., Ghosh, Y., Hollingsworth, J., Sandoghdar, V., Götzinger, S. (2014). Experimental realization of an optical antenna designed for collecting 99% of photons from a quantum emitter. *Optica*, 1, 203–208.

de Torres, J., Ghenuche, P., Moparthi, S. B., Grigoriev, V., Wenger, J. (2015). FRET enhancement in aluminum zero-mode waveguides. *Chem. Phys. Chem.*, 16, 782–788.

de Torres, J., Mivelle, M., Moparthi, S. B., Rigneault, H., Van Hulst, N. F., Garcia-Parajo, M. F., Margeat, E., Wenger, J. (2016). Plasmonic nanoantennas enable forbidden Förster dipole-dipole energy transfer and enhance the FRET efficiency, *Nano Lett.*, 16, 6222–6230.

Garcia-Etxarri, A., Gomez-Medina, R., Froufe-Perez, L. S., Lopez, C., Chantada, L., Scheffold, F., Aizpurua, J., Nieto-Vesperinas, M., Saenz, J. J. (2011). Strong magnetic response of submicron silicon particles in the infrared. *Opt. Express*, 19, 4815–4826.

Edel, J. B., Kornyshev, A. A., Urbakh, M. (2013). Self-assembly of nanoparticle arrays for use as mirrors, sensors, and antennas. *ACS Nano*, 7, 9526–9532.

Eggeling, C., Ringemann, C., Medda, R., Schwarzmann, G., Sandhoff, K., Polyakova, S., Belov, V. N., Hein, B., von Middendorff, C., Schönle, A., Hell, S. W. (2009). Direct observation of the nanoscale dynamics of membrane lipids in a living cell. *Nature*, 457, 1159–1162.

Eid, J., Fehr, A., Gray, J., Luong, K., Lyle, J., Otto, G., Peluso, P., Rank, D., Baybayan, P., Bettman, B., Bibillo, A., Bjornson, K., Chaudhuri, B., Christians, F., Cicero, R., Clark, S., Dalal, R., deWinter, A., Dixon, J., Foquet, M., Gaertner, A., Hardenbol, P., Heiner, C., Hester, K., Holden, D., Kearns, G., Kong, X., Kuse, R., Lacroix, Y., Lin, S., Lundquist, P., Ma, C., Marks, P., Maxham, M., Murphy, D., Park, I., Pham, T., Phillips, M., Roy, J., Sebra, R., Shen, G., Sorenson, J., Tomaney, A., Travers, K., Trulson, M., Vieceli, J., Wegener, J., Wu, D., Yang, A., Zaccarin, D., Zhao, P., Zhong, F., Korlach, J., Turner, S. (2009). Real-time DNA sequencing from single polymerase molecules. *Science*, 323, 133–138.

Estrada, L. C., Aramendia, P. F., Martinez, O. E. (2008). 10000 times volume reduction for fluorescence correlation spectroscopy using nanoantennas. *Opt. Express*, 16, 20597–20602.

Evlyukhin, A. B., Novikov, S. M., Zywietz, U., Eriksen, R. L., Reinhardt, C., Bozhevolnyi, S. I., Chichkov, B. N. (2012). Demonstration of magnetic dipole resonances of dielectric nanospheres in the visible region. *Nano Lett.*, 12, 3749–3755.

Fan, J. A., He, Y., Bao, K., Wu, C., Bao, J., Schade, N. B., Manoharan, V. N., Shvets, G., Nordlander, P., Liu, D., Capasso, F. (2011). DNA-enabled self-assembly of plasmonic nanoclusters. *Nano Lett.*, 11, 4859–4864.

Fan, J. A., Wu, C., Bao, K., Bao, J., Bardhan, R., Halas, N. J., Manoharan, V. N., Nordlander, P., Shvets, G., Capasso, F. (2010). Self-assembled plasmonic nanoparticle clusters. *Science*, 328, 1135–1138.

Flauraud, V., Regmi, R., Winkler, P. M., Alexander, D. T. L., Rigneault, H., van Hulst, N. F., García-Parajo, M. F., Wenger, J., Brugger, J. (2017). In-plane plasmonic antenna arrays with surface nanogaps for giant fluorescence enhancement. *Nano Lett.*, 17, 1703–1710.

Flauraud, V., van Zanten, T. S., Mivelle, M., Manzo, C., García-Parajo, M. F., Brugger, J. (2015). Large-scale arrays of bowtie nanoaperture antennas for nanoscale dynamics in living cell membranes. *Nano Lett.*, 15, 4176–4182.

Flynn, J. D., Haas, B. L., Biteen, J. S. (2017). Plasmon-enhanced fluorescence from single proteins in living bacteria. *J. Phys. Chem. C*, 120, 20512–20517.

Ghenuche, P., de Torres, J., Moparthi, S. B., Grigoriev, V., Wenger, J. (2014). Nanophotonic enhancement of the Förster resonance energy transfer rate with single nanoapertures. *Nano Lett.*, 14, 4707–4714.

Ghenuche, P., Mivelle, M., de Torres, J., Moparthi, S. B., Rigneault, H., Van Hulst, N. F., Garcia-Parajo, M. F., Wenger, J. (2015). Matching nanoantenna field confinement to FRET distances enhances Förster energy transfer rates. *Nano Lett.*, 15, 6193–6201.

Gill, R., Tian, L., Somerville, W. R. C., Le Ru, E. C., van Amerongen, H., Subramaniam, V. (2012). Silver nanoparticle aggregates as highly efficient plasmonic antennas for fluorescence enhancement. *J. Phys. Chem. C*, 116, 16687–16693.

Grzelczak, M., Vermant, J., Furst, E. M., Liz-Marzán, L. M. (2010). Directed self-assembly of nanoparticles. *ACS Nano*, 4, 3591–3605.

Halas, N. J., Lal, S., Chang, W. S., Link, S., Nordlander, P. (2011). Plasmons in strongly coupled metallic nanostructures. *Chem. Rev.*, 111, 3913–3961.

Hoang, T. B., Akselrod, G. M., Argyropoulos, C., Huang, J. N., Smith, D. R., Mikkelsen, M. H. (2015). Ultrafast spontaneous emission source using plasmonic nanoantennas. *Nat. Commun.*, 6, 7788.

Hoang, T. B., Akselrod, G. M., Mikkelsen, M. H. (2016). Ultrafast room-temperature single photon emission from quantum dots coupled to plasmonic nanocavities. *Nano Lett.*, 16, 270–275.

Holzmeister, P., Acuna, G. P., Grohmann, D., Tinnefeld, P. (2013). Breaking the concentration limit of optical single-molecule detection. *Chem. Soc. Rev.*, 43, 1014–1028.

Khatua, S., Paulo, P. M. R., Yuan, H., Gupta, A., Zijlstra, P., Orrit, M. (2014). Resonant plasmonic enhancement of single-molecule fluorescence by individual gold nanorods. *ACS Nano*, 8, 4440–4449.

Khatua, S., Yuan, H., Orrit, M. (2015). Enhanced-fluorescence correlation spectroscopy at micro-molar dye concentration around a single gold nanorod. *Phys. Chem. Chem. Phys.*, 17, 21127–21132.

Kinkhabwala, A., Yu, Z. F., Fan, S. H., Avlasevich, Y., Mullen, K., Moerner, W. E. (2009). Large single-molecule fluorescence enhancements produced by a bowtie nanoantenna. *Nat. Photonics*, 3, 654–657.

Kinkhabwala, A. A., Yu, Z. F., Fan, S. H., Moerner, W. E. (2012). Fluorescence correlation spectroscopy at high concentrations using gold bowtie nanoantennas. *Chem. Phys.*, 406, 3–8.

Koenderink, A. F. (2017). Single-photon nanoantennas. *ACS Photonics*, 4, 710–722.

Kuznetsov, A. I., Miroshnichenko, A. E., Fu, Y. H., Zhang, J., Luk'yanchuk, B. (2012). *Magentic light. Sci. Rep.*, 2, 492.

Lakowicz, J. R. (2006). *Principles of Fluorescence Spectroscopy*, 3rd ed. (Springer, Berlin/New York).

Lakowicz, J., Fu, Y. (2009). Modification of single molecule fluorescence near metallic nanostructures. *Laser Photonics Rev.*, 3, 221–232.

Lee, K. G., Chen, X. W., Eghlidi, H., Kukura, P., Lettow, R., Renn, A., Sandoghdar, V., Götzinger, S. (2011). A planar dielectric antenna for directional single-photon emission and near-unity collection efficiency. *Nat. Photonics*, 5, 166–169.

Levene, M. J., Korlach, J., Turner, S. W., Foquet, M., Craighead, H. G., Webb, W. W. (2003). Zero-mode waveguides for single-molecule analysis at high concentrations. *Science*, 299, 682–686.

Lohmüller, T., Iversen, L., Schmidt, M., Rhodes, C., Tu, H. L., Lin, W. C., Groves, J. T. (2012). Single molecule tracking on supported membranes with arrays of optical nanoantennas. *Nano Lett.*, 12, 1717–1721.

Lu, G. W., Liu, J., Zhang, T. Y., Li, W. Q., Hou, L., Luo, C. X., Lei, F., Manfait, M., Gong, Q. H. (2012). Plasmonic near-field in the vicinity of a single gold nanoparticle investigated with fluorescence correlation spectroscopy. *Nanoscale*, 4, 3359–3364.

Medintz, I. L., Clapp, A. R., Mattoussi, H., Goldman, E. R., Fisher, B., Mauro, J. M. (2003). Self-assembled nanoscale biosensors based on quantum dot FRET donors. *Nat. Materials*, 2, 630–638.

Moran-Mirabal, J. M., Craighead, H. G. (2008). Zero-mode waveguides: Subwavelength nanostructures for single molecule studies at high concentrations. *Methods*, 46, 11–17.

Novotny, L., Hecht, B. (2006). *Principles of Nano-Optics*. Cambridge University Press, Cambridge.

Novotny, L., van Hulst, N. (2011). Antennas for light. *Nat. Photonics*, 5, 83–90.

Pradhan, B., Khatua, S., Gupta, A., Aartsma, T., Canters, G., Orrit, M. (2016). Gold-nanorod-enhanced fluorescence correlation spectroscopy of fluorophores with high quantum yield in lipid bilayers. *J. Phys. Chem. C*, 120, 25996–26003.

Puchkova, A., Vietz, C., Pibiri, E., Wünsch, B., Sanz Paz, M., Acuna, G. P., Tinnefeld, P. (2015). DNA origami nanoantennas with over 5000-fold fluorescence enhancement and single-molecule detection at 25 μM. *Nano Lett.*, 15, 8354–8359.

Punj, D., de Torres, J., Rigneault, H., Wenger, J. (2013). Gold nanoparticles for enhanced single molecule fluorescence analysis at micromolar concentration. *Opt Express*, 21, 27338–27343.

Punj, D., Ghenuche, P., Moparthi, S. B., de Torres, J., Grigoriev, V., Rigneault, H., Wenger, J. (2014). Plasmonic antennas and zero-mode waveguides to enhance single molecule fluorescence detection and fluorescence correlation spectroscopy toward physiological concentrations. *WIREs Nanomed. Nanobiotechnol.*, 6, 268–282.

Punj, D., Mivelle, M., Moparthi, S. B., van Zanten, T. S., Rigneault, H., van Hulst, N. F., García-Parajo, M. F., Wenger, J. (2013). A plasmonic 'antenna-in-box' platform for enhanced single-molecule analysis at micromolar concentrations. *Nat. Nanotechnol.*, 8, 512–516.

Punj, D., Regmi, R., Devilez, A., Plauchu, R., Moparthi, S. B., Stout, B., Bonod, N., Rigneault, H., Wenger, J. (2015). Self-assembled nanoparticle dimer antennas for plasmonic-enhanced single-molecule fluorescence detection at micromolar concentrations. *ACS Photonics*, 2, 1099–1107.

Regmi, R., Berthelot, J., Winkler, P. M., Mivelle, M., Proust, J., Bedu, F., Ozerov, I., Begou, T., Lumeau, J., Rigneault, H., Garcia-Parajo, M. F., Bidault, S., Wenger, J., Bonod, N. (2016). All-dielectric silicon nanogap antennas to enhance the fluorescence of single molecules. *Nano Lett.*, 16, 5143–5151.

Regmi, R., Winkler, P. M., Flauraud, V., Borgman, K. J. E., Manzo, C., Brugger, J., Rigneault, H., Wenger, J., García-Parajo, M. F. (2017). Planar optical nano-antennas resolve cholesterol-dependent nanoscale heterogeneities in the plasma membrane of living cells. *Nano Lett.*, 17(10), 6295–6302.

Schuller, J. A., Barnard, E. S., Cai, W. S., Jun, Y. C., White, J. S., Brongersma, M. L. (2010). Plasmonics for extreme light concentration and manipulation. *Nat. Materials*, 9, 193–204.

Sezgin, E., Levental, I., Mayor, S., Eggeling, C. (2017). The mystery of membrane organization: Composition, regulation and roles of lipid rafts. *Nat. Rev. Mol. Cell Biol.*, 18, 361–374.

Uemura, S., Aitken, C. E., Korlach, J., Flusberg, B. A., Turner, S. W., Puglisi, J. D. (2010). Real-time tRNA transit on single translating ribosomes at codon resolution, *Nature*, 464, 1012–1017.

Vietz, C., Kaminska, I., Sanz Paz, M., Tinnefeld, P., Acuna, G. P. (2017). Broadband fluorescence enhancement with self-assembled silver nanoparticle optical antennas. *ACS Nano*, 11, 4969–4975.

Wang, Q., Lu, G., Hou, L., Zhang, T., Luo, C., Yang, H., Barbillon, G., Lei, F. E., Marquette, C. A., Perriat, P., Tillement, O., Roux, S., Ouyang, Q., Gong, Q. (2011). Fluorescence correlation spectroscopy near individual gold nanoparticle. *Chem. Phys. Lett.*, 503, 256–261.

Weiss, S. (2000). Measuring conformational dynamics of biomolecules by single molecule fluorescence spectroscopy. *Nat. Struct. Biol.*, 7, 724–729.

Wenger, J. (2012). Fluorescence enhancement factors on optical antennas: Enlarging the experimental values without changing the antenna design. *Int. J. Optics*, 2012, 828121.

Wenger, J., Rigneault, H. (2010). Photonic methods to enhance fluorescence correlation spectroscopy and single molecule fluorescence detection. *Int. J. Mol. Sci.*, 11, 206–221.

Winkler, P. M., Regmi, R., Flauraud, V., Brugger, J., Rigneault, H., Wenger, J., García-Parajo, M. F. (2017). Transient nanoscopic phase separation in biological lipid membranes resolved by planar plasmonic antennas. *ACS Nano*, 11, 7241–7250.

Yuan, H., Khatua, S., Zijlstra, P., Yorulmaz, M., Orrit, M. (2013). Thousand-fold enhancement of single-molecule fluorescence near a single gold nanorod. *Angew. Chem.*, 125, 1255–1259.

Zander, C., Enderlein, J., Keller, R. A. (2002). (Eds.), *Single-Molecule Detection in Solution: Methods and Applications* (VCH-Wiley, Berlin/New York).

Zhang, T., Gao, N., Li, S., Lang, M. J., Xu, Q. H. (2015). Single-particle spectroscopic study on fluorescence enhancement by plasmon coupled gold nanorod dimers assembled on DNA origami, *J. Phys. Chem. Lett.*, 6, 2043–2049.

Chapter 7

Plasmonic-Based SERS-Traceable Drug Nanocarriers in Cancer Theranostics

Monica Potara,[a] Timea Nagy-Simon,[a] Sorina Suarasan,[a] and Simion Astilean[a,b]

[a]*Nanobiophotonics and Laser Microspectroscopy Center, Interdisciplinary Research Institute in Bio-Nano-Sciences, Babes-Bolyai University, T. Laurian Str. 42, Cluj-Napoca 400271, Romania*
[b]*Department of Biomolecular Physics, Faculty of Physics, Babes-Bolyai University, M. Kogalniceanu Str. 1, Cluj-Napoca 400084, Romania*

monica.potara@phys.ubbcluj.ro, timea.simon@ubbcluj.ro, sorina.suarasan@ubbcluj.ro, simion.astilean@phys.ubbcluj.ro

The development of drug nanocarriers which integrate sensitive diagnostic, optical imaging and targeted therapeutic abilities into single theranostic nanoparticle heralds a new paradigm in the fight against cancer. Owing to their biocompatibility, reduced size, feasible surface modification and unique optical properties, plasmonic nanoparticles appear as one of the most promising building blocks able to merge the specific targeting and inhibition of tumor progression with simultaneous drug localization and real-time monitoring of the therapeutic efficiency. Indeed, due

Plasmonics in Chemistry and Biology
Edited by Marc Lamy de la Chapelle and Nordin Felidj

ISBN 978-981-4800-03-7 (Hardcover), 978-0-429-45875-0 (eBook)
www.jennystanford.com

to localized surface plasmon resonance (LSPR) supported by plasmonic nanoparticles, surface-enhanced Raman scattering (SERS) can be implemented not only as an ideal tool to perform the diagnostic with ultrahigh sensitivity and multiplexing capability but also as imaging contrast for the tracking of drug nanocarriers inside cells and tumors. This chapter reviews the recent progress in literature related to the development of such SERS-traceable drug nanocarriers in cancer theranostics, with a particular focus on chemotherapeutic drugs and photosensitizers delivery systems. The prospect of plasmonic-based SERS-traceable drug nanocarriers is relevant for future personalized diagnostic and treatment with improved specificity and therapeutic effect.

7.1 Introduction

The emergence of nanotechnology for medicine holds great promise to achieve a new paradigm and revolutionary advances in the fight against cancer by providing new solutions from earlier diagnostics to better, more efficient, and more targeted therapy as well as accurate imaging. Nowadays, a hot research topic is developing at the interface between physics, chemistry and materials science from one side and biology and medicine from other side, aiming to provide novel nanotools for cancer diagnostic and treatment. Several innovative nanomaterials integrating good biocompatibility with sensitive diagnostic, optical imaging and targeted therapeutic abilities have been fabricated and tested [12]. Among them, noble-metal (gold, silver) nanoparticles (plasmonic nanoparticles, PNPs) represent a distinct class of nanoobjects with enormous potential in cancer diagnostic, therapy and treatment monitoring due to their highly attractive physical, chemical, optical and electronic properties. In particular, PNPs exhibit unique optical properties due to their strong interaction with light. Actually the incoming light couples with the oscillation frequency of the conduction electrons in PNPs and a so-called localized surface plasmon resonance (LSPR) arises, which is manifested as a strong UV-visible-NIR absorption band, depending of their size and shape [24]. Strong light absorption by PNPs at LSPR with subsequent energy dissipation in heat has promoted plasmonic-induced photothermal cancer therapy [56]. Another consequence of plasmon excitation is the high electromagnetic field created at the surface of the PNPs under resonant excitation. The effect is currently exploited in sensing and biodetection via a number of surface-enhanced spectroscopies such as surface-enhanced Raman spectroscopy (SERS), metal-enhanced fluorescence (MEF), etc. [69]. However, some of these techniques have several limitations in investigating biological systems. For example, MEF does not provide information about the nature of the adsorbed molecules, thus being unsuitable for the identification of unknown molecules. In addition, the spectral overlapping of fluorescence has limited applications in multiplex analysis. In contrast to fluorescence, SERS is a highly specific spectroscopic technique which allows identifying molecular species and providing structural information from the unique vibrational Raman fingerprint [76]. Beyond ultrasensitive analytical detection by SERS, an increasing interest is devoted nowadays to the development of a novel class of nanoprobes consisting of PNPs encoded with strong Raman active molecules (named Raman reporter molecules). This kind of optical nanoprobes called "SERS nanotags" are commonly used for targeted detection of cancer biomarkers, noninvasive imaging of living cells and tumors and monitoring the response to therapy [82].

Apart from their unique optical properties, PNPs exhibit several other characteristics which make them attractive candidates for implementation in nanomedicine. For example, the nanometric size of PNPs allows them to be easy integrated in biological systems. In addition, nanosized particles can selectively accumulate into tumor tissues, by a mechanism termed as enhanced permeability and retention (or EPR) effect [71]. This is possible due to the altered anatomy of tumor vessels which are irregular in shape, dilated, leaky and can present fenestrations in endothelial cells. Moreover, the surface chemistry of PNPs allows the conjugation with specific biomolecules, which are able to endow stability and biocompatibility and to make them receptive to specific target organs or pathologic sites. These properties have led to new nanotechnological solutions in cancer treatment by utilizing PNPs as targeted drug nanocarriers in order to achieve therapeutic dosing in the site of disease, thus resulting in a more effective treatment with minimum side effects.

There are a lot of examples in the recent literature where PNPs have been integrated either in cancer diagnostic as image-guided agents [41] or in cancer therapy as drug carriers [71]. However, the use of separate agents for diagnosis and treatment can lead to undesirable complications in reading the biodistribution and selectivity of designed agents. Nowadays, the research is advancing toward the integration of several functionalities as diagnostic, therapeutic and imaging capabilities into a single nanoparticle system. A new term is created, theranostics (therapy + diagnostics), to describe this promising generation of new medical nanotools based upon the fusion of therapeutics and diagnostics. The use of theranostic nanoparticles will enable simultaneous targeted drug delivery, noninvasive imaging of the diseased tissue and treatment monitoring, all in one system, resulting in a more effective and personalized medical treatment [35, 61].

Aiming at the enhancement of the potency and safety of future cancer treatment, innovative multifunctional theranostic PNPs have been fabricated and tested in vitro with cancer cells and the research field is steadily progressing toward demonstration in preclinical animal cancer models. Among those PNPs employed in cancer theranostics, SERS-traceable drug nanocarriers based on SERS nanotags loaded with anticancer drugs possesses excellent functionalities of noninvasive, real-time quantitative evaluation of the dynamic process of drug delivery and controlled release inside cells and tumors, combined with targeted therapeutic abilities.

In this chapter, we highlight the recent progress in the development of SERS-traceable drug nanocarriers for cancer theranostics. We start with a brief discussion on the fabrication of SERS nanotags, followed by presenting some examples related to their applications in detection and imaging of cancer cells and tumors. Then, we review the contributions in the field of theranostic agents suitable for targeted delivery of chemotherapeutic drugs and photosensitizers and SERS-guided diagnosis and treatment.

7.2 SERS Encoded Plasmonic Nanoparticles for Cancer Detection and Imaging

Typically, a SERS nanotag consists of at least three components: (i) a metal nanoparticle, (ii) a Raman reporter molecule anchored onto the surface of nanoparticles (a molecule with a large Raman cross section), and (iii) a coating layer to render the tag with stability and biocompatibility. Thereby a "core–shell" structure is formed with the outer shell potentially open to be conjugated with capture molecules such as antibodies, peptides, etc., for specific targeting. Within this particular architecture, the SERS-encoded nanoparticles are easily identified by the specific SERS signal of reporter molecules (spectral fingerprint) [82].

The development of SERS nanotags has spawned new opportunities for a wide range of applications in oncology, such as traceable drug delivery in cancer therapy for image-guided therapy, indirect and highly sensitive detection of cancer biomarkers for early cancer diagnostic, etc. [41]. However, these applications strongly depend on the optical stability, sensitivity and biocompatibility of SERS nanotags. According to the electromagnetic mechanism (EM), the enhancement in SERS arises from Raman laser excitation and emission coupling with the nanoparticle LSPR modes, and a relationship between the LSPR spectrum and the SERS spectrum collected from a SERS-active substrate results [5, 14, 25, 26, 93]. It was demonstrated that the greatest Raman enhancement can be achieved when the LSPR λ_{max} falls between the excitation wavelength and the wavelength of the scattered photon [26, 49, 93]. In other words, there is a complementary nature of LSPR and SERS and several efforts have been invested for tuning the plasmonic response of nanoparticles in order to optimize the SERS signals [23, 25, 36, 65]. Moreover, for in vivo applicability is highly desirable to design SERS nanotags that exhibit strong SERS signal under near infrared (NIR) excitation where the intrinsic tissue chromophores and water molecules exhibit minimal absorption and scattering for the excitation light, providing maximum irradiation penetration through tissue. Therefore, choosing the metal core, size, structure, and coating agent is a key issue for fabricating SERS nanotags with desired optical properties. The most commonly used metal cores are individual nanoparticles, especially colloidal gold and silver due to their small size and consistent SERS signal. A number of SERS tags for targeted cancer imaging in live cells have been developed lately based on noble-metal nanoparticles, starting from the simplest spherical shapes to the more complicated anisotropic or bimetallic/hybrid nanostructures. Although early SERS studies were performed with silver nanoparticles to monitor the intracellular distribution of drugs, gold nanoparticles (AuNPs) have been considered most suitable as bioimaging platforms due to their good biocompatibility and long-term stability [2]. A systematic study performed on the amplification efficiency of the spherical AuNPs of different sizes revealed that 40–60 nm nanoparticles provide the highest signal intensity [33]. Moreover, it was also shown that the internalization efficiency into cells of AuNPs is the highest in this size range [13], which made them the most adequate for cellular imaging applications. Successful SERS probing and imaging of individual cells using spherical AuNPs conjugated with rose bengal or crystal violet as reporter molecules was performed by Kneipp et al. [40]. The authors demonstrated that the prepared labels present good stability, were not toxic to cells, and did not suffer photobleaching. Moreover, once internalized by cells, the nanotags also provided sensitive spectral information on their immediate molecular environment enabling imaging of the native constituents of the cell. However, spherical nanoparticles exhibit plasmonic bands in the visible region which limit their in vivo applicability. In addition, these individual nanoparticles show a rather small Raman enhancement and their intracellular SERS efficiency is mostly attributed to the aggregation of nanoparticles inside the endosomal cellular compartments which results in the formation of hot-spots between interconnecting nanoparticles thus providing high SERS enhancement [39]. However, this aggregation process can hardly be controlled, which can result in undesired effects. Instead, controlled aggregation and stabilization of the nanoparticles for intracellular SERS imaging application could be a good option to facilitate the formation of hot spots between nanoparticles. On the other hand, the plasmon coupling between interconnected nanoparticles results in a shift of the plasmon bands toward the NIR spectral region, desired for excitation of biological samples. Several strategies for controlled aggregation of PNPs in solution have been reported in literature and also in our group, mostly using polymers to encapsulate and maintain the formed structures [9, 66, 75]. For instance, NIR responsive small ensembles of AuNPs (dimers and trimers) enveloped in thiolated PEG operating as SERS nanotags were recently reported in our group [9]. The peculiarity of these tags is that no further encoding with reporters was necessary as a fraction of the ascorbic acid molecules used as reducing agent remained attached to the nanoparticles, providing persistent and strong SERS to be detectable inside human adult retinal pigment epithelial (D407) cells. In another study, we presented the fabrication of close-packed assemblies of AuNPs using salt as aggregation agent in the presence of methanol, which were stabilized by Pluronic F127 [66]. Conjugated with methylene blue as reporter molecule, these nano-aggregates provided a robust SERS signal inside cancer cells which enabled intracellular nanoparticles localization and cellular mapping at a 3D level at 633 nm laser excitation.

The necessity of the excitation in the biological window for in vivo studies has raised the research interest toward NIR-absorbing Raman reporters. Resonance excitation of SERS tags, referred as surface-enhanced resonance Raman scattering (SERRS), can further increase the SERS signals by 10- to 100-fold [50] which can be sensitive enough for single cancer cells studies. As an example, Qian et al. fabricated NIR-active SERRS nanotags using an infrared dye, diethylthiatricarbocyanine (DTTC), as Raman reporter conjugated to spherical AuNPs stabilized by thiolated PEG [60]. By anchoring with single-chain variable fragment (ScFv) antibodies, the conjugated nanoparticles were able to actively target both cancer cells and xenograft tumors in animal models. Furthermore, the obtained NIR-active nanotags were more than 200 times brighter than NIR emitting quantum dots and allowed spectroscopic detection of small tumors (0.03 cm^3) at a penetration depth of 1–2 cm. In another study, ultrasensitive targeted NIR nanoprobes for tumor detection have been developed based on 60 nm diameter gold nanospheres conjugated with newly designed NIR reporters [63]. The synthesis and screening of an 80-member tricarbocyanine library led to the identification of CyNAMLA-381 as an optimal NIR SERS reporter surpassing the sensitivity of the DTTC standard by 12-fold. The scFv antiHER2 antibody-targeted nanoconjugates showed good selectivity toward HER2-positive cancer cells and xenograft models and displayed good SERRS imaging capability both in vitro and in vivo at NIR laser excitation (785 nm).

Alternatively to spherical or aggregated nanoparticles, anisotropic PNPs, due to the localized hot-spots at their corners and edges offer the major advantage that can operate individually as a highly sensitive and reproducible SERS tags [8, 57, 59]. Furthermore, the tunability of their LSPR and their multiple plasmonic resonances allow the use of a wide range of laser excitation from visible to NIR. Among various anisotropic nanoparticles, gold nanorods (AuNRs) are probably the most intensely exploited nanoparticles in SERS bio-imaging applications due to the large and controllable tunability across the visible and NIR region, good chemical stability and biocompatibility under biological conditions [32, 48].

For instance, mercaptopyridine-tagged antibody-conjugated AuNRs have been successfully used by Park et al. for targeting and imaging MCF7 cells overexpressing breast cancer marker HER2 24. The authors confirmed the superior SERS imaging performance of GNRs compared to gold nanospheres, attributed to the strong longitudinal surface plasmon oscillation and the creation of a higher density of junction points (Fig. 7.1).

Multi-branched AuNPs, such as nanostars, urchin- and flower-like shapes, etc., are another class of anisotropic nanoparticles with tunable LSPR in the NIR optical window, also presenting multiple sharp "hot-spots", which make this type of nanoparticles widely exploited in intracellular SERS imaging applications [7, 84]. As an example, PEG-protected and basic-fucsin encoded flower-shaped AuNPs were fabricated in our group as SERS-active tags for cellular spectral detection [8]. The reliability of the SERS tags was demonstrated on epithelial cells from human retina (D407), which showed a low level of toxicity for the concentrations required for SERS investigation.

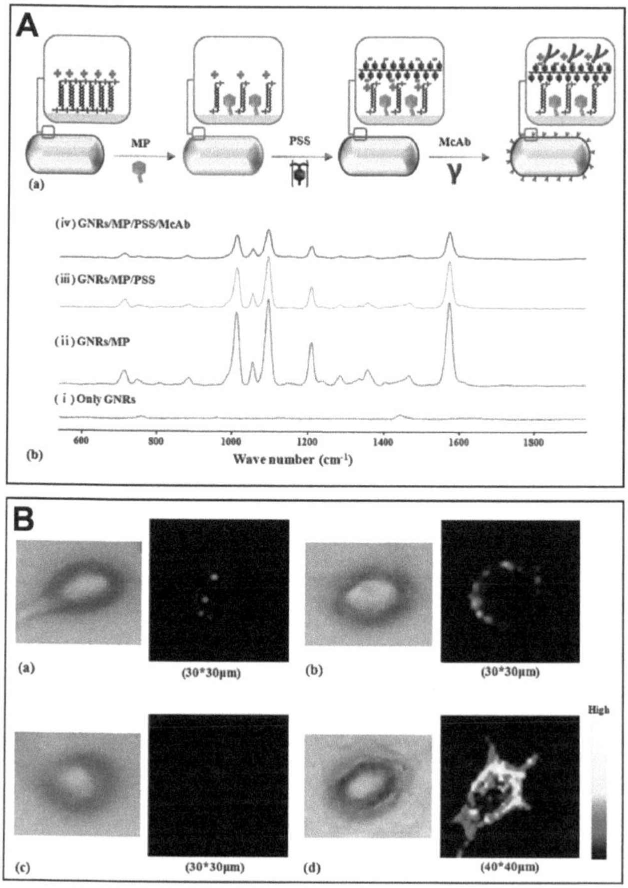

Figure 7.1 (A) Schematic illustration of the layer-by-layer process of assembling antibodies onto the nanorod surface through electrostatic interactions, (b) corresponding SERS spectra for those processes: (i) GNRs/MP; (ii) PSS powder only; (iii) GNPs/MP/PSS; (iv) GNPs/MP/PSS/McAb. Notes: GNRs = gold nanorods; MP = 4-mercaptopyridine; PSS = poly(sodium 4-styrene-sulfonate); McAb = monoclonal anti-rabbit antibody. (B) SERS mapping images of (a) MCF7/WILD using GNSs, (b) MCF7/HER2 using GNSs, (c) MCF7/WILD using GNRs and (d) MCF7/HER2 using GNRs. In (a), (b), (c), and (d), the left pictures are bright-field images, while the right images are Raman mapping images at the 1095 cm^{-1} peak of Raman reporter MP. All of the mapping images were scanned with an interval of 1 μm, and the numbers in parentheses denote the mapping pixel sizes. Reproduced from reference [57] with permission of The Royal Society of Chemistry.

A multi-wavelength responsive high-sensitivity SERS nanotag based on chitosan-coated silver nanotriangles (Chit-AgNTs) labeled with para-aminothiophenol (p-ATP) has been also reported in our group [59]. In addition to preventing nanoparticle aggregation, the chitosan layer demonstrated its role in conferring biocompatibility to the silver nanoparticles toward lung cancer cells (A549) without interfering with cell proliferation. The intracellular SERS performance of the obtained nanotags was demonstrated under three laser lines excitation (532, 633 and 785 nm) which besides the signature of the reporter molecule, provided sensitive molecular information about cellular components offering in this way multiple information, both about nanoparticle localization and specific interaction inside cells.

Other structures, such as gold–silver bimetallic nanoparticles have attracted considerable interest in the development of SERS nanotags due to their composition dependent characteristics [62, 86].

Besides the size and shape control, the usefulness of SERS nanotags in different biomedical applications requires stable and well-defined resonances. For example, a shift of plasmon resonance can reduce the SERS intensity by orders of magnitude. This, in turn, masks the relationship between the unique, size and shape-dependent properties of nanoparticles and the Raman enhancement of the reporter molecules. In biological fluids, colloidal nanoparticles stabilized by electrostatic repulsion are likely to undergo significant modifications such as aggregation, etching, and surface neutralization. Moreover, the proteins present in biological system can be adsorbed on the particle surface, leading to the detachment of the Raman reporter molecules. The biocompatibility of nanoparticles to in vitro and in vivo environments is another important requirement to make SERS encoded nanoparticles suitable for biomedical applications. Therefore, choosing an appropriate protecting agent with excellent biocompatibility to coat the nanoparticles is a key issue that should be properly addressed. The use of (bio)polymers represents an attractive solution to achieve the goal of stabilization of SERS nanotags, especially toward their safety use and further biofunctionalization. The most commonly used coating agents is polyethylene glycol (PEG), a polymer that serves to multiple purposes for SERS nanotags such as biocompatibility, steric stabilization, weak affinity to interfering molecules and excellent abilities to further functionalization for specific cellular targeting [8, 20]. Chitosan, a well-known biopolymer is attracting increasingly more attention in construction of SERS nanotags due to its unique biological and physicochemical characteristics such as excellent biocompatibility, biodegradability, and nontoxicity and excellent affinities to targeting agents via electrostatic or covalent conjugation [59, 87]. Gelatin, a natural, biocompatible, biodegradable, nontoxic and low-cost biopolymer obtained from denatured collagen represents another option to obtain stable SERS nanotags which can be safely integrated in biological applications [42]. Polyvinylpyrrolidone (PVP) and pluronic have also been successfully used to achieve the goal of SERS nanotags stabilization and biocompatibilization [66, 75]. Silica encapsulation, biomolecule coating or liposome encapsulations are other attractive methods to obtain stable, robust and biocompatible SERS nanotags [16, 73, 91].

For further detailed information on SERS nanotags in view of cancer detection and imaging applications, the reader is directed toward other comprehensive materials that are available in the literature [6, 41].

7.3 Combining SERS Imaging with Therapy for Cancer Theranostics

The combination of therapy and diagnosis, also termed theranostics, has shown immense potential in anti-cancer therapies, still dealing with major problems such as inability to deliver sufficient dose of drugs in a controlled manner, lack of specificity, undesired side effects, etc. Therefore, the physical properties of a theranostic agent must follow certain requirements, such as stability under physiological conditions, ability to penetrate through the biological barriers encountered in the way to the targeted tissue or organ, specificity against the cancer, the lowest possible toxicity against healthy tissue and reduced side effects.

The concept of imaging-guided therapy is particularly valuable in case of anticancer drug delivery especially when targeting moieties are used to specifically address certain cancer types. First of all, targeted imaging could provide useful information about tumor localization and dimension, as well as the connection and delimitation of the tumor from surrounding healthy tissues. Tracking of the therapeutic agent could be extremely helpful to determine the maximum level accumulation at the tumor site, to monitor the release of the drug from the nanocarrier and also could give precious indications in the case of photosensitizer drug delivery regarding the irradiation timing and dosimetry. Moreover, the imaging function of theranostic nanoparticles could be exploited to monitor the progression and outcome of diseases after therapy.

Extensive research interest has been recently paid to the development of novel theranostic agents integrating selective targeted cancer cell and tumor imaging by SERS and anticancer treatment using cytotoxic agents such as chemotherapeutic or photosensitizing drugs. The general construction/structure of a theranostic agent for SERS-imaging guided drug delivery is schematically presented in Fig. 7.2 and includes several key components: (i) a component for noninvasive imaging (a PNPs decorated with a Raman reporter molecule), (ii) a polymer coating to impart nanoparticle stability, biocompatibility and provide functional groups for bioconjugation, (iii) a therapeutic agent (anticancer drug), and (iv) a targeting moiety (antibody, folic acid, peptides, etc.) for specific cellular drug delivery.

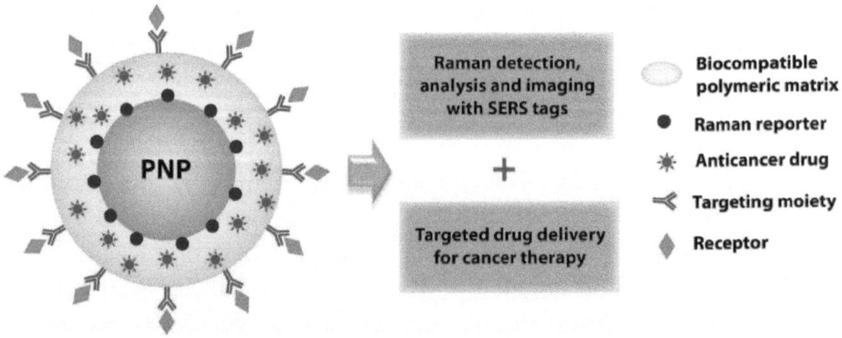

Figure 7.2 Schematic representation of a SERS-imaging guided drug delivery theranostic nano-construct.

7.3.1 SERS-Traceable Plasmonic Nanoparticles in Chemotherapeutic Drug Delivery Applications

Although chemotherapeutic treatment of cancer has experienced some progress over the past decades, the efficiency of conventional chemotherapy in the management of cancer is still limited by several factors, including the low drug levels at the diseased site, nonspecific toxicity, drug resistance mechanisms, and low solubility of drugs [79, 85]. In an attempt to improve chemotherapeutic outcomes in cancer, several efforts have been invested on developing drug delivery systems that can preferentially localize therapeutic concentration of drug to the tumor sites with minimum side effects. Owing to their biocompatibility, reduced size and feasible surface modification, nanoparticles appear as promising building blocks for fabricating efficient drug nanocarriers. Loading of drugs onto nanoparticles can have beneficial impact in cancer chemotherapy, such as (i) prolonging blood retention time, enhancing the drug concentration at the tumor sites and providing sustained drug release; (ii) overcoming the drug resistance mechanisms; (iii) reduction of the side effects associated with conventionally used pharmaceutical excipients; (iv) lowering toxicity to healthy tissues; (v) the high surface to volume ratio of nanoparticles allows loading more drug molecules on a single particle than conventional carrier, and even several types of drugs simultaneously. A key goal for designing efficient drug delivery systems is the ability to control the drug release at pathological sites. To this end, stimuli-responsive nanocarriers have been fabricated as triggered delivery systems [46, 83]. These nanocarriers can release drugs in response to an internal stimulus such as pH, temperature, enzymatic action, redox potential, or an external stimulus such as ultrasound, light, applied magnetic, and electrical field [46, 83]. Owing to their unique optical properties, PNPs provide important opportunities to build traceable drug nanocarriers that allow the real-time monitoring of stimuli-triggered drug release. Due to the benefit of ultrahigh sensitivity and multiplexing capability, surface-enhanced Raman scattering technique has emerged as an ideal diagnostic tool for SERS tracking of drug nanocarriers inside cells and tumors. A good example is the work of Ock et al. By exploiting the ability of purine or pyrimidine chemotherapeutic drugs to generate SERS signals when loaded onto spherical AuNPs, the authors have monitored the intracellular glutathione-triggered release of anticancer drugs using label-free Raman spectroscopic technique [53]. In this system, initial SERS signals were recorded from drug molecules loaded on AuNPs. The adsorption of glutathione onto AuNPs surface causes the detachment of anticancer drugs leading to the decrease of the SERS signal of drugs. By monitoring the variation of SERS intensities, the glutathione-induced release of anticancer drugs was estimated inside mammalian cells. The same approach was applied for in vitro and in vivo monitoring of glutathione-induced release of thiopurine anticancer drug loaded on AuNPs [52]. Gold nanocages loaded with doxorubicin have been further used for SERS monitoring of photothermal-induced release of drug molecules from plasmonic nanocages [78]. Later, Hossain et al. have demonstrated the SERS monitoring of doxorubicin release from biohybrid nanoparticles inside human breast cancer cells [27]. The biohybrid nanoparticles reported here consist of AuNPs modified with cell-penetrating peptide and cancer targeting antibody to facilitate the uptake by the cancer cells and specific cell targeting. Such biohybrid nanoparticles were proved to operate as excellent Raman signal enhancers of doxorubicin molecules loaded onto their surface. By monitoring the time-dependent change of intracellular SERS signal of doxorubicin, the dynamic process of drug delivery and glutathione-mediated release inside cancer cells was successfully investigated. In rather the same time, Lee et al. have reported the fabrication of a multifunctional nanocarrier based on PEGylated AuNPs and doxorubicin with potential use for chemotherapy and SERS detection of cancer cells [43].

Another very efficient strategy to directly track the delivery and release dynamics of doxorubicin from AuNPs in single living cells was proposed by Kang et al. and is based on combination of Raman spectroscopy with fluorescence imaging [34]. In this procedure, a pH-responsive nanocarrier was first designed through the conjugation of doxorubicin to the nanoparticle surface via a pH-sensitive hydrazone linkage. Initially, a strong SERS signal was recorded from doxorubicin molecules loaded on

AuNPs, while the fluorescence signal was quenched. Once the drug molecules were detached from the AuNPs, the Raman signal of doxorubicin molecules was significantly reduced, while their fluorescence signal was recovered. By monitoring the selective plasmonic enhancement or quenching of the Raman and the fluorescence intensity from the doxorubicin–AuNP conjugates, the tracking of the drug delivery and pH-mediated release in single living cells was accomplished.

Recently, our group has successfully investigated the dynamic process of drug delivery and controlled release inside breast cancer cells (MCF-7) by combining confocal Raman microscopy with fluorescence and fluorescence lifetime imaging (FLIM) [70]. Here, a new pH- and temperature-responsive delivery nanochemotherapeutic system was designed based on doxorubicin noncovalently bound to biosynthesized gelatin-coated AuNPs. The gelatin coating endows the nanochemotherapeutic system not only the property of biocompatibility, but also the abilities to respond and release doxorubicin in tumor microenvironments. The dynamic process of delivery and intracellular release and accumulation of doxorubicin was monitored noninvasively via wide-field fluorescence microscopy and scanning confocal microspectroscopy by recording either time-resolved fluorescence or SERS signals of drug molecules. In particular, FLIM recorded under 485 nm pulsed diode laser excitation revealed the bimodal distribution of doxorubicin in cells (Fig. 7.3B). The shorter fluorescence lifetime of doxorubicin localized in nuclei (1.52 ns) than in the cytoplasm (2.4 ns) is consistent with the cytotoxic mechanism induced by doxorubicin-DNA intercalation. Remarkably, the few doxorubicin molecules captured between nanoparticles ("electromagnetic hotspots") after most drug is released act as SERS reporters for the localization of plasmonic nanocarriers in MCF-7 cells (Fig. 7.3C,D). The combined functionalities of high capacity of drug loading via noncovalent complexation, pH and temperature-triggered release of doxorubicin for maximum antitumor activity against breast cancer cells and noninvasive quantitative evaluation of cellular uptake and drug release inside cells indicate the potential of the designed nanocarrier for implementation in nanomedicine.

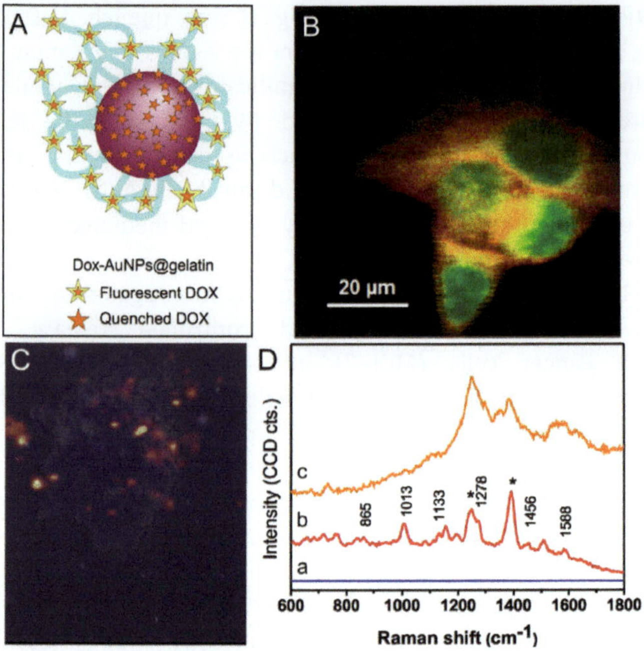

Figure 7.3 (A) Schematic illustration of DOX–AuNPs@gelatin nanochemotherapeutic system; (B) FLIM image of MCF-7 cells treated with 5 μg/mL of DOX loaded onto DOX–AuNPs@gelatin for 24 h at 485 nm excitation wavelength. (C) Overlay of the Raman map of DOX–AuNPs@gelatin incubated cell with the corresponding optical image. (D) SERS spectra collected from different points of a treated cell: (a) nuclear region and (b) extranuclear region. Bands marked with * correspond to DOX molecules. For comparison, the SERS spectrum of DOX–AuNPs@gelatin in colloidal solution is presented in the spectrum (c). Reprinted with permission from reference [70]. Copyright (2016) American Chemical Society.

Yet another strategy for real-time tracking of drug delivery and release inside cells and tumors was recently proposed by Tian et al. and relies on plasmonic-tunable Raman/Fourier transform infrared (FTIR) spectroscopy imaging [77]. In this system, mitoxantrone, a chemotherapeutic drug with significant clinical activity, was conjugated onto the surface of gold nanostars via a thiolated and carboxylated poly(ethylene glycol) (PEG) spacer. Due to its strong Raman scattering signal when loaded onto AuNPs, mitoxantrone was localized inside cells by Raman spectroscopy and in vivo through a plasmonic-tunable Raman/FTIR imaging platform. This novel strategy was proved to be effective in detecting drug accumulation in the heart of healthy mice and in the tumor of lung cancer mice, revealing a strong accumulation in the heart of healthy mice 5 min after administration and infiltration in the tumor site of lung cancer mice 5 h after systemic injection (Fig. 7.4). Thanks to their multiple abilities of drug delivery and controlled release combined with in vivo real-time imaging of drug accumulation, the designed gold nanostars hold great promise for application in image-guided cancer chemotherapy.

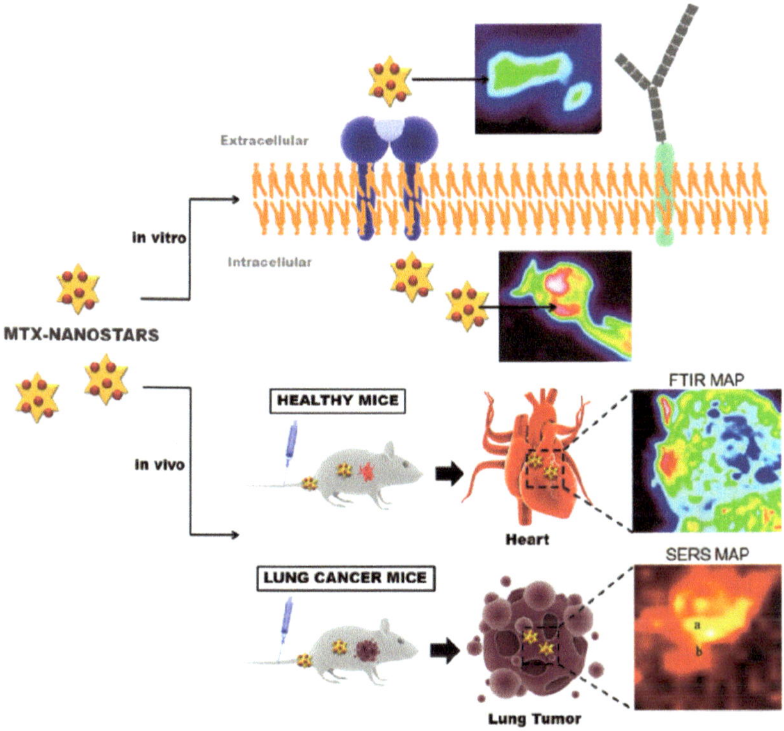

Figure 7.4 Gold nanostars for theranostics: intracellular and in vivo SERS detection combined with real-time drug delivery using plasmonic-tunable Raman/FTIR imaging. This system can directly track in real-time the delivery and release of an anticancer drug (MTX) from gold nanostars in single living cells and in mice (healthy and lung cancer mice models). This in vivo SERS detection method holds a great promise for application in image-guided cancer chemotherapy or as a nonspecific anti-inflammatory therapy for patients with cardiovascular diseases or chronic heart failure. Reprinted from *Biomaterials*, vol. 106, Tian, F., Conde, J., Bao, C., Chen, Y., Curtin, J., and Cui, D., Gold nanostars for efficient in vitro and in vivo real-time SERS detection and drug delivery via plasmonic-tunable Raman/FTIR imaging, pp. 87–97, Copyright (2016), with permission from Elsevier [77].

The real-time monitoring of intracellular dual-drug controlled release was recently reported by Cao et al. and is based on combination of fluorescence resonance energy transfer (FRET) with SERS [10]. In this work, mesoporous silica and gold were separately functionalized with FRET pair coumarin/doxorubicin and 6-mercaptopurine by either pH or redox responsive bond, leading to traceable platform for evaluation of the intracellular dual-drug stimuli-responsive release. By detecting the changes of FRET signals of coumarin/doxorubicin and SERS intensity of 6-mercaptopurine, the pH and glutathione triggered intracellular release of doxorubicin and 6-mercaptopurine was successfully investigated.

Wang et al. have developed a biocompatible nanoplatform that incorporates bimodal fluorescence-SERS imaging abilities and chemotherapeutic effects on breast cancer cells [81]. In this work, a robust and reproducible SERS nanotag was first constructed based on colloidal silver nanoparticles encoded with 4-mercaptopyridine Raman reporter and protected with a thin silica layer. In the second step, the individual silica-protected SERS tags were coated with a mesoporous titania shell with the aim to endow them capacity for fluorescent dye labeling (flavin mononucleotide) and anticancer drug loading (doxorubicin). Easily internalized by breast cancer cells (MCF-7), as demonstrated by both their strong fluorescence and SERS signals, the fabricated drug-loaded nanotags were proved to exhibit comparable or even greater therapeutic effects compared with the free drug, owing to the increased cell-uptake of the anticancer drug.

pH-sensitive self-assembled plasmonic vesicles were exploited as carriers for anticancer drugs and SERS imaging probes to specifically label targeted cancer cells and monitor intracellular drug delivery [68]. In this system, citrate-stabilized AuNPs were first encoded with a Raman reporter and coated with mixed amphiphilic polymer brush for targeted delivery of doxorubicin. In the second step, the strong interparticle coupling gives rise to plasmonic vesicles with significantly different plasmonic properties and greatly enhanced SERS intensity compared with isolated AuNPs. The fabricated plasmonic vesicles exhibit efficient drug loading and release and are easily detectable inside cells by both plasmonic imaging and Raman spectroscopy. The pH-responsive disassembly of the plasmonic vesicles influences the SERS intensity of Raman reporter molecules which can serve as independent feedback mechanism for real-time monitoring of drug release from the vesicles. Later, the same group of authors has developed biodegradable SERS-active plasmonic vesicles, tagged with cancer targeting ligands, which allows for simultaneous SERS detection and synergistic chemo-photothermal therapy of specific cancer cells [67].

A SERS-traceable and stimuli-responsive nanocarrier for controlled intracellular drug delivery was designed by Zong and co-workers [92]. In their work, the authors have used Raman-encoded Au@Ag nanorods as the SERS active core and mesoporous silica as the drug containing shell. By monitoring the SERS signal of the nanocarriers and the fluorescence of the drug molecules, the dynamic process of drug delivery and controlled release inside living cells was successfully investigated.

Later, Fang et al. designed mesoporous silica nanoparticles as the pH-controllable drug carriers that allow for simultaneous SERS/fluorescence imaging and chemotherapy of specific cancer cells [21]. Raman-encoded silver nanoparticles protected with a silica layer were used as the SERS-active core, while a second layer of mesoporous silica is coated on the core for doxorubicin loading and release. To endow this medical platform with pH-responsive releasing characteristics, a supplementary layer of chitosan/poly(methacrylicacid) was introduced outside the mesoporous silica shell. By conjugating transferrin onto the nanocarrier surface, its targeting ability to transferrin receptor overexpressed cancer cells was achieved. This class of carriers was proved to simultaneously fulfill the functionalities of HeLa cells targeting, pH-responsive drug release, SERS-traceable cellular uptake and fluorescence-based monitoring of intracellular release of doxorubicin.

A highly sensitive nanotheranostic platform for in vivo tumor recognition and treatment was designed by Conde et al. [15]. The here reported agents are based on AuNPs encoded with a Raman reporter, encapsulated in polymer and conjugated with cetuximab, a monoclonal antibody drug that specifically targets cancer biomarkers, such as epidermal growth factor receptors. The designed theranostic carriers were proved to target and inhibit tumor progression, leading to tumor size decrease, while simultaneously allowing for spectroscopic detection of relatively small xenograft tumors.

Yolk-shell nanoparticles were proposed by Zhang and co-workers as promising candidates for developing SERS-based theranostic agents [88]. In this system, NIR light absorbing gold nanorods (AuNRs) were first synthesized and embedded in mesoporous silica shell. In the second step, a further coating of mesoporous titania ($mTiO_2$) was applied on multiple silica-protected AuNRs, followed by a controllable etching of the silica shell, in order to obtain AuNR@-void@$mTiO_2$ yolk-shell nanoparticles. Then, Raman reporters and anticancer drug doxorubicin were co-loaded on the fabricated nanoparticles to endow them with chemotherapeutic and imaging abilities. Within this particular architecture, AuNRs operate as both, SERS imaging probes and photothermal agents and modulate the drug release, while the outer $mTiO_2$ shell serves as a chemically inert drug loading material. This class of theranostic agents was successfully used for intracellular SERS imaging and in vivo monitoring and synergistic chemo-photothermal therapy.

Yet another unique nanomedical platform designed for cancer theranostics is based on individual Raman-encoded gold nanocages encapsulated into porous SiO_2 shell and functionalized with a Tat peptide [30]. In their work, Hu et al. have shown that gold nanocages perform as effective photothermal transducers for localized hyperthermia cancer therapy, as well as sensitive SERS nanotags, while the porous SiO_2 shell guarantees the chemical and optical stability of the gold nanocages, enhances the drug loading capability, and vastly improves the cellular uptake efficiency after Tat functionalization.

Small AuNPs with SERS reporters were distributed around silica nanoparticles leading to SERS-traceable plasmonic-based systems [45]. To endow the system with therapeutic characteristics, doxorubicin was then conjugated onto the nanoplatform surface via the covalent pH cleavable drug-carriers links. Owing to the combined functionalities of SERS tracing and targeted cytotoxicity toward cancer cells, this type of theranostic platform showed a great potential to investigate the dynamic delivery of drugs inside living cells.

Recently, SERS monitoring of apoptotic events during therapy was taken into account in the development of theranostic platforms. A good example is the work of Narayanan et al., who have reported the design of a theranostic nanoprobe for simultaneous targeted photothermal-chemotherapy of human fibrosarcoma cell line (HT1080) and SERS evaluation of the therapeutic efficiency [51]. Here, a Raman reporter (squaraine dye) was first attached to a targeting component (matrix metalloproteinases) and then conjugated to gold nanorods to fabricate a targeted photothermal agent with SERS imaging abilities. To endow this medical platform with pH-responsive chemotherapeutic characteristics, an anticancer drug (doxorubicin) was loaded to the gold nanorods surface via an acid labile hydrazone linker. The targeting abilities of the fabricated nanocarriers were evaluated with the SERS "turn on"/"turn off" mechanism. It was found that the enzymatic recognition of cancer cells causes the cleavage of the matrix metalloproteinases peptide substrate, which ultimately detaches the Raman reporter from nanoparticles surface, leading to the gradual decrease in SERS signal intensity. The cellular uptake and intracellular drug release was investigated using the fluorescence properties of doxorubicin, while the structural and functional variations of cellular components during the site-specific cytotoxic action of the theranostic platform were examined through SERS. The results revealed that the combined photothermal-chemotherapy induced the apoptotic DNA fragmentation which can be effectively monitored via SERS.

7.3.2 SERS-Traceable Plasmonic Nanoparticles in Photosensitizer Delivery Applications

Photosensitizers (PSs) are light-sensitive drugs used in photodynamic therapy (PDT), which after delivery to the targeted tumor cells or tissues are (photo)activated by irradiation with light at the proper dose and wavelength, promoting the photochemical conversion of molecular oxygen into singlet oxygen, a cytotoxic agent that destroys the cancer cells [1, 55].

From clinical point of view, an ideal PS should have an intense absorption in the 600–800 nm spectral domain, also called the *phototherapeutic window*, as the penetration depth of the light into tissue increases with its wavelength, thus deeper tissue section can be reached and treated. However, other photophysical and photochemical properties, such as high-absorption cross section, efficient intersystem crossing to the triplet state, high quantum yield of singlet oxygen formation, etc. are also crucial factors in view of efficient PDT [4].

First generation PSs, based on porphyrins, (e.g., photophrin, hematoporphyrin), present some disadvantages, such as low absorbance and long-term photosensitivity of the skin after treatment [3]. To overcome these drawbacks and improve the treatment efficacy, several strategies have been developed to produce novel tumor-selective therapeutic agents based on porphyrin derivatives or synthetic PS with reduced side effects. Hence, numerous *second-generation* PSs have been developed, some of them also used in clinical trials. They present some improved properties such as chemical purity, high extinction coefficient, and strong absorbance in the therapeutic wavelength domain. The majority of the molecular PSs currently used in cancer therapy are aromatic molecules, based on a tetrapyrrole structure which is efficient in the formation of long lived triplet excited state [54]. However, many clinically approved PSs suffer from poor solubility in biological media, poor bioavailability and unfavorable biodistribution [11]. The approach of conjugation with biological targeting moieties (e.g. antibody, peptides) or encapsulation into nanocarriers of the well-designed second generation PSs gave rise to the so-called *third generation PSs,* which may stand a good chance to overcome the above mentioned drawbacks [58]. These nanoformulations can take advantage of the enhanced permeability and retention (EPR) effect to improve the cellular uptake and the relatively high loading capacity and also can be endowed themselves with targeting molecules to facilitate selective accumulation in cancer cells. The most frequently employed nanocarriers for PS delivery, also approved for clinical use, include polymeric micelles and liposomes. Their main advantages consist of their ability to solubilize hydrophobic PSs with high loading capacity and provide long circulation time in the blood stream due to the fact that prevent recognition by proteins and macrophages.

As the efficiency of PDT is highly dependent on the yield of singlet oxygen production and on its successful interplay with the targeted biological substrate, two different approaches employing nanoparticle systems can be followed. On the one hand, one can use biodegradable or stimuli-responsive nanoparticles which upon reaching the tumor site can release the PS which subsequently irradiated can produce singlet oxygen in direct contact with the targeted biological substrate [72]. On the other hand, in the cases when the PS is not released from the carrier the singlet oxygen generated has to diffuse out. This approach can offer significant benefits in case of PSs which have photodynamic properties compromised when interacting with biological macromolecules or enzymes [74] and when the presence of the nanoparticles provides amplification of the photodynamic efficiency, such as the so-called *plasmon-enhanced singlet oxygen generation* effect, frequently addressed when using noble metal nanoparticles as PS nanocarriers [29]. The enhanced electromagnetic field around plasmonic nanoparticles (PNPs), as a result of surface plasmon excitation is known to have considerable effects on the photophysical and photochemical properties of organic molecules located in their closed vicinity, giving rise to a number of remarkable effects in both imaging and therapeutic applications [47]. PSs placed in the close vicinity of metal nanoparticles due to the coupling between the PS's molecular dipole and the electric field near the metal nanoparticles surface exhibit enhanced excitation rate which subsequently contribute to the enhancement of intersystem crossing and excited state triplet yield, resulting in enhancement of the singlet oxygen generation yield. The extent of the plasmon enhanced photodynamic effect is dependent on a variety of factors such as optical properties of the PNPs which are determined by their size, shape and material, spectral match between NP and PS, distance between PS and nanoparticle surface, issues which demand thorough optimization process and that were broadly studied in the recent years [37]. Combination of PS with PNPs not only that can have beneficial impact on their photodynamic activity, but also provides various opportunities to integrate multiple imaging functionalities into the therapeutic nanoconstruct, facilitating the therapy management which can consequently enhance the treatment response.

Due to their intrinsic fluorescence, PS molecules can be used for imaging and locating tumor sites, therefore fluorescence has became the most broadly used spectroscopic technique to monitor PS accumulation. Based on the fact that the degree of PS photobleaching can related to the amount of singlet oxygen generated, the measurement of the fluorescence emission of the PS has been a very convenient method to assess the efficiency of PDT [17]. However, this method is not applicable in some cases, such as for PSs with high singlet oxygen quantum yield, but reduced fluorescence emission. On the other hand, the presence of PNPs can constructively or destructively influence the emission properties of fluorescent molecules located in their close vicinity, which can compromise the direct correlation of emission intensity with the PS concentration, impeding in this way the accurate evaluation of PS accumulation.

Another high-sensitivity spectroscopic technique based on plasmonic nanostructures, SERS, has been recently recognized as an alternative to fluorescence imaging during PDT. Due to the well-resolved spectral fingerprint of Raman-active molecules in contrast to the broad fluorescence emission of fluorophores, SERS nanotags are more suitable for multiplex detection, while photobleaching effects are also minimalized [80]. Two approaches are mainly addressed when designing SERS-nanotags for imaging guided PDT, namely using distinct molecules as Raman reporter and PS or using a single molecule, a Raman-active PS, for both imaging and therapeutic purposes, both options having their own advantages and limitations. For instance, using different molecules as imaging and therapeutic moiety offers the possibility to uncouple the two mechanisms in order to independently optimize both of them. On the other hand, monitoring the spectral signature of the PS during PDT could also provide useful information about molecular conformations, could reveal the undesirable impairment of PS, which all can influence the efficiency and outcome of the treatment.

The combination of the two powerful imaging techniques, SERS and fluorescence, could be also a promising solution by exploiting their complementarity, therefore compensating their limitations by each other. For instance, as fluorescence imaging offers a fast and wide-area detection tool, it can be used as an immediate indicator of the target position, meanwhile the outstanding multiplex capability and specificity of SERS detection can be exploited as an accurate and high-contrast sensing tool for imaging and localization of nanocarriers, biomarker distributions, drug release monitoring, etc.

The first theranostic nanoparticle construct combining SERS imaging and photosensitization was reported by Fales et al. in 2011 [19]. They designed a SERS-active nanoparticles platform based on silica coated gold nanostars loaded with MB, a common PS which served as photodynamic agent and also as fluorescent label. 3,3′-Diethylthiatricarbocyanine iodide (DTTC) NIR dye used as Raman reporter provided a strong signal at 785 nm laser excitation attributed to the so-called SERRS effect, which further amplified the imaging efficiency. The enhanced rate of singlet oxygen generation in the case of MB-loaded gold nanostars compared to the physical mixture of the two components was attributed, on the one hand, to the protective effect of the silica matrix preventing PS aggregation and, on the other hand, to the plasmonic effect provided by the gold nanostars. Consequently, MB-loaded gold nanostars acted as efficient photodynamic nanoplatforms against BT549 cells upon 633 nm laser irradiation.

Later on, Zhang et al. reported multilayer silica coated gold nanorods doped with two different dyes for dual modal in vivo tumor detection by fluorescence and SERS [89]. Additionally, they have shown that porphyrin PS can also be co-encapsulated, able to induce PDT treatment after tumor imaging. Although multiple functionalities were integrated into a single system, imaging and therapeutic properties could be activated by different excitation channels, NIR excitation for imaging and 532 nm line excitation for PDT. In another study, the same group reported PEGylated gold nanostars tagged with DTCD as Raman reporter and encapsulated in a silica shell which was loaded with the PS Protoporphyrin IX [20]. In this case, a targeting moiety, transactivator of transcription (TAT) peptide, was conjugated to the nanoparticles which facilitated specific uptake by BT-549 breast cancer cells, as demonstrated by Raman mapping at 633 nm excitation. Photodynamic efficiency of the nanoparticles was demonstrated under UV irradiation of the nanoparticle-treated cells.

NIR-active silica encapsulated gold nanorods were also investigated by Seo et al. as SERS imaging guided combined therapeutic nanoparticles [64]. MB as both reporter and PS was embedded in the silica shell, and the obtained nanoconjugates showed very good Raman performance (enhancement factor of 3.0×10^{10}), enabling the detection of a single cancer cell.

Moreover, the nanoparticles exhibited a synergistic effect of PDT and PTT therapies of cancer under single-wavelength NIR laser irradiation.

In the meantime, Raghavan et al. synthesized branched gold nanostars by a green chemistry approach which were coated with the Raman reporter, DTTCI (3,3′-Diethylthiatricarbocyanine iodide) and the PS, hypericin, a naturally occurring plant-based perylene-quinone dye [31]. Denaturated BSA used as stabilizing agent also served as spacer upon PS conjugation in order to prevent its fluorescence quenching. Endowed with multimodal optical bioimaging capabilities through SERS and fluorescence, the fabricated gold nanosensitizer demonstrated the ability to destroy cancer cells by both PDT and PTT under 543 nm and 675 nm laser excitation, respectively.

SERS-active gold nanochains as a potential theranostic system for multiplex detection and PDT of cancer were reported by Zhao et al. [90]. The nanochains were prepared by the assembly citrate-capped AuNPs using hyaluronic acid and hydrocaffeic acid conjugates as template. In view of multiplex detection by Raman spectroscopy combined with PDT, 2-naphthalenethiol (NPT) reporter molecule and thiolated pheophorbide a (Pheo) PS were conjugated onto the surface of the nanochains. It was demonstrated that these chainlike nanostructures were well-uptaken by HeLa cells via the endocytic pathway, and that chain length did not considerably influence the cellular uptake behavior. PDT efficiency of the nanoconjugate under He–Ne laser irradiation (670 nm) was noticeably higher than that of the free photosensitizer, ascribed to the improved solubility of the hydrophobic PS in the case of the nanoconjugate which provided increased singlet oxygen generation quantum yield.

The theranostic nanoparticles designed by Farhadi et al. based on spherical AuNPs and liposome-like palladium-pyrolipid nanostructures offers the possibility to use the same excitation wavelength (638 nm) to simultaneously activate both the therapeutic and imaging function of the nanoparticles [22]. This was the first report on simultaneous SERS mapping and PDT using the same molecule, a palladium containing porphysome, as reporter and PS. On the one hand, this approach greatly facilitated the real-time monitoring of the PS, and on the other hand the effective PDT dose could be monitored through photobleaching-induced loss of the SERS signal.

Shortly thereafter, our group proposed a novel class of theranostic agents serving as multimodal contrast agents via SERS and FLIM in cell imaging with plasmon-enhanced photodynamic effect against murine colone carcinoma (C-26) cells [66]. The design of these theranostic nanoparticles is based on the close-packed nano-assemblies of AuNPs stabilized using Pluronic and labeled with the common PS, Methylene Blue (GNPA-F127-MB), serving as both imaging and therapeutic moiety. Uptake and accurate localization of the nanoparticles inside cells were performed by dark-field and differential interference contrast (DIC) microscopies. SERS imaging performance of our nanoparticles was demonstrated by their strong, distinct signal inside C-26 cells which could be accurately detected at a 3D level.

The nanoparticle localization maps at different Z-depths (left column, Fig. 7.5A) were obtained at 633 nm excitation by integrating the peak ascribed to the C–C ring stretching vibrations of MB centered at 1623 cm^{-1}, meanwhile the basic cellular structure was visualized using the peak at 2800–3100 cm^{-1} originating from the C–H stretching vibrations of lipids (right column in Fig. 7.5A). At the same time, we showed that MB molecules separated from the metal surface by the polymer spacer can provide fluorescence signal allowing the detection of the nanoparticles inside cells via FLIM. The in vitro evaluation of the photodynamic performance of the theranostic agents against C-26 cancer cells demonstrated improved therapeutic efficiency relative to the free PS under low-dose LED irradiation at 660 nm wavelength without showing dark cytotoxicity (Fig. 7.5D).

At the same time, Kim et al. developed a novel multifunctional core–shell nanoplatform by combining the unique properties of AuNPs and graphene oxide with the photodynamic performance of the FDA-approved second generation zinc phthalocyanine PS [38]. Stability of the nanoplatform and low dark cytotoxicity were assured by PEGylation and the applicability as a Raman probe in living cells was demonstrated on HeLa cells based on the intrinsic Raman signal of graphene oxide. Combinational treatment by sequential PTT and PDT on HeLa cells incubated with the obtained nanosystem showed promising enhanced therapeutic effect, resulting in a significantly lower cell viability (6.8%) than that with exposure to only PTT (29.3%) or PDT (11.4%) alone.

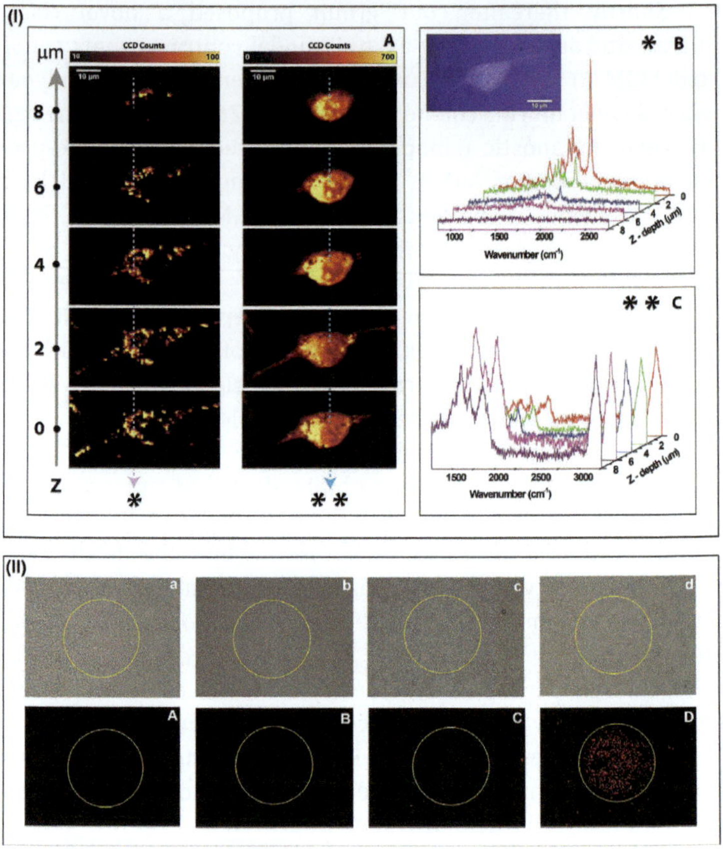

Figure 7.5 (I) (A) SER(R)S maps of C-26 cells incubated with GNPA-F127-MB generated by plotting the intensity of a characteristic Raman peak over the scanned area. The nanoparticle localization (left column) was obtained at excitation with the laser line at 633 nm using the peak centered at 1623 cm^{-1} (C–C ring stretching vibrations of MB). The basic cellular structure (right column) was visualized at excitation with the laser line at 532 nm using the peak at 2800–3100 cm^{-1} C–H stretching vibrations of lipids). (B) Correspondent extracted SERRS spectra from each Z at the spot marked with the arrow and * in (A). The inset shows the bright field image of the scanned cell. (C) Correspondent extracted SERS spectra from each Z at the spot marked with the arrow and ** in (A). (II) Transmitted light and immunofluorescent staining images of C-26 cell: not treated and irradiated (a, A), treated with GNPA-F127 and irradiated (b, B), treated with MB and irradiated (c, C), and treated with GNPA-F127-MB and irradiated. Irradiation performed by LED emitting at 660 nm for 40 min with a power of 50 mW (17.7 W/cm^2). The yellow circles represent the irradiated areas. Reprinted with permission from reference [66]. Copyright (2016) American Chemical Society.

Very recently, Fales et al. reported a silver-embedded gold nanostar based folate-targeted theranostic nanocomposite [18]. The diagnostic component consists of a SERS nanotag built by embedding para-mercaptobenzoic acid (pMBA) Raman reporter between the gold core and silver shell, meanwhile the therapeutic function is provided by the PS protoporphyrin IX (PpIX) loaded onto a silica shell encapsulating the whole nanoconstruct. Promising SERS-based selective detection and targeted PDT treatment were conducted on two lines of the folate receptor positive cells (HeLa cervical cancer cells and SK-BR-3 breast cancer cells), while MDA-MB-468 breast cancer cells were used as a negative control.

Another Raman imaging-guided and folic acid-targeted PDT agent against HeLa cancer have been recently reported by Huang et al. using Au@Ag nanohollow nanoparticles labeled with MB which served as both Raman reporter and PS [28]. Their results indicated a robust protection of MB against oxidant species, which facilitated the tracking of nanoparticles inside cells using Raman mapping (Fig. 7.6B,D). LED-activated PDT at 660 nm of HeLa cells treated with the MB-loaded nanohollows showed an increasing cell death rate in function of incubation period which reached 77% for 6 h treatment. To improve the therapeutic efficacy, an additional illumination with 808 nm light was applied in order to induce a synergistic way of destroying cancer cells through combined PDT and PTT (Fig. 7.6C).

A complex nanocomposite for tri-modal SERS/dark-field/fluorescence imaging and synergistic PDT/PTT have been newly reported by Li et al. [44]. This nanostructure is based on silica-coated AuNRs tagged with the Raman reporter 4-MBA and decorated with gold nanoclusters as fluorescent label which were further enriched with the photosensitizer chlorine e6. The fabricated nanoconstruct showed enhanced photophysical stability, high drug loading efficiency (up to 55%), long-term stability and good biocompatibility. A novel targeting approach using two types of targeting molecules was employed (FA and a cell penetrating peptide), which synergistically enhanced the efficiency of the uptake of nanocomposites by HeLa cells (up to 70.7%). In vitro imaging assays demonstrated the successful targeting of cancer cell using high-resolution and noninvasive spectroscopic techniques. Moreover, the combined PDT/PTT treatment using two different laser wavelengths to perform PDT

(633 nm) and PTT (808 nm) induced synergistic therapeutic efficiency on cancer cells compared to the individual therapies destroying only a fraction of the irradiated cells.

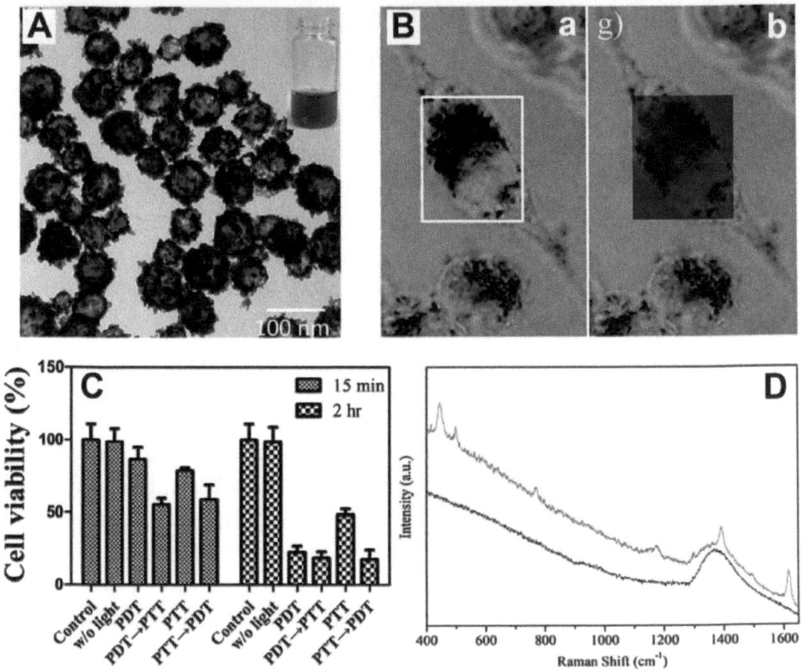

Figure 7.6 (A) TEM image of 3-D Au@Ag nanohollows; (B) Bright field image of HeLa cells incubated with the nanohollows for 6 h (a) and corresponding overlaid image with SERS map constructed from the Raman peak at 1619 cm^{-1} (b); (C) MTT assay of the Au@Ag nano-hollows@PS@FA-treated HeLa cells received LED light at 660 nm (51 mW), laser light at 808 nm for PTT (115 mW), 660/808 nm, and 808/660 nm. (D) Raman spectra collected from cells treated with Au@Ag nanohollows (red spectrum) and measured outside of the cells (black spectrum). Reproduced from reference [28] with permission of The Royal Society of Chemistry.

7.4 Conclusions

In summary, significant progress has been recently made toward exploiting the plasmonic nanoparticles as drug nanocarriers to create theranostic nanoplatforms for cancer targeting, detection, and therapy. There are numerous examples in the recent literature where SERS active plasmonic nanoparticles have been successfully applied in cancer diagnostic and imaging with a tremendous effort invested behind application in the thorough fabrication and optimization process. As an emerging approach, the development of novel targeted theranostic agents integrating SERS-traceable plasmonic nanoparticles as nanocarriers in the delivery applications of anticancer agents such as chemotherapeutic or photosensitizing drugs has become a hot research topic.

With the aim to precisely control the targeted delivery and release of chemotherapeutic drugs at pathological sites, a number of efficient drug delivery systems based on SERS-traceable plasmonic nanoparticles have been recently designed and successfully applied in vitro as well as in vivo. The SERS performance of the reported theranostic nanoparticles greatly facilitated the monitoring of drug accumulation and allowed the direct tracking of the drug delivery and release dynamics. Current efforts have been also dedicated to the design of theranostic nanoplatforms combining SERS-imaging with the cytotoxic effect of photosensitizing drugs. Besides conferring imaging capabilities and generous surface for drug loading, the amplified electromagnetic field around plasmonic nanoparticles has been demonstrated to provoke the enhancement of the singlet oxygen generation yield, thus improving therapeutic efficiency. Furthermore, the intrinsic fluorescence of both chemotherapeutic drug molecules and photosensitizers endows these theranostic agents with multiple imaging abilities, facilitating in this way the therapy management through the possibility of real-time monitoring of the drug accumulation and monitoring the therapy outcome in a complementary manner. In order to enhance the therapeutic efficiency, a few SERS guided dual modal therapeutic nanoparticles have been also reported based on NIR absorbing plasmonic nanoparticles acting also as photothermal agents upon laser irradiation, evoking this way a synergistic therapeutic effect.

The recent advances in the field of SERS-traceable plasmonic nanoparticles based drug nanocarriers show promising perspectives both in vitro and in vivo, however their success in the clinical applications is not yet realized. We believe that the promising integration of SERS probes with complementary imaging techniques and conventional therapeutic modalities represents a big step toward the translation of plasmonic nanoparticles based theranostic nanosystems to the clinical environment.

Acknowledgments

This work was supported by a grant from the Romanian National Authority for Scientific Research, CNCS–UEFISCDI, project number PN-II-RU-TE-2014-4-1988 and a grant of Ministry of Research and Innovation, CNCS-UEFISCDI, project number PN-III-P4-ID-PCE-2016-0837, within PNCDI III. Nagy-Simon T. acknowledges the financial support from Babes-Bolyai University, project number GTC_31803/2016.

References

1. Agostinis, P., Berg, K., Cengel, K. A., Foster, T. H., Girotti, A. W., Gollnick, S. O., Hahn, S. M., Hamblin, M. R., Juzeniene, A., Kessel, D., Korbelik, M., Moan, J., Mroz, P., Nowis, D., Piette, J., Wilson, B. C., and Golab, J. (2011). Photodynamic therapy of cancer: An update, *CA Cancer J. Clin.*, **61**, 250–281.
2. Alkilany, A. M., and Murphy, C. J. (2010). Toxicity and cellular uptake of gold nanoparticles: What we have learned so far? *J. Nanoparticle Res.*, **12**, 2313–2333.
3. Allison, R. R., Downie, G. H., Cuenca, R., Hu, X.-H., Childs, C. J., and Sibata, C. H. (2004). Photosensitizers in clinical PDT, *Photodiagnosis Photodyn. Ther.*, **1**, 27–42.
4. Allison, R. R., and Sibata, C. H. (2010). Oncologic photodynamic therapy photosensitizers: A clinical review, *Photodiagnosis Photodyn. Ther.*, **7**, 61–75.
5. Aroca, R. (2006). *Surface Enhanced Vibrational Spectroscopy*. (John Wiley & Sons, Ltd, Chichester, UK).
6. Asiala, S., Barrett, L., Mabbott, S., and Graham, D. (2016). Advances in biofunctional SERS-active nanoparticles for future clinical diagnostics and therapeutics. In: *Front. Plasmon Enhanc. Spectrosc.* vol. 1. American Chemical Society, pp. 131–161.
7. Barbosa, S., Agrawal, A., Rodríguez-Lorenzo, L., Pastoriza-Santos, I., Alvarez-Puebla, R. A., Kornowski, A., Weller, H., and Liz-Marzán, L. M. (2010). Tuning size and sensing properties in colloidal gold nanostars, *Langmuir*, **26**, 14943–14950.

8. Boca, S., Rugina, D., Pintea, A., Barbu-Tudoran, L., and Astilean, S. (2011). Flower-shaped gold nanoparticles: Synthesis, characterization and their application as SERS-active tags inside living cells, *Nanotechnology*, **22**, 055702.

9. Boca, S., Rugina, D., Pintea, A., Leopold, N., and Astilean, S. (2012). Designing gold nanoparticle-ensembles as surface enhanced Raman scattering tags inside human retinal cells, *J. Nanotechnol.*, **2012**, 1–10.

10. Cao, H., Yang, Y., Chen, X., and Shao, Z. (2016). Intelligent Janus nanoparticles for intracellular real-time monitoring of dual drug release, *Nanoscale*, **8**, 6754–6760.

11. Castano, A. P., Demidova, T. N., and Hamblin, M. R. (2005). Mechanisms in photodynamic therapy: Part three—Photosensitizer pharmacokinetics, biodistribution, tumor localization and modes of tumor destruction, *Photodiagnosis Photodyn. Ther.*, **2** 91–106.

12. Chen, G., Roy, I., Yang, C., and Prasad, P. N. (2016). Nanochemistry and nanomedicine for nanoparticle-based diagnostics and therapy, *Chem. Rev.*, **116** 2826–2885.

13. Chithrani, B. D., Ghazani, A. A., and Chan, W. C. W. (2006). Determining the size and shape dependence of gold nanoparticle uptake into mammalian cells, *Nano Lett.*, **6**, 662–668.

14. Colas, F. J., Cottat, M., Gillibert, R., Guillot, N., Djaker, N., Lidgi-Guigui, N., Toury, T., Barchiesi, D., Toma, A., Di Fabrizio, E., Gucciardi, P. G., and Lamy de la Chapelle, M. (2016). Red-shift effects in surface enhanced Raman spectroscopy: Spectral or intensity dependence of the near-field? *J. Phys. Chem. C*, **120**, 13675–13683.

15. Conde, J., Bao, C., Cui, D., Baptista, P. V., and Tian, F. (2014). Antibody–drug gold nanoantennas with Raman spectroscopic fingerprints for in vivo tumour theranostics, *J. Control. Release*, **183**, 87–93.

16. Doering, W. E., and Nie, S. (2003). Spectroscopic tags using dye-embedded nanoparticles and surface-enhanced Raman scattering, *Anal. Chem.*, **75**, 6171–6176.

17. Dysart, J. S., Singh, G., and Patterson, M. S. (2005). Calculation of singlet oxygen dose from photosensitizer fluorescence and photobleaching during mTHPC photodynamic therapy of MLL cells, *Photochem. Photobiol.*, **81**, 196–205.

18. Fales, A. M., Crawford, B. M., and Vo-Dinh, T. (2016). Folate receptor-targeted theranostic nanoconstruct for surface-enhanced Raman

scattering imaging and photodynamic therapy, *ACS Omega,* **1**, 730–735.

19. Fales, A. M., Yuan, H., and Vo-Dinh, T. (2011). Silica-coated gold nanostars for combined surface-enhanced Raman scattering (SERS) detection and singlet-oxygen generation: A potential nanoplatform for theranostics, *Langmuir,* **27**, 12186–12190.
20. Fales, A. M., Yuan, H., and Vo-Dinh, T. (2013). Cell-penetrating peptide enhanced intracellular Raman imaging and photodynamic therapy, *Mol. Pharm.,* **10**, 2291–2298.
21. Fang, W., Wang, Z., Zong, S., Chen, H., Zhu, D., Zhong, Y., and Cui, Y. (2014). pH-controllable drug carrier with SERS activity for targeting cancer cells. *Biosens Bioelectron* **57**, 10–15.
22. Farhadi, A., Roxin, Á., Wilson, B. C., and Zheng, G. (2015). Nano-enabled SERS reporting photosensitizers, *Theranostics,* **5**, 469–476.
23. Félidj, N., Aubard, J., Lévi, G., Krenn, J. R., Salerno, M., Schider, G., Lamprecht, B., Leitner, A., and Aussenegg, F. R. (2002). Controlling the optical response of regular arrays of gold particles for surface-enhanced Raman scattering, *Phys. Rev. B,* **65**, 075419.
24. Garcia, M. A. (2011). Surface plasmons in metallic nanoparticles: fundamentals and applications, *J. Phys. Appl. Phys.,* **44**, 283001.
25. Guillot, N., and Lamy de la Chapelle, M. (2012). The electromagnetic effect in surface enhanced Raman scattering: Enhancement optimization using precisely controlled nanostructures, *J. Quant. Spectrosc. Radiat. Transf.,* **113**, 2321–2333.
26. Guillot, N., Shen, H., Frémaux, B., Péron, O., Rinnert, E., Toury, T., and Lamy de la Chapelle, M. (2010). Surface enhanced Raman scattering optimization of gold nanocylinder arrays: Influence of the localized surface plasmon resonance and excitation wavelength, *Appl. Phys. Lett.,* **97**, 023113.
27. Hossain, M. K., Cho, H.-Y., Kim, K.-J., Choi, J.-W. (2015). In situ monitoring of doxorubicin release from biohybrid nanoparticles modified with antibody and cell-penetrating peptides in breast cancer cells using surface-enhanced Raman spectroscopy, *Biosens. Bioelectron.,* **71**, 300–305.
28. Huang, C.-C., Lin, P.-H., and Lee, C.-W. (2016). OFF/ON galvanic replacement reaction for preparing divergent AuAg nano-hollows as a SERS-visualized drug delivery system in targeted photodynamic therapy, *RSC Adv.,* **6**, 64494–64498.
29. Huang, X., Tian, X.-J., Yang, W., Ehrenberg, B., and Chen, J.-Y. (2013). The conjugates of gold nanorods and chlorin e6 for enhancing the

fluorescence detection and photodynamic therapy of cancers, *Phys. Chem. Chem. Phys.*, **15**, 15727–15733.

30. Hu, F., Zhang, Y., Chen, G., Li, C., and Wang, Q. (2015). Double-walled Au nanocage/SiO_2 nanorattles: Integrating SERS imaging, drug delivery and photothermal therapy, *Small*, **11**, 985–993.

31. Raghavan, V., Connolly, J. M., Fan, H. M., Dockery, P., Wheatley, A., Keogh, I., and Olivo, M. (2014). Gold nanosensitisers for multimodal optical diagnostic imaging and therapy of cancer, *J. Nanomed. Nanotechnol.*, **5**, 238.

32. Jiang, L., Qian, J., Cai, F., and He, S. (2011). Raman reporter-coated gold nanorods and their applications in multimodal optical imaging of cancer cells, *Anal. Bioanal. Chem.*, **400**, 2793.

33. Joseph, V., Matschulat, A., Polte, J., Rolf, S., Emmerling, F., and Kneipp, J. (2011). SERS enhancement of gold nanospheres of defined size, *J. Raman Spectrosc.*, **42**, 1736–1742.

34. Kang, B., Afifi, M. M., Austin, L. A., and El-Sayed, M. A. (2013). Exploiting the nanoparticle plasmon effect: Observing drug delivery dynamics in single cells *via* Raman/fluorescence imaging spectroscopy, *ACS Nano*, **7**, 7420–7427.

35. Kelkar, S. S., and Reineke, T. M. (2011). Theranostics: Combining imaging and therapy, *Bioconjug. Chem.*, **22**, 1879–1903.

36. Kessentini, S., Barchiesi, D., D'Andrea, C., Toma, A., Guillot, N., Di Fabrizio, E., Fazio, B., Maragó, O. M., Gucciardi, P. G., and Lamy de la Chapelle, M. (2014). Gold dimer nanoantenna with slanted gap for tunable LSPR and improved SERS, *J. Phys. Chem. C*, **118**, 3209–3219.

37. Oo, M. K. K., Yang, Y., Hu, Y., Gomez, M., Du, H., and Wang, H. (2012). Gold nanoparticle-enhanced and size-dependent generation of reactive oxygen species from protoporphyrin IX, *ACS Nano*, **6**, 1939–1947.

38. Kim, Y.-K., Na, H.-K., Kim, S., Jang, H., Chang, S.-J., and Min, D.-H. (2015). One-pot synthesis of multifunctional Au@graphene oxide nanocolloid core@shell nanoparticles for Raman bioimaging, photothermal, and photodynamic therapy, *Small*, **11** 2527–2535.

39. Kneipp, J., Kneipp, H., McLaughlin, M., Brown, D., and Kneipp, K. (2006). In vivo molecular probing of cellular compartments with gold nanoparticles and nanoaggregates, *Nano Lett.*, **6**, 2225–2231.

40. Kneipp, J., Kneipp, H., Rajadurai, A., Redmond, R. W., and Kneipp, K. (2009). Optical probing and imaging of live cells using SERS labels, *J. Raman Spectrosc.*, **40**, 1–5.

41. Lane, L. A., Qian, X., and Nie, S. (2015). SERS nanoparticles in medicine: From label-free detection to spectroscopic tagging, *Chem. Rev.*, **115** 10489–10529.

42. Lee, C., and Zhang, P. (2014). Highly stable gelatin layer-protected gold nanoparticles as surface-enhanced Raman scattering substrates, *J. Nanosci. Nanotechnol.*, **14**, 4325–4330.

43. Lee, K. Y. J., Wang, Y., Nie, S. (2015). In vitro study of a pH-sensitive multifunctional doxorubicin–gold nanoparticle system: therapeutic effect and surface enhanced Raman scattering, *RSC Adv.*, **5**, 65651–65659.

44. Li, N., Li, T., Liu, C., Ye, S., Liang, J., and Han, H. (2016). Folic acid-targeted and cell penetrating peptide-mediated theranostic nanoplatform for high-efficiency tri-modal imaging-guided synergistic anticancer phototherapy, *J. Biomed. Nanotechnol.*, **12**, 878–893.

45. Liu, L., Tang, Y., Dai, S., Kleitz, F., and Qiao, S. Z. (2016). Smart surface-enhanced Raman scattering traceable drug delivery systems, *Nanoscale*, **8**, 12803–12811.

46. Liu, M., Du, H., Zhang, W., and Zhai, G. (2017). Internal stimuli-responsive nanocarriers for drug delivery: Design strategies and applications, *Mater. Sci. Eng. C*, **71**, 1267–1280.

47. Li, Y., Wen, T., Zhao, R., Liu, X., Ji, T., Wang, H., Shi, X., Shi, J., Wei, J., Zhao, Y., Wu, X., and Nie, G. (2014). Localized electric field of plasmonic nanoplatform enhanced photodynamic tumor therapy, *ACS Nano*, **8**, 11529–11542.

48. von Maltzahn, G., Centrone, A., Park, J.-H., Ramanathan, R., Sailor, M. J., Hatton, T. A., and Bhatia, S. N. (2009). SERS-coded gold nanorods as a multifunctional platform for densely multiplexed near-infrared imaging and photothermal heating, *Adv. Mater. Deerfield Beach. Fla.*, **21**, 3175–3180.

49. McFarland, A. D., Young, M. A., Dieringer, J. A., and Van Duyne, R. P. (2005). Wavelength-scanned surface-enhanced Raman excitation spectroscopy, *J. Phys. Chem. B*, **109** 11279–11285.

50. McNay, G., Eustace, D., Smith, W. E., Faulds, K., and Graham, D. (2011). Surface-enhanced Raman scattering (SERS) and surface-enhanced resonance Raman scattering (SERRS): A review of applications, *Appl. Spectrosc.*, **65** 825–837.

51. Narayanan, N., Nair, L. V., Karunakaran, V., Joseph, M. M., Nair, J. B., N, R. A., Jayasree, R. S., and Maiti, K. K. (2016). Investigation of apoptotic events at molecular level induced by SERS guided targeted theranostic nanoprobe, *Nanoscale*, **8**, 11392–11397.

52. Ock, K., Jeon, W. I., Ganbold, E. O., Kim, M., Park, J., Seo, J. H., Cho, K., Joo, S.-W., and Lee, S. Y. (2012). Real-time monitoring of glutathione-triggered thiopurine anticancer drug release in live cells investigated by surface-enhanced Raman scattering, *Anal. Chem.*, **84**, 2172–2178.

53. Ock, K.-S., Ganbold, E. O., Park, J., Cho, K., Joo, S.-W., and Lee, S. Y. (2012). Label-free Raman spectroscopy for accessing intracellular anticancer drug release on gold nanoparticles, *Analyst*, **137**, 2852.

54. O'Connor, A. E., Gallagher, W. M., and Byrne, A. T. (2009). Porphyrin and nonporphyrin photosensitizers in oncology: Preclinical and clinical advances in photodynamic therapy, *Photochem. Photobiol.*, **85**, 1053–1074.

55. Ogilby, P. R. (2010). Singlet oxygen: There is indeed something new under the sun, *Chem. Soc. Rev.*, **39**, 3181–3209.

56. O'Neal, D. P., Hirsch, L. R., Halas, N. J., Payne, J. D., and West, J. L. (2004). Photo-thermal tumor ablation in mice using near infrared-absorbing nanoparticles, *Cancer Lett.*, **209**, 171–176.

57. Park, H., Lee, S., Chen, L., Lee, E. K., Shin, S. Y., Lee, Y. H., Son, S. W., Oh, C. H., Song, J. M., Kang, S. H., and Choo, J. (2009). SERS imaging of HER2-overexpressed MCF7 cells using antibody-conjugated gold nanorods, *Phys. Chem. Chem. Phys.*, **11**, 7444.

58. Paszko, E., Ehrhardt, C., Senge, M. O., Kelleher, D. P., and Reynolds, J. V. (2011). Nanodrug applications in photodynamic therapy, *Photodiagnosis Photodyn. Ther.*, **8**, 14–29.

59. Potara, M., Boca, S., Licarete, E., Damert, A., Alupei, M.-C., Chiriac, M. T., Popescu, O., Schmidt, U., and Astilean, S. (2013). Chitosan-coated triangular silver nanoparticles as a novel class of biocompatible, highly sensitive plasmonic platforms for intracellular SERS sensing and imaging, *Nanoscale*, **5**, 6013.

60. Qian, X., Peng, X.-H., Ansari, D. O., Yin-Goen, Q., Chen, G. Z., Shin, D. M., Yang, L., Young, A. N., Wang, M. D., Nie, S. (2007). In vivo tumor targeting and spectroscopic detection with surface-enhanced Raman nanoparticle tags, *Nat. Biotechnol.*, **26**, 83–90.

61. Roy Chowdhury, M., Schumann, C., Bhakta-Guha, D., and Guha, G. (2016). Cancer nanotheranostics: Strategies, promises and impediments, *Biomed. Pharmacother.*, **84**, 291–304.

62. Samal, A. K., Polavarapu, L., Rodal-Cedeira, S., Liz-Marzán, L. M., Pérez-Juste, J., and Pastoriza-Santos, I. (2013). Size tunable Au@Ag core–shell nanoparticles: Synthesis and surface-enhanced Raman scattering properties, *Langmuir*, **29**, 15076–15082.

63. Samanta, A., Maiti, K. K., Soh, K.-S., Liao, X., Vendrell, M., Dinish, U. S., Yun, S.-W., Bhuvaneswari, R., Kim, H., Rautela, S., Chung, J., Olivo, M., and Chang, Y.-T. (2011). Ultrasensitive near-infrared Raman reporters for SERS-based in vivo cancer detection, *Angew. Chem. Int. Ed.,* **50**, 6089–6092.

64. Seo, S.-H., Kim, B.-M., Joe, A., Han, H.-W., Chen, X., Cheng, Z., and Jang, E.-S. (2014). NIR-light-induced surface-enhanced Raman scattering for detection and photothermal/photodynamic therapy of cancer cells using methylene blue-embedded gold nanorod@SiO_2 nanocomposites, *Biomaterials,* **35**, 3309–3318.

65. Shen, H., Guillot, N., Rouxel, J., de la Chapelle, M. L., and Toury, T. (2012). Optimized plasmonic nanostructures for improved sensing activities, *Opt. Express,* **20**, 21278–21290.

66. Simon, T., Potara, M., Gabudean, A.-M., Licarete, E., Banciu, M., and Astilean, S. (2015). Designing theranostic agents based on pluronic stabilized gold nanoaggregates loaded with methylene blue for multimodal cell imaging and enhanced photodynamic therapy, *ACS Appl. Mater. Interfaces,* **7**, 16191–16201.

67. Song, J., Pu, L., Zhou, J., Duan, B., and Duan, H. (2013). Biodegradable theranostic plasmonic vesicles of amphiphilic gold nanorods, *ACS Nano,* **7**, 9947–9960.

68. Song, J., Zhou, J., and Duan, H. (2012). Self-assembled plasmonic vesicles of SERS-encoded amphiphilic gold nanoparticles for cancer cell targeting and traceable intracellular drug delivery, *J. Am. Chem. Soc.,* **134**, 13458–13469.

69. Stewart, M. E., Anderton, C. R., Thompson, L. B., Maria, J., Gray, S. K., Rogers, J. A., and Nuzzo, R. G. (2008). Nanostructured plasmonic sensors, *Chem. Rev.,* **108**, 494–521.

70. Suarasan, S., Focsan, M., Potara, M., Soritau, O., Florea, A., Maniu, D., and Astilean, S. (2016). Doxorubicin-incorporated nanotherapeutic delivery system based on gelatin-coated gold nanoparticles: Formulation, drug release, and multimodal imaging of cellular internalization, *ACS Appl. Mater. Interfaces,* **8**, 22900–22913.

71. Sun, T., Zhang, Y. S., Pang, B., Hyun, D. C., Yang, M., and Xia, Y. (2014). Engineered nanoparticles for drug delivery in cancer therapy, *Angew. Chem. Int. Ed.,* **53**, 12320–12364.

72. Svenskaya, Y., Parakhonskiy, B., Haase, A., Atkin, V., Lukyanets, E., Gorin, D., and Antolini, R. (2013). Anticancer drug delivery system based on calcium carbonate particles loaded with a photosensitizer, *Biophys. Chem.,* **182**, 11–15.

73. Tam, N. C. M., Scott, B. M. T., Voicu, D., Wilson, B. C., and Zheng, G. (2010). Facile synthesis of Raman active phospholipid gold nanoparticles, *Bioconjug. Chem.*, **21**, 2178–2182.

74. Tang, W., Xu, H., Park, E. J., Philbert, M. A., and Kopelman, R. (2008). Encapsulation of methylene blue in polyacrylamide nanoparticle platforms protects its photodynamic effectiveness, *Biochem. Biophys. Res. Commun.*, **369**, 579–583.

75. Tan, X., Wang, Z., Yang, J., Song, C., Zhang, R., and Cui, Y. (2009). Polyvinylpyrrolidone- (PVP-) coated silver aggregates for high performance surface-enhanced Raman scattering in living cells, *Nanotechnology*, **20**, 445102.

76. Taylor, J., Huefner, A., Li, L., Wingfield, J., and Mahajan, S. (2016). Nanoparticles and intracellular applications of surface-enhanced Raman spectroscopy, *Analyst*, **141**, 5037–5055.

77. Tian, F., Conde, J., Bao, C., Chen, Y., Curtin, J., and Cui, D. (2016). Gold nanostars for efficient in vitro and in vivo real-time SERS detection and drug delivery via plasmonic-tunable Raman/FTIR imaging, *Biomaterials*, **106**, 87–97.

78. Tian, L., Gandra, N., and Singamaneni, S. (2013). Monitoring controlled release of payload from gold nanocages using surface enhanced Raman scattering, *ACS Nano*, **7**, 4252–4260.

79. Vinogradov, S., and Wei, X. (2012). Cancer stem cells and drug resistance: the potential of nanomedicine, *Nanomedicine*, **7**, 597–615.

80. Vo-Dinh, T., Liu, Y., Fales, A. M., Ngo, H., Wang, H.-N., Register, J. K., Yuan, H., Norton, S. J., and Griffin, G. D. (2015). SERS nanosensors and nanoreporters: golden opportunities in biomedical applications: SERS nanosensors and nanoreporters, *Wiley Interdiscip. Rev. Nanomed. Nanobiotechnol.*, **7**, 17–33.

81. Wang, Y., Chen, L., and Liu, P. (2012). Biocompatible triplex Ag@SiO_2@mTiO_2 core-shell nanoparticles for simultaneous fluorescence-SERS bimodal imaging and drug delivery, *Chem. Eur. J.*, **18**, 5935–5943.

82. Wang, Y., Yan, B., and Chen, L. (2013). SERS tags: Novel optical nanoprobes for bioanalysis, *Chem. Rev.*, **113**, 1391–1428.

83. Yao, J., Feng, J., and Chen, J. (2016). External-stimuli responsive systems for cancer theranostic, *Asian J Pharm Sci.*, **11**, 585–595.

84. Yuan, H., Fales, A. M., Khoury, C. G., Liu, J., and Vo-Dinh, T. (2013). Spectral characterization and intracellular detection of surface-enhanced Raman scattering (SERS)-encoded plasmonic gold nanostars, *J. Raman Spectrosc.*, **44**, 234–239.

85. Zahreddine, H., and Borden, K. L. B. (2013). Mechanisms and insights into drug resistance in cancer, *Front Pharmacol.*, **4**, 28.

86. Zeng, L., Pan, Y., Wang, S., Wang, X., Zhao, X., Ren, W., Lu, G., and Wu, A. (2015). Raman reporter-coupled Ag_{core}@Au_{shell} nanostars for in vivo improved surface enhanced Raman scattering imaging and near-infrared-triggered photothermal therapy in breast cancers, *ACS Appl. Mater. Interfaces*, **7**, 16781–16791.

87. Zhang, G., Li, J., Shen, A., and Hu, J. (2015). Synthesis of size-tunable chitosan encapsulated gold-silver nanoflowers and their application in SERS imaging of living cells, *Phys. Chem. Chem. Phys. PCCP*, **17**, 21261–21267.

88. Zhang, W., Wang, Y., Sun, X., Wang, W., and Chen, L. (2014). Mesoporous titania based yolk–shell nanoparticles as multifunctional theranostic platforms for SERS imaging and chemo-photothermal treatment, *Nanoscale*, **6**, 14514–14522.

89. Zhang, Y., Qian, J., Wang, D., Wang, Y., and He, S. (2013). Multifunctional gold nanorods with ultrahigh stability and tunability for in vivo fluorescence imaging, SERS detection, and photodynamic therapy, *Angew. Chem. Int. Ed.*, **52**, 1148–1151.

90. Zhao, L., Kim, T.-H., Kim, H.-W., Ahn, J.-C., and Kim, S. Y. (2015). Surface-enhanced Raman scattering (SERS)-active gold nanochains for multiplex detection and photodynamic therapy of cancer, *Acta Biomater.*, **20**, 155–164.

91. Zheng, X.-S., Hu, P., Cui, Y., Zong, C., Feng, J.-M., Wang, X., and Ren, B. (2014). BSA-coated nanoparticles for improved SERS-based intracellular pH sensing, *Anal. Chem.*, **86**, 12250–12257.

92. Zong, S., Wang, Z., Chen, H., Yang, J., and Cui, Y. (2013). Surface enhanced Raman scattering traceable and glutathione responsive nanocarrier for the intracellular drug delivery, *Anal. Chem.*, **85**, 2223–2230.

93. Félidj, N., Aubard, J., Lévi, G., Krenn, J. R., Hohenau, A., Schider, G., Leitner, A., and Aussenegg, F. R. (2003). Optimized surface-enhanced Raman scattering on gold nanoparticle arrays, *Appl. Phys. Lett.*, **82**, 3095–3097.

Chapter 8

Label-Free SERS Detection of Heme-Proteins with Porous Silver Nanocubes

Maximilien Cottat,[a,c] Marella de Angelis,[a] Elizaveta Panfilova,[b] Nikolai Khlebtsov,[b] Roberto Pini,[a] and Paolo Matteini[a]

[a]*Institute of Applied Physics "Nello Carrara", National Research Council, via Madonna del Piano 10, 50019, Sesto Fiorentino, Italy*

[b]*Institute of Biochemistry and Physiology of Plants and Microorganisms, Russian Academy of Sciences, 13 ProspektEntuziastov, 410049, Saratov, Russia*

[c]*Laboratoire des Technologies et de la Microélectronique CNRS, 17 Avenue des Martyrs, 38054 Grenoble Cedex 9, France*

p.matteini@ifac.cnr.it

8.1 Introduction

The first observation of the surface-enhanced Raman scattering (SERS) effect was done in 1974 by Fleischman et al. [1], who described an increase in Raman intensity of adsorbed pyridine on a rough silver electrode. This growth of Raman signal was attributed to the larger surface provided by the electrode roughness, causing a larger number of molecules to contribute to the Raman intensity. A few years later, the relation between SERS phenomenon and excitation of the plasmons generated by

Plasmonics in Chemistry and Biology
Edited by Marc Lamy de la Chapelle and Nordin Felidj

ISBN 978-981-4800-03-7 (Hardcover), 978-0-429-45875-0 (eBook)
www.jennystanford.com

nanostructured metal surfaces, inducing a localized enhancement of the electromagnetic field (E-field), was proposed [2–4]. Afterwards, it has been demonstrated that the SERS effect is composed of two distinct mechanisms: (1) a chemical contribution due to direct interaction of the analyte with the metallic surface that induces a charge transfer between the adsorbed molecule and the metal [5, 6], and (2) an electromagnetic contribution related to the excitation of the localized surface plasmon resonances (LSPRs) of the nanostructures that occurs when the wavelength of excitation light is close to the that of a particular plasmonic mode [7, 8]. The electromagnetic enhancement has been demonstrated to be several orders of magnitude higher than the chemical effect and can account for a factor of 10^8 [9] or more as a function of type and geometry of the nanostructures.

Initially, SERS substrates were constituted of simple metal surfaces with a nano-roughness. Nowadays, nanoparticles of noble metals obtained by chemical or physical methods are mainly used thanks to the development of fabrication techniques ensuring a fine control on the geometric parameters on the nanometer scale. Different SERS responses are associated with the chosen metal (silver, gold, copper, etc.), but also with the shape and size of the nanoparticles [10, 11]. The choice of an excitation wavelength matching the LSPRs position, as well as of a suitable polarized light in the case of anisotropic nanostructures, are critical parameters to maximize the SERS efficiency [12–15]. The sharpness of the nanostructure plays also an important role in the SERS intensity, due to the "tip effect" that is produced at its extremities. Thus, for similar size and optical properties, sharp nanoparticles usually provide a stronger enhancement of local electromagnetic fields than spherical ones and therefore a higher SERS signal [16–18]. Various shapes of nanoparticles are commonly used as SERS substrates, such as nanospheres [19, 20], nanorods [17, 21, 22], nanostars [23, 24], and nanocubes [16, 25, 26]. Recently, a new kind of substrates has been proposed for SERS analysis, consisting of porous and hollow nanoparticles produced by galvanic replacement [27, 28].

The improvement in the fabrication of the SERS substrates has enabled to move the limit of detection down to the single molecule [29]. On account of the super sensitivity offered by SERS, this technique has been proposed for several applications in the analysis and sensing of biomolecules [30–36]. However, the large majority of biomolecules, like proteins, shows a low Raman cross-section and unspecific interactions with a metallic surface, which limits their direct and label-free detection. To overcome these limitations, in the most cases of protein detection, nanoparticles are aggregated by adding salts or alkanethiols and then dried, in order to create a very intense electromagnetic field in the tiny gaps formed at the interface between aggregated nanoparticles, designated as "hot-spots", in turn increasing the sensitivity. However, chemical additives and drying can alter the native protein conformation while the presence of highly localized spots of huge electromagnetic field irregularly generated throughout the aggregated SERS substrate can jeopardize signal reproducibility as well as induce photo-damages to the protein structure, unless illumination conditions are carefully adjusted from point to point.

In this chapter, we describe a novel strategy to detect intense SERS signals of proteins in physiological buffer by using isolated nonspherical silver nanoparticles. Two different types of nanoparticles were tested: nanocubes and porous nanocubes as obtained by galvanic replacement. A theoretical simulation based on the finite element method (FEM) was performed to estimate the E-field of both substrates. SERS spectra of heme-proteins including myoglobin and cytochrome C were achieved and the impact on the protein structure caused by the excitation of the nanoparticles was considered. The comparison between SERS signals of standard and porous nanocubes has shown that the latter are the most efficient for sensitive detection of proteins. Using this type of nanoparticles, we were able to measure spectra of the protein in physiological buffer with a limit of detection in the nanomolar range without the need of a Raman tag.

8.2 Protein Detection Using Standard and Porous Nanocubes

8.2.1 Experimental Section

8.2.1.1 Fabrication of the nanoparticles

Standard and porous nanocubes were fabricated by following a sulfide-mediated polyol method, which involves silver reduction with ethylene glycol in the presence of polyvinylpyrrolidone (PVP) [37]. A solution of ethylene glycol (30 mL) was mixed with 350 μL of 3 mM sodium sulfide in ethylene glycol, followed by the addition of 150 mg of PVP. The solution was heated at 150°C during 30 min under argon flow. Then, 120 mg of silver nitrate ($AgNO_3$) was added, which immediately changed the solution color, from yellow to green ocher. The reaction was stopped by rapid cooling in ice. Porous nanocubes were prepared by using a galvanic replacement method. For this purpose, 200 μL of the as-prepared Ag nanocubes were injected in 10 mL of a heated solution (100°C) containing PVP (1 g/L). Then, a controlled volume of 1 mM tetrachloroauric acid ($HAuCl_4$) was added drop by drop. The $HAuCl_4$ volume permits to adjust the size of the nanopores and thus to control the LSPR position. Nanocubes and porous nanocubes were washed twice by centrifugation (10000× g, 30 min) to remove excess of all reagents. Finally, nanocubes and porous nanocubes were resuspended in ethanol and in pure water, respectively.

8.2.1.2 SERS analysis

SERS measurements were carried out in a 7 mM phosphate buffer prepared by mixing Na_2HPO_4 and NaH_2PO_4 in the proper amount to reach a pH of 7.4. Cytochrome C and myoglobin were dissolved in water to obtain a stock solution of 10^{-4} M, and then diluted in the phosphate buffer to obtain several final concentration values (from 10^{-6} down to 10^{-8} M) in the analyzed sample. For both nanoparticle batches, the initial concentration was around 10^{11} particles/mL. Typically, 10 μL of nanoparticles were mixed with 440 μL of phosphate buffer followed by 2 min of sonication. Then 50 μL of stock protein solution were added for incubation during 2 h at room temperature followed by centrifugation to remove protein excess (10000× g, 10 min). Extinction analysis of the as-synthesized nanoparticle batches was achieved with a UV-Vis Jasco V-560 spectrophotometer while extinction spectra of the nanoparticles after incubation with the proteins were acquired by means of a bench spectrometer (EPP2000 by Stellarnet Inc.) connected to a Leica DM2500 microscope operated in transmission mode. SERS measurements were carried out on suspended drops of 7 μL as previously described in [38] by means of a Horiba Xplora micro-Raman equipped with two lasers (532 and 638 nm) and with a spectral resolution of about 10 cm^{-1}. Back scattering light was collected by a 10× microscope objective with a NA of 0.25 providing a 7 μm-size laser spot. Laser power at the sample was 4.8 mW at 532 nm and 2.1 mW at 638 nm. Acquisition time of each spectrum was 10 s with two accumulations. The SERS spectra presented consist of an average of 10 spectra followed by baseline correction as obtained by LabSpec 6 spectroscopy software.

8.2.2 Results and Discussion

8.2.2.1 Characterization of the nanoparticles

The shape and size of standard and porous nanocubes were characterized by transmission electron microscopy (TEM) (Figs. 8.1A and 8.1B). The nanoparticles shape appeared highly homogeneous with narrow size dispersion for both nanoparticle types.

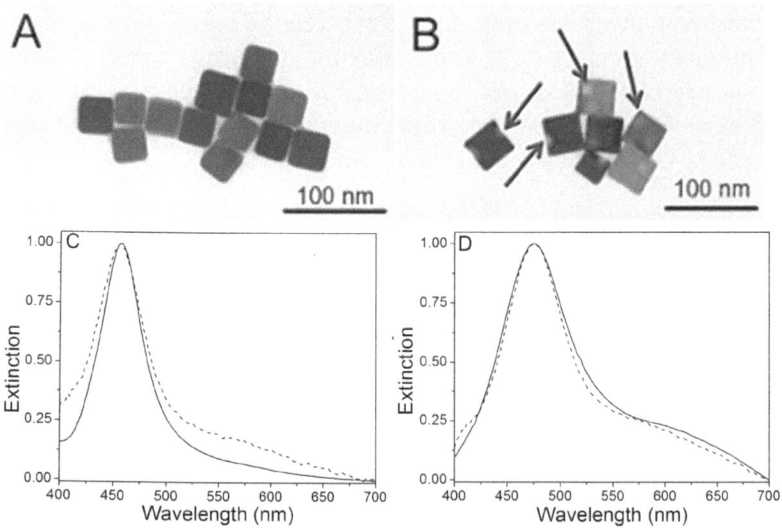

Figure 8.1 TEM images of standard (A) and porous (B) nanocubes. Extinction spectra of standard (C) and porous (D) nanocubes before (solid line) and after (dashed line) incubation with 10^{-6} M myoglobin.

Both particle types exhibited a size of around 50 nm and a regular cubic shape, as a consequence of using PVP as stabilizer in their synthesis [39, 40]. Indeed, PVP has a strong affinity for the {100} crystalline facets of the initial silver seeds [41], inducing a specific orientation in the nanoparticle growth. Final nanocubes have faces with a {100} crystalline orientation and which are fully covered by PVP, and rounded corners with variable crystalline orientations with lacking or scarce PVP presence. During the galvanic replacement, the oxidation of silver by gold ions leads to deposition of one gold atom mainly occurring on the cube corners at the expense of three silver atoms removed from the cube faces [42]. In turn, the nanoparticles become porous with LSPRs moving from blue to red as the reaction proceeds [43].

In Fig. 8.1B, the presence of local nanopores (indicated by arrows) on the cube's faces confirms that silver oxidation occurs at specific locations and mostly on three to five sides of the nanoparticle. Nanocubes and porous nanocubes show LSPR bands located at 457 nm (Fig. 8.1C) and at 478 nm (Fig. 8.1D), respectively. Moreover, while nanocubes exhibit a narrow band, a broader peak with a shoulder is noticeable for porous nanocubes. These observations have been previously reported and associated to plasmon damping [28]. After incubation with the protein (dashed spectra of Fig. 8.1C and 8.1D), the maximum peak position and shape are not modified, revealing that our experiments were effectively carried out on isolated nanoparticles.

8.2.2.2 Simulation of the E-field distribution

Simulation of the induced E-field was performed using a 3D numerical FEM under Comsol Multiphysics 5.1 software and operating in the scattering mode of the wave-optics-module to solve the Helmholtz equation. Our geometry approximated the nanoparticle shapes detected via TEM and shown in Fig. 8.1A for the standard nanocubes and in Fig. 8.1B for the porous nanocubes. We included a silver nanocube with 50 nm size, with rounded edges at a 6 nm radius of curvature [38]. We then adopted a model with an 8 nm-size pore on one of the six cube faces located in the proximity of one of the corner. In the calculation, the nanoparticles were surrounded by water and the silver was considered as unique material for the nanoparticles. This choice is supported by considering that gold mass fraction is less than 0.76% and it bedecks only small PVP-free zones on the nanoparticle surface thus changing very little the response of the nanoparticle to the incoming E-field [28]. Although the model may look simplified, it can capture the essential features in terms of E-field distribution.

The local field, E-field, was expressed in the form of normalized electric field intensity $|E|/|E_0|$ at the laser wavelength, where E_0 is the amplitude of the electric field incident onto the nanoparticle. The metal was described through its measured dielectric function [44]. The domains were delimited by perfectly matched layers (PML) in order to reach a close contact at the outer boundaries. The incident field was assumed to be an electromagnetic plane wave propagating along the z axis and linear polarization along the y direction. The E-field distribution of the porous nanocube is presented in Fig. 8.2A.

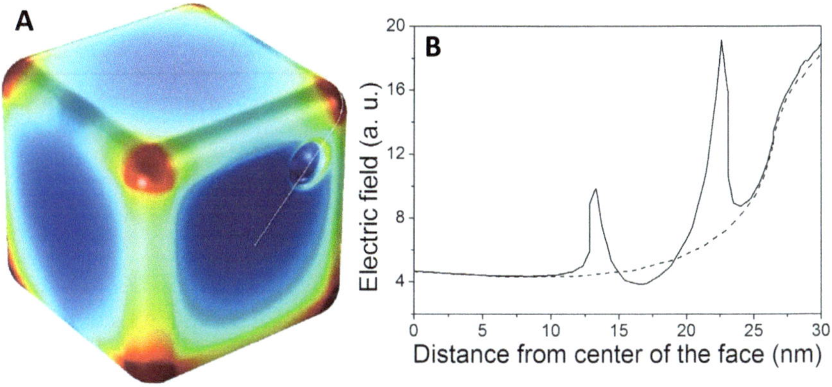

Figure 8.2 (A) FEM simulation of the distribution of the E-field over the surface of a 50 nm silver nanocube containing one nanopore. (B) Profile of the E-field distribution along the white line shown in (A) connecting the apex of a corner to the center of a face and passing through the nanopore (solid line) compared to a similar cube region without nanopores (dashed line).

We have calculated the E-field along a curve from the apex of the cube corner to the center of one of the six cube faces. The curve passes through the nanopore and shows two peaks of E-field intensity localized at the nanopore edges (Fig. 8.2B, solid line). When compared to the value of a corresponding cube region without nanopore (Fig. 8.2B, dashed line), the E-field peak is twice larger and of the same magnitude as the electric field on a cube corner. Since the surface area influenced by this increase (Fig. 8.2A) is of the same extent of a single hot spot on a cube corner, we can conclude that each pore of the cube acts as an additional hot spot.

8.2.2.3 SERS detection

SERS properties of standard and porous nanocubes were investigated using 532 and 638 nm excitation wavelengths. We considered two heme-proteins, i.e., myoglobin and cytochrome C, because of their well-characterized SERS properties [45–50]. It is worth to note that the use of a 532 nm excitation is advantageous for both proteins due to a resonance Raman effect with the Q band of the heme-group lying in the 500–575 nm range (Fig. 8.3). Myoglobin shows also a smaller heme band at 640 nm that is in resonance with the red excitation (Fig. 8.3 inset, dashed line).

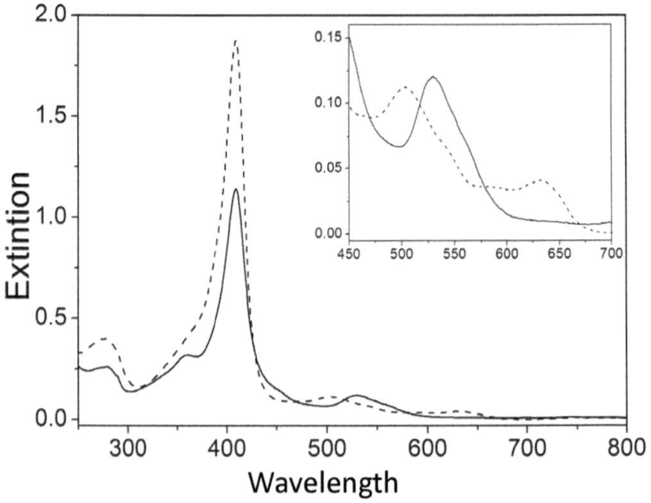

Figure 8.3 Extinction spectra of cytochrome C (solid line) and myoglobin (dashed line). Inset: magnification of the 450–700 nm range.

Figure 8.4 shows SERS spectra of 10^{-6} M myoglobin acquired at λ_{ex} = 532 nm (Fig. 8.4A) and λ_{ex} = 638 nm (Fig. 8.4B). The bottom spectrum corresponds to the Raman profile obtained with a protein concentration of 1 mM. The Raman signal is mainly characterized by vibration modes assigned to the heme group, especially when using the 532 nm laser excitation (Fig. 8.4A), as reported in Table 8.1. The presence of amide III bands, which are sensitive to the protein structure (α-helix and β-sheet), is also noticed (1235 and 1300 cm^{-1}) [51].

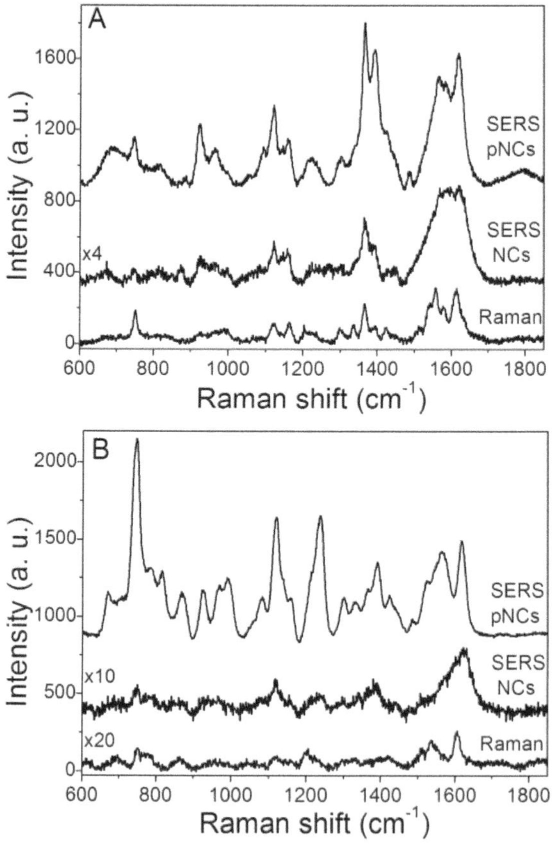

Figure 8.4 SERS spectra of myoglobin (1 μM) measured with porous (SERS pNCs) and standard (SERS NCs) nanocubes at λ_{ex} = 532 nm (A) and at λ_{ex} = 638 nm (B). The Raman spectrum of 1 mM myoglobin is also reported in (A) and in (B) as reference.

In general, SERS signals provided by porous nanocubes are sharper and more intense as compared to the case of standard nanocubes. Accordingly, SERS spectra obtained with the latter have been multiplied by 4 and 10 to reach similar intensity values. The low signals measured with standard nanocubes produced poorly resolved SERS spectra with low signal-to-noise ratio. Nevertheless, modes at 750, 1122, 1164, 1366 and 1395 cm^{-1} clearly identify the possibility to detect micromolar concentrations of myoglobin. In the case of porous nanocubes, all the vibration modes of the protein are visible and closely match those of the Raman spectrum. Similar Raman and SERS patterns suggest that the protein structure was not affected by the interaction with the nanoparticles. Then we repeated the SERS analysis on cytochrome C, which is presented in Fig. 8.5. As for the myoglobin, we detected several vibration modes that are assigned to the heme group (747, 1124, 1164, 1225, 1312, 1367, 1400, 1558, 1581, and 1632 cm^{-1}), as indicated in Table 8.1.

Table 8.1 Peaks position and assignment of Raman spectra of myoglobin and cytochrome C measured at λ_{ex} = 532 nm

Raman peak position		Assignment [46–49, 52–54]	
Myoglobin	**Cytochrome C**	**Mode**	**Symmetry**
750	747	υ_{15}	B_{1g}
926		υ_{46}	E_u
1122	1124	υ_{22}	A_{2g}
1164	1164	υ_{30}	B_{2g}
1220	1225	υ_{13}	B_{1g}
1235		Amide III (β-sheet)	
1300		Amide III (α-helix)	
	1312	υ_{21}	A_{2g}
1334		υ_{41}	E_u
1366	1367	υ_{4}	A_{1g}
1395	1400	υ_{20}	A_{2g}
1424		υ_{28}	B_{2g}
1512		υ_{38}	E_u
1539		υ_{11}	B_{1g}
1558	1558	υ_{11}	B_{1g}
1579	1581	υ_{19}	A_{2g}
1613		υ_{10}	B_{1g}
1637	1632	υ_{10}	B_{1g}

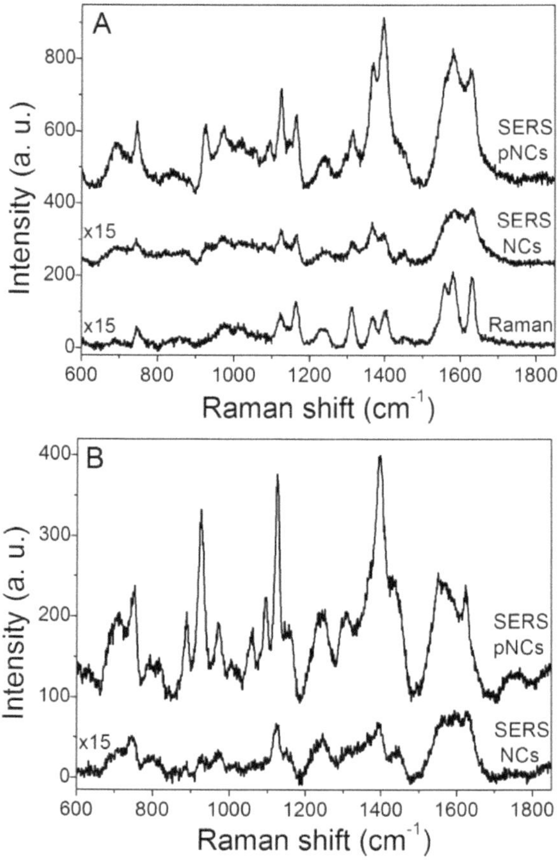

Figure 8.5 SERS spectra of 10^{-6} M cytochrome C measured with porous (SERS pNCs) and standard (SERS NCs) nanocubes at λ_{ex} = 532 nm (A) and at λ_{ex} = 638 nm (B). The Raman spectrum of 1 mM cytochrome C obtained at λ_{ex} = 532 nm is also displayed in (A).

Similar to the above, a large variation in SERS intensity between porous and standard nanocubes is apparent in spite of the laser excitation used. Again, the SERS spectrum of cytochrome C obtained with porous nanocubes is optimally resolved and exhibits the same bands compared with the Raman spectrum. Conversely, only few peaks in the SERS spectra with standard nanocubes can be discerned for identifying the protein.

For both proteins, SERS spectra acquired with silver porous nanocubes showed more intense signals and better resolution than observed with standard nanocubes. Thus, we may conclude that the SERS efficiency of porous nanocubes is higher than the one provided by using the standard counterpart. We can tentatively explain these observations on account of a larger surface of the SERS active area in the presence of nanopores. In fact, the SERS signal of standard nanocubes is predicted to be mostly generated by the hot spots localized on the cube's corners (Fig. 8.2A) where the E-field is very intense due to the tip effect and the PVP is lacking or scarce [38]. Cube's sides have been demonstrated to minimally contribute to the SERS signal because of the presence of a PVP shielding layer on the surface [38], which prevents direct protein interactions to occur [55]. On the other side, the presence of nanopores creates new SERS-active sites (see Figs. 8.2A and 8.2B), which can also participate to the final signal. Similarly to corner sites, the multi-crystalline features in the pore cavities is expected to discourage the formation of a dense PVP coating favoring direct protein–metal interactions and, in turn, effective SERS detection of adsorbed proteins. Overall, the presence of nanopores confers an extensive area available for E-field enhancement, causing a large number of protein molecules to be detected.

8.2.2.4 SERS sensitivity

Figure 8.6 shows SERS spectra obtained with porous nanocubes recorded for myoglobin and cytochrome C at various concentrations, i.e., from 10^{-6} M to 10^{-8} M.

The SERS intensity of myoglobin (Figs. 8.6A and 8.6B) decreases regularly as the protein concentration is reduced, which suggests the possibility to perform a quantitative detection of this protein. Concerning cytochrome C (Figs. 8.6C and 8.6D), the signal decrease is less significant passing from 10^{-6} M to 10^{-7} M, indicating that almost all SERS-active sites are already occupied at the latter concentration value. This behavior is possibly in accordance with the fact that the interaction of cytochrome C with the nanoparticles should be driven by electrostatic attractions. In fact, the protein is positively charged at our operating pH because of an isoelectric point close to 10, which favors a physical adsorption of protein molecules on the PVP-decorated nanoparticle surfaces and a saturation of the hot spots at lower concentration values.

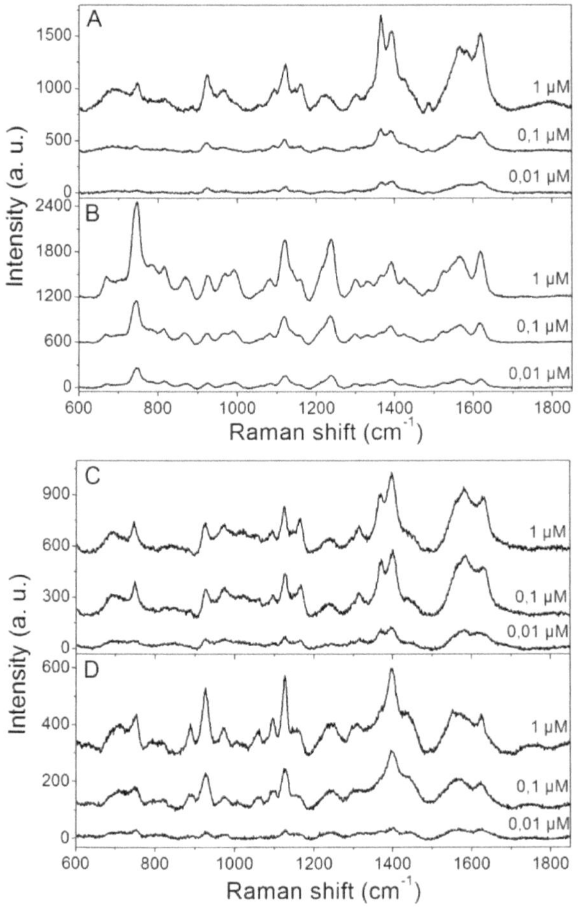

Figure 8.6 SERS spectra of myoglobin (A and B) and cytochrome C (C and D) measured with porous nanocubes at different concentration, using λ_{ex} = 532 nm (A and C) and λ_{ex} = 638 nm (B and D).

Finally, for both proteins and at both excitations considered, we were able to observe the characteristic signals of the proteins at 10^{-8} M with a nice signal to noise ratio, which points out the high sensitivity offered by the porous nanocubes. The strong SERS signals obtained in both resonant and nonresonant excitation conditions combined with a high sensitivity demonstrate that protein detection with silver porous nanocubes is feasible in the nanomolar range without using any Raman tag or specific surface functionalization.

8.3 Conclusions

We have demonstrated that silver porous nanocubes fabricated by galvanic replacement offer a better SERS efficiency than standard silver nanocubes. Detection of heme-proteins, in resonant and nonresonant excitation conditions, has been achieved at nanomolar concentration, proving the high sensitivity of isolated porous nanocubes at physiological pH. The acquired SERS spectra closely resemble the Raman spectra, suggesting that modification in the protein native state is not occurring, which is of interest for a conformational analysis of the protein structure. These results show the ability of silver porous nanocubes to be used for sensitive and label-free detection of proteins by SERS.

Acknowledgements

P.M., M.deA., and R.P. acknowledge the financial support from the European Community within the EuroNanoMed3 ERANET cofund (H2020) SPEEDY project (ID 221). The work by E.P. and N.K. was supported by RFBR grants 16-02-00054, 17-02-00075. The work by N.K. was supported by RFBR grant 18-62-7803 (RFBR-CNR collaborative project).

References

1. Fleischmann, M., Hendra, P. J., and McQuillan, A. J. (1974). Raman spectra of pyridine adsorbed at a silver electrode, *Chem. Phys. Lett.*, **26**(2), 163–166.
2. Albrecht, M. G. and Creighton, J. A. (1977). Anomalously intense Raman spectra of pyridine at a silver electrode, *J. Am. Chem. Soc.*, **99**(15), 5215–5217.
3. Moskovits, M. (1978). Surface roughness and the enhanced intensity of Raman scattering by molecules adsorbed on metals, *J. Chem. Phys.*, **69**(9), 4159–4161.
4. Jeanmaire, D. L. and Van Duyne, R. P. (1977). Surface Raman spectroelectrochemistry: Part I. Heterocyclic, aromatic, and aliphatic amines adsorbed on the anodized silver electrode, *J. Electroanal. Chem. Interfacial Electrochem.*, **84**(1), 1–20.

5. Otto, A. (1984). *Light Scattering in Solids IV: Electronics Scattering, Spin Effects, SERS, and Morphic Effects*, eds., Cardona, M., and Güntherodt, G., (Berlin, Heidelberg). pp. 289–418.
6. Campion, A., and Kambhampati, P. (1998). Surface-enhanced Raman scattering, *Chem. Soc. Rev.*, **27**(4), 241–250.
7. Wokaun, A. (1984). *Solid State Phyics*, eds., Ehrenreich, H., Turnbull, D., and Seitz, F., pp. 223–294.
8. Schatz, G. C., and Van Duyne, R. P. (2006). *Handbook of Vibrational Spectroscopy*, Chapter "Electromagnetic mechanism of surface enhanced spectroscopy".
9. Aroca, R. F., Alvarez-Puebla, R. A., Pieczonka, N., Sanchez-Cortez, S., and Garcia-Ramos, J. V. (2005). Surface-enhanced Raman scattering on colloidal nanostructures, *Adv. Colloid Interface Sci.*, **116**(1–3), 45–61.
10. Hanarp, P., Käll, M., and Sutherland, D. S. (2003). Optical properties of short range ordered arrays of nanometer gold disks prepared by colloidal lithography, *J. Phys. Chem. B*, **107**(24), 5768–5772.
11. Jensen, T. R., Malinsky, M. D., Haynes, C. L., and Van Duyne, R. P. (2000). Nanosphere lithography: Tunable localized surface plasmon resonance spectra of silver nanoparticles, *J. Phys. Chem. B*, **104**(45), 10549–10556.
12. Grand, J., de la Chapelle, M. L., Bijeon, J. L., Adam, P. M., Vial, A., and Royer, P. (2005). Role of localized surface plasmons in surface-enhanced Raman scattering of shape-controlled metallic particles in regular arrays, *Phys. Rev. B*, **72**(3), 033407.
13. Haynes, C. L., and Van Duyne, R. P. (2003). Plasmon-sampled surface-enhanced Raman excitation spectroscopy, *J. Phys. Chem. B*, **107**(30), 7426–7433.
14. Emory, S. R., Haskins, W. E., and Nie, S. (1998). Direct observation of size-dependent optical enhancement in single metal nanoparticles, *J. Am. Chem. Soc.*, **120**(31), 8009–8010.
15. McFarland, A. D., Young, M. A., Dieringer, J. A., and Van Duyne, R. P. (2005). Wavelength-scanned surface-enhanced raman excitation spectroscopy, *J. Phys. Chem. B*, **109**(22), 11279–11285.
16. Rycenga, M., Kim, M. H., Camargo, P. H. C., Cobley, C., Li, Z.-Y., and Xia, Y. (2009). Surface-enhanced Raman scattering: Comparison of three different molecules on single-crystal nanocubes and nanospheres of silver, *J. Phys. Chem. A*, **113**(16), 3932–3939.

17. Nikoobakht, B., and El-Sayed, M. A. (2003). Surface-enhanced Raman scattering studies on aggregated gold nanorods, *J. Phys. Chem. A*, **107**(18), 3372–3378.
18. Tian, F., Bonnier, F., Casey, A., Shanahan, A. E., and Byrne, H. J. (2014). Surface enhanced Raman scattering with gold nanoparticles: Effect of particle shape, *Anal. Methods*, **6**(22), 9116–9123.
19. Li, W., Camargo, P. H. C., Lu, X., and Xia, Y. (2009). Dimers of silver nanospheres: facile synthesis and their use as hot spots for surface-enhanced Raman scattering, *Nano Lett.*, **9**(1), 485–490.
20. Njoki, P. N., Lim, I. I. S., Mott, D., Park, H.-Y., Khan, B., Mishra, S., Sujakumar, R., Luo, J., and Zhong, C.-J. (2007). Size correlation of optical and spectroscopic properties for gold nanoparticles, *J. Phys. Chem. C*, **111**(40), 14664–14669.
21. Fazio, B., D'Andrea, C., Foti, A., Messina, E., Irrera, A., Donato, M. G., Villari, V., Micali, N., Maragò, O. M., and Gucciardi, P. G. (2016). SERS detection of biomolecules at physiological pH via aggregation of gold nanorods mediated by optical forces and plasmonic heating, *Sci. Rep.*, **6**, 26952.
22. Orendorff, C. J., Gearheart, L., Jana, N. R., and Murphy, C. J. (2006). Aspect ratio dependence on surface enhanced Raman scattering using silver and gold nanorod substrates, *PCCP*, **8**(1), 165–170.
23. Khoury, C. G., and Vo-Dinh, T. (2008). Gold nanostars for surface-enhanced Raman scattering: Synthesis, characterization and optimization, *J. Phys. Chem. C*, **112**(48), 18849–18859.
24. Hrelescu, C., Sau, T. K., Rogach, A. L., Jäckel, F., and Feldmann, J. (2009). Single gold nanostars enhance Raman scattering, *Appl. Phys. Lett.*, **94**(15), 153113.
25. McLellan, J. M., Siekkinen, A., Chen, J., and Xia, Y. (2006). Comparison of the surface-enhanced Raman scattering on sharp and truncated silver nanocubes, *Chem. Phys. Lett.*, **427**(1–3), 122–126.
26. Banchelli, M., Tiribilli, B., de Angelis, M., Pini, R., Caminati, G., and Matteini, P. (2016). Controlled veiling of silver nanocubes with graphene oxide for improved surface-enhanced Raman scattering detection, *ACS Appl. Mater. Interfaces*, **8**(4), 2628–2634.
27. Yang, Y., Zhang, Q., Fu, Z.-W., and Qin, D. (2014). Transformation of Ag nanocubes into Ag–Au hollow nanostructures with enriched Ag contents to improve SERS activity and chemical stability, *ACS Appl. Mater. Interfaces*, **6**(5), 3750–3757.

28. Rycenga, M., Hou, K. K., Cobley, C. M., Schwartz, A. G., Camargo, P. H. C., and Xia, Y. (2009). Probing the surface-enhanced Raman scattering properties of Au-Ag nanocages at two different excitation wavelengths, *PCCP*, **11**(28), 5903–5908.

29. Ru, E. C. L., and Etchegoin, P. G. (2012). Single-molecule surface-enhanced Raman spectroscopy, *Annu. Rev. Phys. Chem.*, **63**(1), 65–87.

30. Han, X. X., Huang, G. G., Zhao, B., and Ozaki, Y. (2009). Label-free highly sensitive detection of proteins in aqueous solutions using surface-enhanced raman scattering, *Anal. Chem.*, **81**(9), 3329–3333.

31. David, C., Guillot, N., Shen, H., Toury, T., and de la Chapelle, M. L. (2010). SERS detection of biomolecules using lithographed nanoparticles towards a reproducible SERS biosensor, *Nanotechnology*, **21**(47), 475501.

32. Matteini, P., de Angelis, M., Ulivi, L., Centi, S., and Pini, R. (2015). Concave gold nanocube assemblies as nanotraps for surface-enhanced Raman scattering-based detection of proteins, *Nanoscale*, **7**(8), 3474–3480.

33. Chen, J., Jiang, J., Gao, X., Liu, G., Shen, G., and Yu, R. (2008). A new aptameric biosensor for cocaine based on surface-enhanced raman scattering spectroscopy, *Chem. Eur. J.*, **14**(27), 8374–8382.

34. Cottat, M., D'Andrea, C., Yasukuni, R., Malashikhina, N., Grinyte, R., Lidgi-Guigui, N., Fazio, B., Sutton, A., Oudar, O., Charnaux, N., Pavlov, V., Toma, A., Di Fabrizio, E., Gucciardi, P. G., and Lamy de la Chapelle, M. (2015). High sensitivity, high selectivity SERS detection of MnSOD using optical nanoantennas functionalized with aptamers, *J. Phys. Chem. C*, **119**(27), 15532–15540.

35. Domenici, F., Bizzarri, A. R., and Cannistraro, S. (2012). Surface-enhanced Raman scattering detection of wild-type and mutant p53 proteins at very low concentration in human serum, *Anal. Biochem.*, **421**(1), 9–15.

36. Siddhanta, S., Karthigeyan, D., Kundu, P. P., Kundu, T. K., and Narayana, C. (2013). Surface enhanced Raman spectroscopy of Aurora kinases: Direct, ultrasensitive detection of autophosphorylation, *RSC Adv.*, **3**(13), 4221–4230.

37. Khlebtsov, B. N., Khanadeev, V. A., Maksimova, I. L., Terentyuk, G. S., and Khlebtsov, N. G. (2010). Silver nanocubes and gold nanocages: Fabrication and optical and photothermal properties, *Nanotechnol. Russia*, **5**(7), 454–468.

38. Matteini, P., Cottat, M., Tavanti, F., Panfilova, E., Scuderi, M., Nicotra, G., Menziani, M. C., Khlebtsov, N., de Angelis, M., and Pini, R. (2017). Site-selective surface-enhanced Raman detection of proteins, *ACS Nano*, **11**(1), 918–926.

39. Xia, X., Zeng, J., Oetjen, L. K., Li, Q., and Xia, Y. (2012). Quantitative analysis of the role played by poly(vinylpyrrolidone) in seed-mediated growth of Ag nanocrystals, *J. Am. Chem. Soc.*, **134**(3), 1793–1801.

40. Chen, J., McLellan, J. M., Siekkinen, A., Xiong, Y., Li, Z.-Y., and Xia, Y. (2006). Facile synthesis of gold–silver nanocages with controllable pores on the surface, *J. Am. Chem. Soc.*, **128**(46), 14776–14777.

41. Al-Saidi, W. A., Feng, H., and Fichthorn, K. A. (2012). Adsorption of polyvinylpyrrolidone on Ag surfaces: Insight into a structure-directing agent, *Nano Lett.*, **12**(2), 997–1001.

42. Sun, Y., and Xia, Y. (2004). Mechanistic study on the replacement reaction between silver nanostructures and chloroauric acid in aqueous medium, *J. Am. Chem. Soc.*, **126**(12), 3892–3901.

43. Panfilova, E., Shirokov, A., Khlebtsov, B., Matora, L., and Khlebtsov, N. (2012). Multiplexed dot immunoassay using Ag nanocubes, Au/Ag alloy nanoparticles, and Au/Ag nanocages, *Nano Res.*, **5**(2), 124–134.

44. Rakić, A. D., Djurišić, A. B., Elazar, J. M., and Majewski, M. L. (1998). Optical properties of metallic films for vertical-cavity optoelectronic devices, *Appl. Opt.*, **37**(22), 5271–5283.

45. Salim, A., Christian, J., Jesper, N., and Thomas, N. (2007). Resonance surface enhanced Raman optical activity of myoglobin as a result of optimized resonance surface enhanced Raman scattering conditions, *J. Phys.: Condens. Matter*, **19**(28), 285205.

46. Bizzarri, A. R., and Cannistraro, S. (2002). Surface-enhanced resonance Raman spectroscopy signals from single myoglobin molecules, *Appl. Spectrosc.*, **56**(12), 1531–1537.

47. Dick, L. A., Haes, A. J., and Van Duyne, R. P. (2000). Distance and orientation dependence of heterogeneous electron transfer: A surface-enhanced resonance Raman scattering study of cytochrome c bound to carboxylic acid terminated alkanethiols adsorbed on silver electrodes, *J. Phys. Chem. B*, **104**(49), 11752–11762.

48. Keating, C. D., Kovaleski, K. M., and Natan, M. J. (1998). Protein: Colloid conjugates for surface enhanced Raman scattering: Stability and control of protein orientation, *J. Phys. Chem. B*, **102**(47), 9404–9413.

49. Niaura, G., Gaigalas, A. K., and Vilker, V. L. (1996). Non-resonant SERS study of the adsorption of cytochrome c on a silver electrode, *J. Electroanal. Chem.*, **416**(1–2), 167–178.

50. Sevilla, P., Sánchez-Cortés, S., García-Ramos, J. V., and Feis, A. (2014). Concentration-controlled formation of myoglobin/gold nanosphere aggregates, *J. Phys. Chem. B*, **118**(19), 5082–5092.

51. Das, G., Chirumamilla, M., Toma, A., Gopalakrishnan, A., Zaccaria, R. P., Alabastri, A., Leoncini, M., and Di Fabrizio, E. (2013). Plasmon based biosensor for distinguishing different peptides mutation states, *Sci. Rep.*, **3**, 1792.

52. Kitahama, Y., Egashira, M., Suzuki, T., Tanabe, I., and Ozaki, Y. (2014). Sensitive marker bands for the detection of spin states of heme in surface-enhanced resonance Raman scattering spectra of metmyoglobin, *Analyst*, **139**(24), 6421–6425.

53. Yu, Q., and Golden, G. (2007). Probing the protein orientation on charged self-assembled monolayers on gold nanohole arrays by SERS, *Langmuir*, **23**(17), 8659–8662.

54. Xu, L.-J., Zong, C., Zheng, X.-S., Hu, P., Feng, J.-M., and Ren, B. (2014). Label-free detection of native proteins by surface-enhanced Raman spectroscopy using iodide-modified nanoparticles, *Anal. Chem.*, **86**(4), 2238–2245.

55. Treuel, L., Malissek, M., Grass, S., Diendorf, J., Mahl, D., Meyer-Zaika, W., and Epple, M. (2012). Quantifying the influence of polymer coatings on the serum albumin corona formation around silver and gold nanoparticles, *J. Nanopart. Res.*, **14**(9), 1102.

Chapter 9

Observation of Biomolecules and Their Dynamics in SERS

Jean Emmanuel Clément, Thibault Brulé, Aymeric Leray, and Eric Finot

Laboratoire Interdisciplinaire Carnot de Bourgogne, UMR 6303 CNRS, Université Bourgogne Franche Comté, 21000 Dijon, France

eric.finot@u-bourgogne.fr

9.1 Introduction

Biomolecules include both small molecules such as metabolites, vitamins, and hormones and large macromolecules such as amino acids and proteins. In the case of proteins of high molecular weight, their three-dimensional arrangements enable them to fold into one or more specific spatial conformations driven by noncovalent interactions. The structural biology is a new transverse topic of molecular biochemistry and biophysics that is concerned with the protein conformations linked to its biological functions. There is then a clear call for developing new physical tools to determine the protein structure and its dynamics with the final

Plasmonics in Chemistry and Biology
Edited by Marc Lamy de la Chapelle and Nordin Felidj

ISBN 978-981-4800-03-7 (Hardcover), 978-0-429-45875-0 (eBook)
www.jennystanford.com

goal of our better understanding of the function of biomolecule or the design of new protein. From a biological point of view, the protein has three main levels of structure. The primary structure consists of a sequence of 20 natural amino acids connected to each other by peptide bonds. The secondary structure describes the local folding of the main protein chain. At this level, there are two main structures: the α-helix and β-sheet. The third structure is the result of complex interactions relevant for the three-dimensional conformation of proteins. To put it more bluntly, the protein might be seen as a sentence in which amino acids are the letters, the secondary structures are the word and finally the third structure is the sentence. From a physicochemical point of view, the protein is a linear polymer having thermodynamic stability. Phase transitions between unfolded and native states are activated by a difference in free energy of few kcal/mol) and driven also by surface properties, including surface charge, surface free energy, and hydrophilicity.

This issue was first approach by simulation techniques in the time scale up to the microsecond. Protein's identity is usually obtained using biological proteomics approaches such as immunoassays based on antibodies (Elisa, Western Blot, SDS-PAGE). Native mass spectrometry plays an important role in structural biology (Siuzdak, 1994) providing details about the complex composition, topology, stability. Limitation of those technique remains in the needs of the homogeneity of large sample volume. Higher resolution structural information down to atomic resolution over minute time series can be determined by techniques requiring heavy infrastructures such as X-ray crystallography, neutron or nuclear magnetic resonance (NMR) spectroscopy (Mittermaier and Kay, 2009). Unfortunately, proteins above a few tens of kDa represent a severe challenge for solution NMR methods. Förster resonance energy transfer (FRET) (Stevens et al., 2010) has been used to monitor transient changes in distance or angle, but the needs of dye donors and acceptors limit can pose serious limitations. In biomedical diagnostics, Raman spectroscopy in its imaging mode has been recognized as successful for identifying bacteria, sorting single cells, and diagnosing tissue sections (Krafft and Popp, 2015). The heterogeneity of the microbial cell populations has already been observed at the single-cell level (Hermelink et al., 2009).

Discrimination between lipids, solvent and high dense proteins can be imaged at the micronic resolution using nonlinear Raman microscopy (coherent anti-Stokes Raman scattering (CARS) and stimulated Raman scattering (SRS)). The ultrafast energy transfer (Ferrante et al., 1991) provides the analysis of the vibrational landscape below the nanosecond time scale.

There is clearly a demand for physical tools to investigate the dynamics of single protein in the time slot (10^{-6} s, 1 s). Predicting reaction rates in complex protein dynamics remains a fundamental challenge to a wide range of biophysical processes. How is the coupling between environment fluctuations and the amino acid interactions put into work in flexible proteins? And we still fail to model the local molecular fluctuating environment of a protein that has potentially significant effects on motions relevant to its function.

9.2 Raman Spectroscopy of Proteins

Raman spectroscopy gives information about the selectivity, the sensitivity and the conformation of molecules.

Raman scattering is an inefficient process and typical scattering cross sections are only 10^{-6} to 10^{-8} m^2. Major requirements to obtain Raman spectra are a bright, highly monochromatic excitation source with spectral line width of the order of 1 cm^{-1}.

Spectral library was generated for proteins (Rygula et al., 2013). Specific bands relevant to the secondary structure have been identified using proteins with known a-helix and b-sheet elements. The interpretation is frequently based on changes in peak shape rather than on individual band assignments. Usually, the focus is on the visual interpretation of amide I and amide III bands (Tuma, 2005). The band position determines the selectivity of the detection. However, the spectral deconvolution based on curve-fitting introduces a certain subjectivity of the approach which is due to the number of parameters involved in the band-narrowing.

The low-frequency range between 10 and 250 cm^{-1} is typical of a disordered molecular system characterized by the overlap of relaxation vibration of large amplitude having a quasi harmonic behavior. The vibrational dynamics of the protein is highly related to the structural organization of the solvent.

The frequency range between 500 and 1800 cm^{-1} allows the analysis of different types of amide I to III and mainly the amide I which determines the composition in sheet or helical form of the secondary structure (Clément, 2017). The most intense amide I is generally between 1600 and 1700 cm^{-1}. The unfolded state is usually characterized compared to the folded state by sliding the band of amide I toward the higher frequencies. The vibration bands of amide II are generally between 1500 and 1580 cm^{-1} corresponding to angular deformation between NH and CN bonds which are generally less sensitive to protein folding. Amide III bands between 1200 and 1400 cm^{-1} mainly correspond to the combination (between the stretching vibrations of the CN and the deformation of the NH bonds (Barth, 2002).

The intensity of spectral features in solution is directly proportional to the concentration of the particular species, the time acquisition, the laser power and the EM gain. In principle, the band intensity can be used to assess the sensitivity of the molecular concentration. Based on this principle, first applications include the medical diagnosis by investigating the variation of lipids, protein density in single cell following in vivo the advanced stages of disease (Krafft and Popp, 2015; Qian et al., 2008; Lussier et al., 2016).

Although the Raman technique has the unique feature of quantifying the degree of freedom of molecule, namely the entropic effect, it is difficult to distinguish in practice from the heterogeneous Raman band broadening between the flexible molecular structures and the conformers for an ensemble of molecules. Investigation of the structural changes of protein requires detailed analysis on individual structure (Brulé, 2016). The analysis of the role played by the solvent on molecules is an additional asset for understanding complicated patterns.

9.3 Detection of Single Structures

In recent years, single-molecule techniques (Deniz et al., 2008) such as fluorescence microscopy and atomic force microscopy (Ando et al., 2013) allow for the direct observation of the dynamics of proteins. This avoids the complexities of averaging over heterogeneous dynamics in bulk biochemical assays. The detection of single structure by Surface Enhanced Raman Spectroscopy has been demonstrated for almost 20 years, at the molecular level (Kneipp et al., 1997) and the protein level (Xu et al., 1999). Emerging temporal signal fluctuations were spotted very early (Weiss and Haran, 2001). Indeed various effects influence the SERS signal, namely the position and intensity of vibrational modes:

(a) The selection rules have first an electromagnetic origin, thereby the orientation of the molecule towards the surface introduces some polarization effect; Raman modes perpendicular to the surface are preferably enhanced and are thus dominating the SERS spectrum.
(b) Due to the evanescent field character of the electric field close to the metal surface, only Raman modes in close vicinity up to roughly 10 nm distance are enhanced by several orders of magnitude.
(c) Vibrations of chemical groups which bind preferably to the metal (sulfur, nitrogen and oxygen showing a high affinity to silver and gold) are dominating the SERS spectrum. The work function (or charge density) of the metal can be modified by a charge transfer.
(d) The surface of nanoparticle composed of 40% of diffusing adatoms is also metastable (Margueritat et al., 2012).
(e) The migration of single molecule has been also imaged within the hot spot at a super-resolution (Stranahan et al., 2010).

Despite many efforts of research in plasmonics to elaborate reproducible SERS active substrate (Margueritat et al., 2011), temporal fluctuations of SERS spectra still exist. It is now admitted that these temporal fluctuations arise from diffusion of single molecule which are likely to occupy several conformations and orientations on the probing site.

9.3.1 Complexity: Sorting of Molecules

The most sophisticated and effective methods for classifying SERS spectra are based on predetermined library of spectra. They are called supervised methods and include linear discriminant analysis (LDA) and partial least squares discriminant analysis (PLS-DA). However, building a valid classification model requires the acquisition of a training set of samples. Furthermore, these classification models are valid only for retrieving one known molecule with one particular SERS based sensor. An alternative sorting tool exists to group SERS spectra by similarities into unknown classes without the need of prerequisite models.

This multivariate analysis called the principal component analysis (PCA) is used to reduce the dimensionality of the measured SERS spectra into a few principal components (Fig. 9.1d). The advantage of this strategy is that no a priori knowledge is required for data analysis. PCA has thus been largely used to regroup similar SERS spectra into clusters (Brulé, 2014). However, this method is limited to disentangle relatively simple mixture because it is graphically limited to three principal components. The Fourier polar analysis (FPA) (Fig. 9.1e) is another nonfitting method based on simple Fourier sine and cosine transforms which produces a fast and graphical representation giving quantitative information (Leray et al., 2016). Hierarchical clustering analysis (HCA) has also been implemented to build a hierarchy of clusters represented in a dendogram (Fig. 9.1c. The appropriate choice between the metrics (number or intensity of Raman Bands, pattern recognition, ...) determines the distance between the clusters. Nevertheless, these methods face the issue of complexity. A complicated problem can be solved over time or with large amounts of data using correlation model. However, there is no causal relationship between parameters in a complex problem. In other words, some complex issue in nanoassembly can only be settled partially and not with an overall picture. Investigating larger chemical structure such as protein having more than 500 amino acids, namely with close to 10,000 atoms and 30,000 normal modes will bring another bigger challenge faced for biological applications.

At the amino acid level, we have shown that multiple adsorption conformations onto a SERS substrate can be determined accurately. For the case of cystein, the complicated analysis shows that the spectrum dynamics can be reasonably reduced to two principal spectra that are identified to different forms of the amino acid absorbed on the gold nanoparticles (Brulé, 2014).

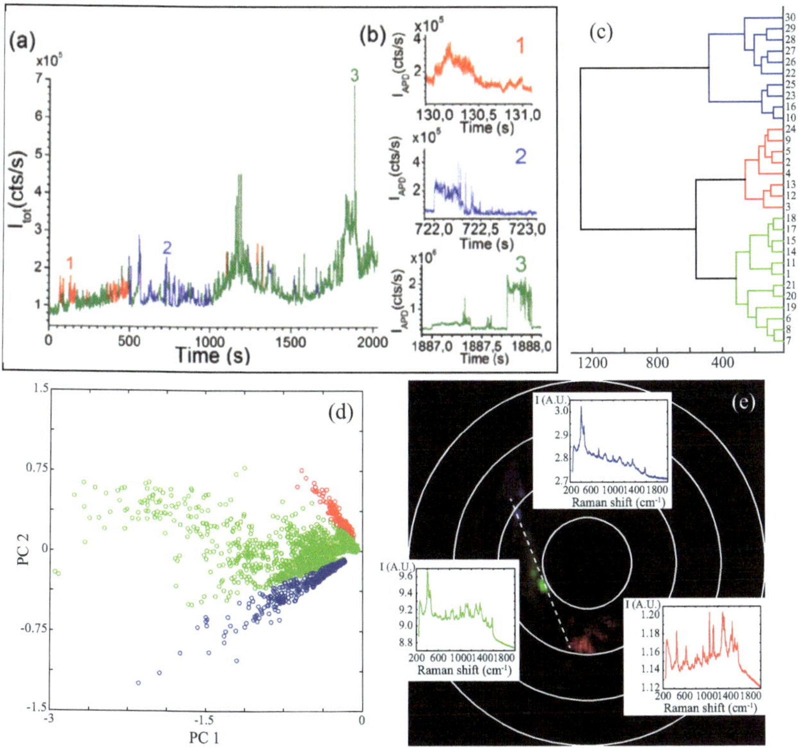

Figure 9.1 Sorting methods: (a, b) Photon counting showing the amplitude fluctuations with time of the spectrally integrated SERS response of the cystein on raspberry-like GNPs. (c) Hierarchical clustering analysis. (d) Principal component analysis. (e) Fourier polar analysis.

9.3.2 Submolecular Resolution: Spectral Pointillism

As already pointed out by several groups, large molecules or proteins cannot fit physically into small active volumes (around nm^3) where enhancement factors are highest. In this case, SERS measurements may probe only a part of the protein whose building blocks are amino acids, leading to unavoidable spectral fluctuations. New analytical methodologies based on the signal processing of fast acquisition of single-molecule signatures are then requested to gain information on the large molecule conformation. Actually the average Raman spectrum is composed of the temporal accumulation of individual spectra or element spectra (that we named spexels). In practice, a large number of spexels is acquired and reliable ranking methods are thus essential for analyzing all these data.

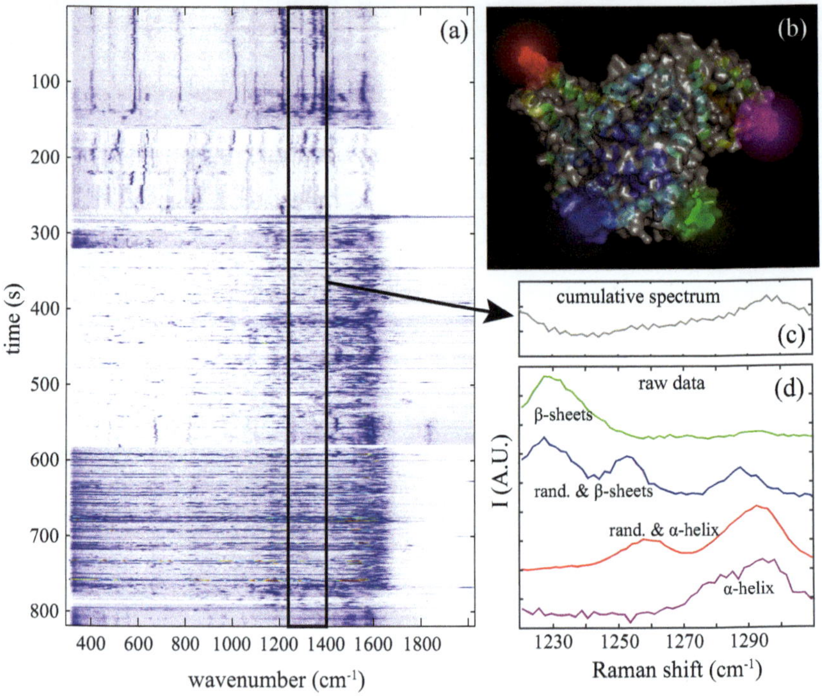

Figure 9.2 Spectral pointillism: (a) Spectrogram showing the large temporal fluctuations of Raman spectra (spexel) of the protein BSA. (b) Schematic view of the protein BSA illustrating several hot spots placed on the protein subdomains. (c) Cumulative spectrum of the ensemble of spexels for the amide II in the (d) extraction of the principal spexels representative of the α-helix and the β-sheet whose proportions determine the protein conformation.

At the protein level, the analysis was found to be more complex (Fig. 9.2a). The spexels stem from single protein subdomains, composed of approximately four distinct amino acids (Fig. 9.2b). The SERS analysis is then similar to the pointillism where only local spots of protein are shined and the protein landscape is rebuilt from these random spexels. The spectral pointillism avoids tedious spectral decomposition treatment which are not accessible with spontaneous Raman spectroscopy.

By probing successively various areas of a single BSA and by analyzing these numerous spexels, we can rebuild the average spectrum from the principal SERS spectra, thereby giving precious information on the protein structure (i.e., polypeptide backbone) and composition (i.e., amino acids) (Clément, 2017). Surface amino acids such as lysine are preferentially detected in the opened form of the protein. The hydrophilic amino acids (those with free amino or carboxyl groups) are generally found on the outer surface of proteins. We show that the unfolded state of the bovine serum albumin can be identified in the time series using the fluctuations of the Raman bands of some amino acids, such as tryptophan, tyrosine, leucine, and histidine, acting as biomarkers (Brulé, 2016).

9.4 Dynamics

Dynamic light scattering (DLS), fluorescence correlation spectroscopy (FCS), and photon correlation spectroscopy (PCS) use the same analysis of the time trace. Copies of the time trace are compared with various other moments and the auto-correlation function is analyzed as function of the delay between original and copy (Loman et al., 2010). The analysis of the dynamics of Raman spectrum provides information on quantitative information for small molecules such as the concentration. For a complex protein system, the dynamics is then governed by nonlinear process characterized by strong volatility hardly predictable (Fig. 9.3a).

9.4.1 Molecular Counting

For small molecules, statistical parameters were found to be independent of time, demonstrating a stationary process. A one stepwise process was found repeatable for smaller molecules. Macroscopic rate can be obtained if the behavior of a single molecule is followed over long enough time to accumulate sufficient statistical significance. The distribution contains more information than the average value alone. The shape of the probability density function in a single-molecule regime can be considered as meaningful indicator of the solution concentration

(Brulé, 2015). The Fourier analysis has been deemed useful with the huge advantage being that it is not limited by a multipoint precalibration or even the knowledge of the enhancement factor of the transducer. The analysis of pink noise in the frequency domain reveals a subdiffusion motion of individual adsorbed molecules in the hot spot. The cutoff frequency between the white and pink noise is used to determine the inline molecular concentration over 6 decades of concentrations between 10^{-11} and 10^{-6} M (Brulé, 2015).

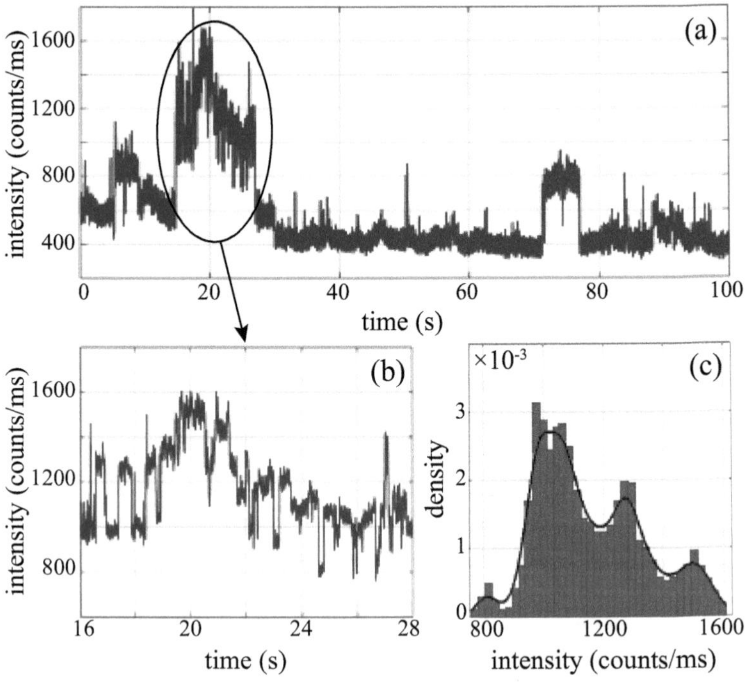

Figure 9.3 Kinetics of molecular adsorption on nanoparticle (a) over 100 s; (b) zoom on a long event showing a multiple step process; (c) multiple peaks or skewed shape of the Raman intensity distribution.

9.4.2 Strong Volatility

Temporal fluctuations observed in large molecules have been poorly exploited in terms of biological output. By definition, fluctuations concern changes in Raman intensity that are rapid with respect to reaction rate that we consider. If we consider a time scale of 0.5 s, corresponding to the acquisition time of one spectra, fluctuations are considered in the time scale below 10 ms and are also relevant to the biological events (Hammes-Schiffer and Benkovic, 2006). Part of the problem is rooted in the fact that due to the hierarchical organization of their conformational landscape, proteins display motions on an extremely wide range of timescales (Frauenfelder et al., 1991). The answer lies in the understanding of how these motions are coupled among each other and which of them are important for function. There is no warranty that ergodicity holds in a protein bathed in a multiwell potential landscape. The time average of a signal at a local spot does not reflect the signal average over the entire volume of the protein. Indeed, the indicators quantifying the strong volatility that we observe are quite difficult to define statistically; the variance cannot apply in a nonstationary signal. However, the shape of the full distribution with multiple peaks or skewed shape is synonymous to the biodiversity or heterogeneity; Levy jump characterized by the Law of Pareto (80–20 rules) used in extreme and rare statistics was found to fit in parts the multiple steps observed. Different topologies of events with different level of fluctuations can be sorted. These individual jumps are generally uncorrelated; thus, the behavior of an ensemble-averaged quantity would show only a broadening and smearing of the Raman band. A molecular memory phenomenon can arise, but, on the average, the basic Michaelis–Menten description seems still to hold (English et al., 2006).

9.5 Conclusion

The complexity highlighted on macromolecules originates from the convolution of at least three parameters: (i) the local enhancement of the electromagnetic field (ii) several Raman fingerprints to describe the same nanostructure, and finally (iii) a nonstationary dynamics observed for large molecules.

Acknowledgments

This work was supported by the Labex ACTION program (ANR-11-LABX-0001-01) and the Burgundy Region Council (PARI

II NANO2BIO, FABER 2014). This work was performed in the context of the European COST Action MP1302 Nanospectroscopy.

References

Ando, T., Uchihashi, T., and Kodera, N. (2013). High-Speed AFM and applications to biomolecular systems, *Annu. Rev. Biophys.*, 42, pp. 393–414.

Barth, A., and Zscherp, C. (2002). What vibrations tell us about proteins, *Q. Rev. Biophys.*, 4, pp. 369–430.

Brulé, T., Bouhelier, A., Dereux, A., and Finot, E. (2016). Discrimination between single protein conformations using Dynamic SERS, *ACS Sens.*, 2, pp. 676–680.

Brulé, T., Bouhelier, A., Yockell-Lelievre, H., Clément, J.-E., Leray, A., Dereux, A., and Finot, E. (2015). Statistical and Fourier Analysis for In-line Concentration Sensitivity in Single Molecule Dynamic-SERS, *ACS Photonics*, 9, 150811080730002.

Brulé, T., Yockell-Lelievre, H., Bouhelier, A., Margueritat, J., Markey, L., Leray, A., Dereux, A., and Finot, E. (2014). Sorting of Enhanced Reference Raman Spectra of a Single Amino Acid Molecule, *J. Phys. Chem. C*, 118, pp. 17975–17982.

Clément, J.-E., Leray, A., Bouhelier, A., and Finot, E. (2017). Spectral pointillism of enhanced Raman scattering for accessing structural and conformational information on single protein, *Phys. Chem. Chem. Phys.*, 19, pp. 458–466.

Deniz, A. A., Mukhopadhyay, S., and Lemke, E. A. (2008). Single-molecule biophysics: at the interface of biology, physics and chemistry, *J. R. S. Interface*, 5, pp. 15–45.

English, B., Min, W., van Oijen, A., Lee, K., Luo, G., Sun, H., Cherayil, B., Kou, S., and Xie, X. (2006). Ever-fluctuating single enzyme molecules: Michaelis–Menten equation revisited, *Nat. Chem. Biol.*, 2, pp. 87–94.

Ferrante, C., Pontecorvo, E., Cerullo, G., Vos, M. H., and Scopigno, T. (2016). Direct observation of subpicosecond vibrational dynamics in photoexcited myoglobin, *Nat. Chem.*, 8, pp. 1137–1143.

Frauenfelder, H., Sligar, S., and Wolynes, P. (1991). The energy landscapes and motions of proteins, *Science*, 254, pp. 1598–1603.

Hammes-Schiffer, S., and Benkovic, S. (2006). Relating protein motion to catalysis, *Ann. Rev. Biochem.*, 75, pp. 519–541.

Hermelink, A., Brauer, A., Lasch, P., and Naumann, D. (2009). Phenotypic heterogeneity within microbial populations at the single-cell level investigated by confocal Raman microspectroscopy, *Analyst*, 134, pp. 1149–1153.

Kneipp, K., Wang, Y., Kneipp, H., Perelman, L., Itzkan, I., Dasari, R. R., and Feld, M. (1997). Single molecule detection using surface-enhanced Raman scattering (SERS), *Phys. Rev. Lett.*, 78, pp. 1667–1670.

Krafft, C., and Popp, J. (2015). The many facets of Raman spectroscopy for biomedical analysis, *Anal. Bioanal. Chem.*, 407, pp. 699–715.

Leray, A., Brulé, T., Buret, M., Colas des Francs, G., Bouhelier, A., Dereux, A., and Finot, E. (2016). Sorting of Single Biomolecules based on Fourier Polar Representation of Surface Enhanced Raman Spectra, *Sci. Rep.*, 6, 20383.

Loman, A., Gregor, I., Stutz, C., Mund, M., and Enderlein, J. (2010). Measuring rotational diffusion of macromolecules by fluorescence correlation spectroscopy, *Photochem. Photobiol. Sci.*, 9, pp. 627–636.

Lussier, F., Brulé, T., Vishwakarma, M., Das, T., Spatz, J. P., and Masson, J. F. (2016). Dynamic-SERS Optophysiology: A Nanosensor for Monitoring Cell Secretion Events, *Nano Lett.*, 16, pp. 3866–3871.

Margueritat, J., Bouhelier, A., Markey, L., and Finot, E. (2012). Discerning the origins of the amplitude fluctuations in dynamic Raman nanospectroscopy, *J. Phys. Chem. C*, 116, pp. 26919–26923.

Margueritat, J., Gehan, H., Félidj, N., Bouhelier, A., Markey, L., and Finot, E. (2011). Influence of the Number of Nanoparticles on the Enhancement Properties of Surface-Enhanced Raman Scattering Active Area: Sensitivity versus Repeatability, *ACS Nano*, 5, pp. 1630–1638.

Mittermaier, A. K., and Kay, L. E. (2009). Observing biological dynamics at atomic resolution using NMR, *Trends Biochem. Sci.*, 34(12), pp. 601–611.

Qian, X. P., Peng, X., Ansari, D., Yin-Goen, Q., Chen, G. Z., Shin, D. M., Yang, L., Young, A. N., Wang, M. D., and Nie, S. (2008). In vivo tumor targeting and spectroscopic detection with surface-enhanced Raman nanoparticle tags, *Nat. Biotech.*, 36, pp. 83–90.

Rygula, A., Majzner, K., Marzec, K. M., Kaczor, A., Pilarczyka, M., and Baranskaa, M. (2013). Raman spectroscopy of proteins: A review, *J. Raman Spectrosc.*, 44, pp. 1061–1076.

Siuzdak, G. (1994). The emergence of mass-spectroscopy in biochemical research, *PNAS*, 91(24), pp. 11290–11297.

Stevens, J., Link, J., Kao, Y., Zang, C., Wang, L., and Zhong, D. (2010). Ultrafast dynamics of resonance energy transfer in myoglobin: Probing local conformation fluctuations, *J. Phys. Chem. B*, 114, pp. 1498–1505.

Stranahan, S. M., and Willets, K. A. (2010). Super-resolution optical imaging of single-molecule SERS hot spots, *Nano Lett.*, 10, pp. 3777–3784.

Tuma, R. (2005). Raman spectroscopy of proteins: from peptides to large assemblies, *J. Raman Spectrosc.*, 36, pp. 307–319.

Weiss, A., and Haran, G. (2001). Time dependent single molecule Raman scattering as a probe of surface dynamics, *J. Phys. Chem. B*, 105, pp. 12348–12354.

Xu, H., Bjerneld, E., K¨all, M., and Börjesson, L. (1999). Spectroscopy of single hemoglobin molecules by surface enhanced Raman scattering, *Phys. Rev. Lett.*, 83, pp. 4357–4360.

Chapter 10

Intracellular Surface-Enhanced Raman Spectroscopy

Jack Taylor,[a] **Anna Huefner,**[a] **Jonathan Wingfield,**[b] **and Sumeet Mahajan**[a]

[a]*Institute for Life Sciences and Department of Chemistry, University of Southampton, SO17 1BJ, United Kingdom*
[b]*Discovery Sciences, Screening and Compound Management, AstraZeneca, Unit 310-Darwin Building, Cambridge Science Park, Milton Road, Cambridge, United Kingdom, CB4 0WG*

s.mahajan@soton.ac.uk

10.1 Intracellular Applications of SERS

Concentrations of biomolecules within cells are typically of the order of nM and therefore a highly sensitive technique is required for intracellular studies of intact and living cells. Fluorescence imaging has been the predominant platform for such biological studies. However, most biomolecules are not fluorescent and therefore, chromophores need to be added or attached to the molecules of interest in order to study them. This labelling process via either staining (addition of dyes) or labelling via endogenously expressed fluorescent proteins through genetic modification is invasive and perturbative, and therefore undesirable in many cases. Raman spectroscopy provides an alternative but signals

Plasmonics in Chemistry and Biology
Edited by Marc Lamy de la Chapelle and Nordin Felidj

ISBN 978-981-4800-03-7 (Hardcover), 978-0-429-45875-0 (eBook)
www.jennystanford.com

are weak, precluding studies that require continuous and rapid monitoring or visualisation. Surface-enhanced Raman spectroscopy (SERS), offering enhancement by factors of 10^6–10^{10} [1, 2] through the use of plasmonic nanomaterials (such as gold and silver nanoparticles), overcomes this primary limitation and thus allows for intracellular studies [3]. SERS retains the advantages of Raman in that it allows fine spectral details to be observed without interference from the vibrational peaks of H_2O observed in IR spectroscopy. SERS also proves advantageous as it is non-destructive and label-free, with simple preparation of samples, utilising an increased depth of penetration by NIR radiation. Sensitivity can be comparable to or greater than fluorescence or biological assays [4, 5], with single molecule detection reported in optimal cases [6–8] and it offers numerous advantages and complimentary information for intracellular analysis.

For successful cellular investigations by SERS, however, the selection of suitable nanoparticles (NPs) is essential. Systems must overcome issues such as NP internalisation and toxicity while maintaining desired optical properties. For *in cellulo* studies, particle diameter must be small enough to penetrate the intracellular matrix yet larger than 15 nm to achieve sufficient SERS enhancement [9]. Spherical AgNPs exhibit stronger plasmonic fields than those of Au, especially in the visible region of the electromagnetic spectrum owing to the partial Au plasmon band overlap with its interband electronic transitions. Notwithstanding this, AuNPs are more widely applied in biological studies due to their well-established and controlled methods of synthesis along with good biocompatibility and chemical stability. The ability to track and detect plasmonic NPs using various analytical tools, especially their localised surface plasmon resonance bands, which can be synthetically tuned into the near-infrared region (the optical transparency window for biological tissues), is an added advantage. Facile surface chemistry allows for easy surface functionalisation, affording not only the binding of specific delivery peptides, but also other applications such as artificial antibodies with binding affinities precisely tuned by varying the density of surface bound ligands. The ability to shield unstable drugs or poorly soluble imaging contrast agents to facilitate their delivery to the otherwise inaccessible cells and tissues of the body is augmented by AuNPs' multivalent nature [10].

From the above it can be seen that the type of NPs, choice of their surface chemistry and consequent interaction with cells (uptake, toxicity) can be critical to their utilisation for intracellular SERS. Through this chapter, we therefore aim to provide an insight into all aspects of intracellular SERS sensing and imaging. We trace the journey of NPs from internalisation to intracellular SERS reporting; focussing on spherical AuNPs as the most widely used and understood SERS probes. AgNPs are also used and alternative NP structures are becoming increasingly popular; therefore, these will be presented where relevant. The crucial roles of NP–cell interactions are discussed, including physical and diffusive routes of internalisation along with biocompatibility factors such as tunability of surface properties to assist uptake and resulting toxicity. The two prevalent approaches, the SERS-reporter and reporter-free approaches, to intracellular SERS studies are then presented. State-of-the-art and recent significant studies are reported and the potential of intracellular SERS for numerous applications in life sciences, therapeutics and drug discovery is discussed.

10.2 Nanoparticle–Cell Interactions

10.2.1 Cellular Internalisation Methods

The key to successful utilisation of NPs inside cells for SERS and other measurements is understanding and manipulating their mechanism of uptake and cellular distribution. Internalisation methods generally fall into three categories: involuntary delivery by physical methods, passive diffusion, and active (or voluntary) uptake.

The basis of physical insertion methods is the theory of applying a force to the cell in order to create localised membrane pores and increase cell permeability. Electroporation and microinjection achieve this with the application of electrical and physical force, respectively [11]. Microinjection has been of particular interest in single cell studies, providing tight control of dosage and timing of delivery but is highly sensitive as inappropriate use (injection location or pressure) can easily damage cellular components. Despite the tight control offered, the complexity of both procedure and instrumentation limits the technique, which is also confined to delivering only large diameter NPs [12]. Electroporation has been compared with diffusive approaches of TATHA2 peptide functionalised and citrate capped AgNPs for facilitated and passive uptake by endocytosis, respectively. Although yeast cells can tolerate high electroporation doses, severe damage was observed when used to deliver AgNPs. Free diffusion of AgNPs resulted in poor uptake which improved with TATHA2-AgNPs. A uniform intracellular distribution was also observed through passive uptake which is a requirement for detection of ubiquitously distributive molecules in cell based biosensors [11].

Voluntary uptake of particles via endocytosis is by far the most commonly used internalisation strategy, whereby metallic species are taken up by the cell's intrinsic machinery. This is owing to the method's relative simplicity and minimal sample preparation as well as tunability by appropriate surface modification [13].

10.2.2 The Endocytotic Pathway

Endocytosis refers to commencement of the vesicular endolysosomal pathway through the cell for the active intake of macromolecules and proteins which are too large to pass directly through the plasma membrane [14] (Fig. 10.1A, white arrow). In its initial step, cargo for internalisation is engulfed by invagination or pit formation of the plasma membrane (Fig. 10.1A–B, black box). Pinching off from the plasma membrane, a small membrane-bound vesicle encapsulates the ingested molecule(s). Many routes of endocytic uptake into the cell exist, divided into phagocytosis, macropinocytosis, endocytosis via non-coated vesicles and energy-dependent, receptor-mediated endocytosis [15–20] (RME, see Fig. 10.1B(i–vi), respectively). The particular intake pathway depends on the kind, size, shape and surface characteristics of the molecules as well as the specific cell type.

A number of endocytotic mechanisms are non-specific, absorptive and/or use processes that have not yet been identified (Fig. 10.3B(iii)). Studied extensively, however, RME relies on the receptor–cargo interaction to initiate the internalisation process [14]. Vesicles emerging from the plasma membrane usually have a size of 50–200 nm depending on the involved receptor and coating molecules. The receptor is internalised with the ligand for gradual recycling back to the plasma membrane (Fig. 10.1B(vi)), thus the process is saturable and used for transmembranal signal transduction [14]. Alongside the disassembly of the coating molecules from the individual vesicles (Fig. 10.1B(vi)), these vesicles fuse to form early endosomes (Fig. 10.1, light green shapes).

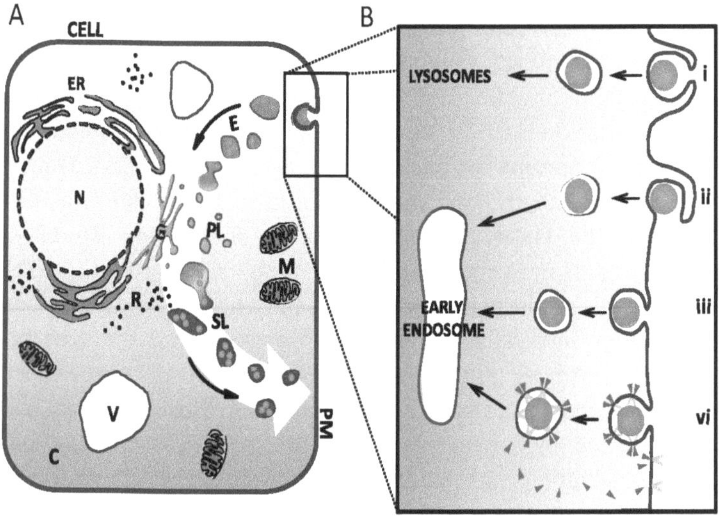

Figure 10.1 (A) A schematic of a cell highlights different compartments. The endocytotic pathway is indicated by the white arrow in the background. After internalisation of the cargo into membrane-bound vesicles inside the cell (light-green, (B) for more details), these vesicles fuse to early endosomes and mature to late endosomes, here shown as green vesicles. Primary lysosomes (abbr. PL, here with light blue colouration) emerge from the Golgi apparatus (G) and merge with endosomes (E, green colouration) to form secondary lysosomes (SL, darker blue colouration). Passing through maturation and degradation processes, they get eventually exocytosed by the cell. (B) Detailed schematic showing the main mechanisms of endocytosis featuring phagocytosis (i), macropinocytosis (ii), non-specific absorptive endocytosis (iii) and receptor-mediated endocytosis (vi) involving receptor (yellow shape) and coating molecules (e.g. clathrin- and caveolin, blue triangle). Abbr.: cytoplasm (C), early and late endosomes (E), endoplasmic reticulum (ER), Golgi apparatus (G), mitochondria (M), nucleus (N), primary lysosome (PL), plasma membrane (PM), ribosome (R), secondary lysosome (SL), vacuole (V). Figure is not to scale. Reprinted with permission from Taylor et al. *Analyst*, 2016-Published by The Royal Society of Chemistry.

Early endosomes are initially relatively small, have a mildly acidic pH and are the main sorting station in the endosomal pathway. While intravesicular fluids are recycled, early endosomes fuse with other incoming cargos from different internalisation pathways (Fig. 10.1A–B) and mature into late endosomes (Fig. 10.1A, green vesicles) [18, 21]. Late endosomes are bigger in size, their pH further decreased and their morphology gradually altered during maturation [18]. At the end of the endosomal pathway, late endosomes and phagosomes eventually fuse with primary lysosomes, forming secondary lysosomes or phagolysosomes which contain components from both fusion partners [18]. Hereafter, the localisation of NP cargos in various cellular compartments depends on parameters such as size and surface coating. Transport to the nuclear membrane and translocation across can be achieved, although the latter has proven to be challenging [9, 13, 22]. Primary lysosomes emerge from the Golgi apparatus (Fig. 10.1A), containing numerous enzymes and various macromolecules from other intracellular, catabolic processes and the membrane turnover through autophagy [18, 23]. Most of the molecular digestion and degradation takes place in these vesicles, which is supported by an increasingly acidic pH. Subsequent dynamic changes in molecular composition and activity of enzymes inside the vesicles also results in modification of the encapsulated molecules [14, 17, 24]. At the end of the lysosomal pathway, vesicles with indigestible material, such as NPs, become residual bodies and are secreted from the cell via exocytosis [23, 25, 26]. Different maturation and fusion steps within this whole pathway are colour-indicated in Fig. 10.1A.

The advantage of the uptake of plasmonic NPs by endocytosis is its simplicity, both in execution and sample preparation. The typical procedure following particle synthesis consists of an incubation period of cells with NPs suspended in culture medium before washing and SERS investigation. Tunability is afforded by altering NP properties such as size, shape and surface coating, which can be manipulated to localise SERS nanoprobes to desired cellular locations as well as to access or escape specific pathways and transport vesicles.

10.2.3 Manipulating Interactions

It is of great importance to understand and quantify the uptake of AuNPs for intracellular SERS. Extensive studies have shown that many factors, including size, shape, surface coating, concentration of NPs, aggregation state of NPs, the type of cell, their environment and the exposure conditions play important roles in their biological interactions [27]. With innovations in nanotechnology, a library of fit-for-purpose AuNPs with different physicochemical properties has been established [1, 28–35].

Several factors contribute to identifying the optimal NP diameter for use. AuNPs larger than 15 nm are of sufficient scattering cross-section and plasmonic activity to provide the required optical contrast and electric field enhancement for SERS interrogation of cells while providing information on their localisation [36]. Additionally, a NP must also be efficiently internalised by the cell. Studies have shown that the optimal diameter for endocytotic uptake of AuNPs is ≤50 nm [37]; although this depends on other factors previously mentioned. Size dependence of uptake can be attributed to 'wrapping time', used to describe how a membrane encloses a particle. This is affected by factors including ratio of adhesion, membrane stretching and membrane bending. AuNPs of 55 nm diameter are calculated to have the fastest wrapping time, with RME receptor binding of a NP producing enough free energy to drive it into the cell. This energy is reduced in NPs of diameters <50 nm, which are therefore required to cluster together in order to achieve internalisation. Conversely, wrapping times of large NPs increase because more RME receptors are required to bind their larger surface area and generate sufficient free energy for uptake. Consequently, the number of AuNPs entering the cell in transport vesicles is reduced; with uptake rate limited by the speed of diffusion of extra receptors to the site of invagination [25, 37]. Particle morphology can also play a key role in determining degrees of cellular uptake, with larger degrees of internalisation by healthy endothelial cells reported in spherical AuNPs and gold nanorods (NRs) than hollow AuNPs and silicon/gold core/shell structures of matching monocarboxy-hexaethyleneglycol coatings [38]. Hence, most intracellular SERS studies conducted utilise spherical AuNPs in the diameter range 40–60 nm.

Cellular uptake via endocytotic pathways may be aided by applying surface modifications to NPs, facilitating their diffusion. A common requirement for stabilisation in order to prevent further growth or aggregation of the NPs also exists. Charged ligands such as citrate are often utilised for this purpose, with coulombic repulsion serving to keep NPs separate. However, the negative charge possessed by citrate and other ligands may produce unfavourable interactions with the similarly negatively charged cell membrane upon contact with the nanoparticle. On the other hand, small, uncharged cargos are favourable for uptake. In cell culture medium, macromolecules such as serum proteins adhere to the noble metal surface by physical or chemical adsorption to produce a double layer structure schematically shown in Fig. 10.2, known as 'protein corona'. These two layers around the colloid consist of an innermost 'hard' layer formed on the Au surface by strong attractive force and a secondary outer layer, comprised of proteins weakly adsorbed to the nanoparticle via intermolecular forces such as the protein–protein interaction which shields the inner layer from the particle environment [39]. The make-up of the protein corona is subject to dynamic changes in composition and thickness (usually measured between 3–15 nm) with adsorbed proteins affecting the hydrodynamic diameter of a NP. Initial formation of the corona occurs rapidly upon contact and over time a stable although dynamic and biocompatible structure in standard culture media is formed due to the presence of serum proteins in such media [39]. This process is of huge synthetic advantage to facilitate NP uptake as the corona layers give the SERS probe its biochemical identity as experienced by the cell. Additionally, it can be combined with functionalisation (e.g. ligands such as localisation signal peptides) for targeting specific organelles inside cells. Ligand selection is crucial in achieving the desired uptake and localisation of a SERS nanoprobe. However, environmental changes such as enzymatic presence or pH are only experienced by the outer corona layers, allowing the use of organic molecules as SERS reporter molecules (discussed later) or fluorescent tags protected within.

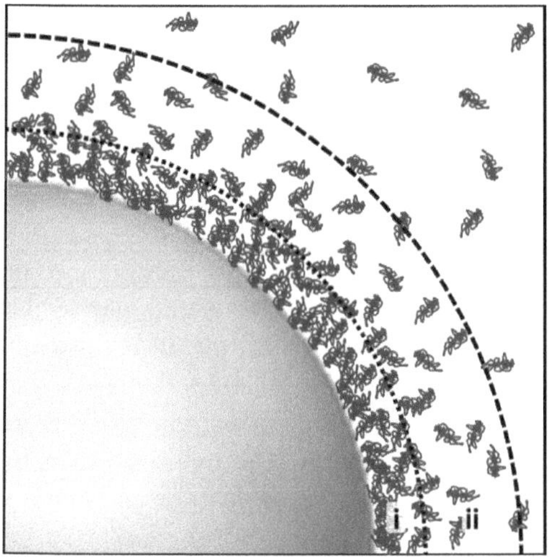

Figure 10.2 Illustration of the protein corona around a spherical gold nanoparticle in a protein-rich environment. Proteins, here shown as blue entangled chains, adhere to the particle's surface forming a 'hard' (i) and 'soft' (ii) layer. While the inner 'hard' layer consists of proteins strongly adsorbed to the particles surface, the outer 'soft' layer is formed by weak protein–protein interactions. Reprinted with permission from Taylor et al. *Analyst*, 2016–Published by The Royal Society of Chemistry.

A SERS NP probe must escape the endolysosomal pathway into the cytosol if the aim is to study the cytoplasm or organelles other than endolysosomal vesicles. This involves biochemical delivery strategies typically based on the conjugation of the particle surface with various types of molecules such as cationic molecules [40], protein transduction domains, cell penetrating peptides (CPPs) [22, 41, 42] and ligands [43]. Escaping endolysosomal vesicles is associated with the ability to penetrate or pass through lipid bilayers. Permeability of the lipid bilayers is achieved through the interaction of cationic molecules with the negatively charged plasma membrane, which is thereby disrupted and forms nanopores. The choice of molecules spans amine-containing polymers [44–46], polypeptides [47] and cationic liposomes [48–50]. For instance, this approach has been applied to achieve cytoplasmic localisation of AuNPs conjugated to cationic liposome alongside plasmid DNA, resulting in improved efficiency of cellular transfection [50].

Furthermore, CPPs have widely been applied to target specific intracellular organelles. Many peptide sequences have been derived from natural systems used by viruses such as the TAT peptide derived from the HIV-1 virus which, conjugated to AuNPs, also potentiates cytoplasmic localisation [41, 51–53]. Even though the exact mechanism of its interaction with biological membranes is not yet known, TAT is thought to mediate cellular import via multiple routes including direct membrane passage [51, 54] and intake through endocytosis [52, 55].

Conversely, ligand molecules facilitate the directed interaction of cargo with specific receptors by binding to them. This can be used for the delivery of AuNPs into cells as shown by Tkachenko and co-workers, who demonstrated that NPs modified with adenoviral RME peptides gain entry into cells via RME [22]. Ligands have particularly been used to target selected cellular components such as the nucleus, by attachment of nuclear localisation signal (NLS) peptides to NPs [56, 57].

10.2.4 Toxicity

A vital consideration to the intracellular SERS experiment is the effect of the administered NPs on its host. That is, any changes or decline in key cellular processes induced by particle–cell interactions which decrease the validity of the experimental system as a representative model of the natural state. Cytotoxicity of NPs is generally attributed to the size of both single particles and clusters thereof, surface modifications, NP concentration and composition of the protein corona in addition to the cell type [26, 58– 60].

Size determines the translocation of NPs across lipid membranes into various subcellular structures possessing size exclusive pores such as nucleopores in the nuclear membrane. Their confinement in cellular vesicles may, however, still present toxicity, when local concentration of noble metal far exceeds its known LD_{50} (the concentration at which a molecule dose is fatally toxic to 50% of a population). The surface area to volume ratio increases with decreasing size and therefore the smaller the size of nanoparticles the higher the levels of interaction they can have with their environment. Indeed single particles of diameters less than 2 nm induce damaging effects on cells [59, 61–63].

In addition to size based effects, homeostatic processes are also disrupted by the presence of NPs within the cell. Oxidative stress results from the production of reactive oxidative species (ROS) upon cellular recognition of 'foreign' inorganic material and is usually compensated for by homeostatic defence mechanisms. However, at high metal concentrations these mechanisms are unable to maintain normal cell function, therefore cause toxicity and cell death [62].

Surface modifications of NPs can also affect cellular toxicity. Oxidative stress, induced by exposure of HeLa cells to 1.4 nm diameter AuNPs was reduced following their pre-treatment with reducing agents/antioxidants N-acetylcysteine, glutathione, and triphenylphosphine monosulfate [62].

AuNPs coated with CTAB, a shape-directing agent utilised in their synthesis, also present toxicity to cells through the interaction of the molecule's strong positive charge with the plasma membrane. Presence of CTAB is reported to disrupt membrane structure at concentrations over 10 nM [58, 61, 64]. Reduction of this toxicity can, however, be achieved by surface modification with capping agents such as PEG [65], PAA [66], PAH [66] and polystyrene sulfonate [67]. For this reason, ligand chemistry is defined as a critical parameter for biocompatibility and toxicity of AuNP SERS probes alongside the earlier discussed size dependence.

In spite of numerous studies, there remains a fundamental requirement for standardisation of investigative procedure before an acceptable consensus on AuNP toxicity can be reached [68]. For instance, the aforementioned toxicity of CTAB has also been attributed to its presence as an unbound molecule in AuNP solutions, which raises the need for tight 'supernatant control' steps to be taken for valid NP toxicity evaluation [58, 66]. Two predominant reasons are attributed to current ambiguity, the first being the complex governance of the phenomenon by a large number of both chemical and physical properties of the NP along with cell types and the degree of NP internalisation. The second then arises from the lack of standardisation in experimental design, execution of procedure and data treatment [68].

To illustrate the latter point, Fratoddi et al. gathered existing data from published studies conducting MTT assays (used to demonstrate cell viability through metabolic activity) of HeLa cells exposed to differently functionalised AuNPs, for analysis by metric numerical particle concentration as opposed to mass concentration or particle size [68]. This produced a simplified view of how parameters such as NP size, concentration and surface coating impact individual cell viability and concluded that size dependence is much less important than the number of particles present per unit volume (cellular concentration). Differently functionalised AuNPs behaved similarly, with the surface coating of the NPs defining the range of particle concentrations at which toxic effects commence.

Thus, there is a need for the standardisation of nanoparticle internalisation protocols and data for advancing the application of SERS to single cells in vitro and beyond; however, promising steps are being taken to achieve this.

10.3 Intracellular SERS

The first application of SERS inside cells was carried out in 1991 when Nabiev et al. detected the presence of antitumor drugs doxorubicin (DOX) and 4′0-tetrahydropyranyl-adriamycin (THP-ADM) in the nucleus and cytoplasm of living cancer cells using AgNPs [69]. Since then, SERS has been successfully applied for detection of a vast array of biological markers and metabolites in turn, for sensing and tracking environmental changes through cellular pathways and processes. These include monitoring cellular functions [70, 71], dynamics [72, 73], enzyme kinetics [74, 75], stress response [76], apoptosis [77] and cell death [78] along with probing specific compartments such as the mitochondria [79] and tracking of drugs released into the cytoplasm by NP carriers [80–82].

The cell-based SERS experiment generally adopts one of the following two methodologies: the SERS-reporter (SERS-label) approach or the reporter-free (label-free) SERS approach, key features of both approaches are summarised in Table 10.1. While the SERS-reporter approach has been more dominant over the last decade, the merits of the label-free approach have begun to emerge with concomitant advancements in computational approaches. This chapter will proceed to explain the fundamental concepts, along with reviewing research advances and developments in each of these two approaches. For general and exhaustive reviews of SERS applications in the biological context, also covering the SERS-reporter approach inside cells, the reader is referred to a few excellent recent publications [1, 83–85].

Table 10.1 Key characteristics of SERS-reporter and reporter-free SERS approaches to intracellular experimentation

SERS-reporter approach	**Reporter-free approach**
SERS probe: 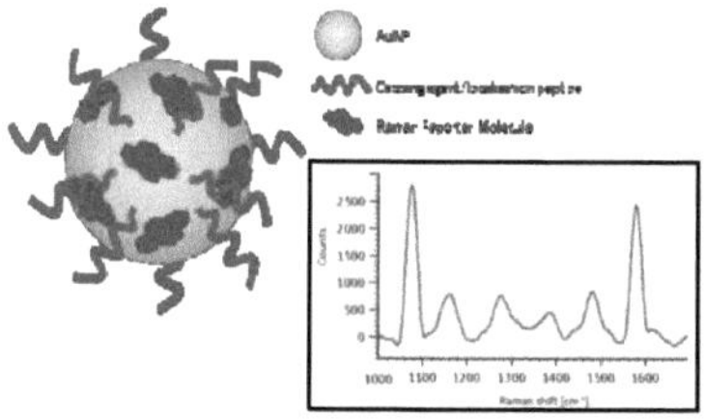	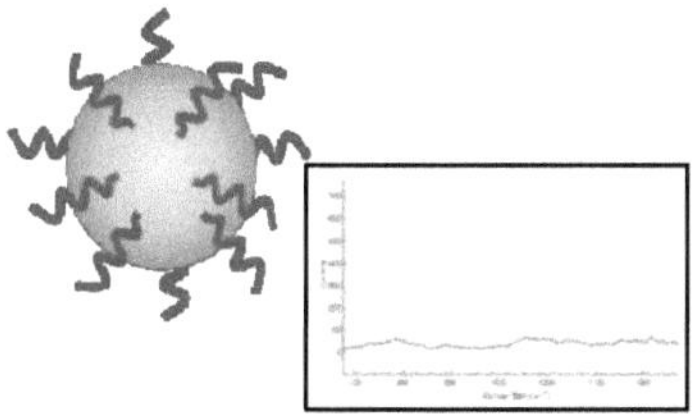
Intracellular SERS: 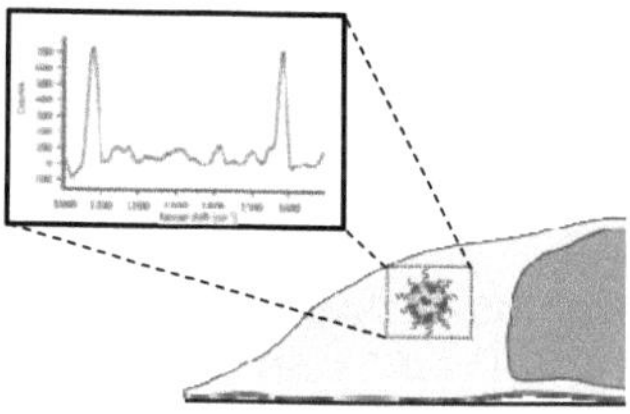	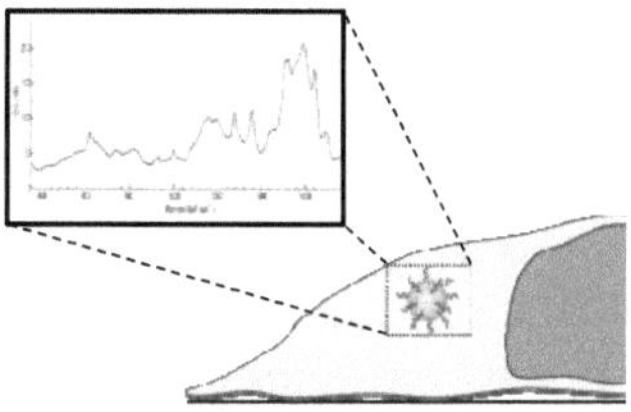
Detects changes to Raman shift/intensities of the reporter molecule spectrum-detection is inferred and specific	Samples the NP's immediate biochemical environment-detection is unspecific, direct and highly localised
Targeted sensing-simple data analysis	Datasets are rich and complex-requires careful analysis
Typically applied for specific sensing applications-cellular environmental factors (e.g. pH), abundance and mapping of biomolecules/markers	More explorative nature-characterising molecular pathways and dynamics, interrogating unknowns

(*Continued*)

Table 10.1 (*Continued*)

SERS-reporter approach	Reporter-free approach
SERS nm-range and narrow band widths give greater spatial and spectral resolution than fluorescence for multiplexed sensing	Wealth of extractable molecular-level information available surpasses that from fluorescence
Increased functionalisation of NPs may complicate cellular interactions, producing toxicity or limiting transport by increased size	Uncontrollable aggregation of spherical NPs *in cellulo* if no capping agent is used-implications for homogeneity of SERS enhancement and reproducibility

Note: Pure and intracellular SERS-reporter spectra were obtained using reporter molecule 4-mercaptobenzoic acid (4-MBA).

10.3.1 Advances in SERS-Reporter Research

The SERS-Reporter approach, whereby NPs are functionalised with covalently bound and strongly Raman-active organic molecules, is the more comprehensively studied approach. The reporter itself, or an additional recognition element, can serve to provide binding specificity to a target. The generated SERS spectrum then reveals the defined signature of the label to facilitate indirect and highly sensitive detection of the target molecule for Raman sensing applications. Examples of this include sensing of reactive oxygen species (ROS), hydrogen peroxide and hypochlorite [86]; and antioxidant glutathione (GSH) [87] in living cells, achieved ratiometrically by comparison of reporter spectral peaks in the presence/absence of target molecules. The Raman-active molecule 4-mercaptobenzoic acid (4-MBA) can be functionalised to NPs as a SERS reporter of cellular pH, which is sensitive to changes between pH 6–8 within the cellular environment [88, 89]. This is owed to spectral differences between the protonated and deprotonated forms (1430 cm^{-1} COO^{-} stretch, specific to deprotonated form in basic conditions) of the molecule, which thus serves as a hydrogen ion (pH) sensor [89]. More recently, this 4-MBA pH sensing was multiplexed with simultaneous measurement of redox potential in live EAhy926 cells. The redox sensitive band at 1666 cm^{-1} (C=P stretch) reports on the oxidation state of SERS reporter molecule N-[2-({2-[(9,10-dioxo-

9,10-dihydroanthracen2yl)formamido]ethyl}disulfanyl)ethyl]-9,10-dioxo-9,10-dihydroanthracene-2-carboxamide (AQ), which was ratiometrically plotted against the redox insensitive vibrational mode at 1606 cm^{-1} (C=C stretch). Cluster analysis then permitted quantitative analysis of redox potential dysregulation, a common indicator for the progressive pathology in neurodegeneration, cardiovascular diseases and cancer [88]. Such application of multivariate data analysis techniques, particularly principal component analysis (PCA), is an incredibly powerful tool for the analysis of both SERS-reporter and reporter-free SERS in sensing and diagnostic applications. PCA has been applied to improve pH sensing by para-MBA SERS reporter NPs in macrophage cells, producing a more robust model than previous ratiometric approaches [90].

The density of SERS labels on the surface of the plasmonic NP can be adjusted to experimental needs. For instance, the co-adsorption of other, undesired molecules as well as their spectral interference can be avoided by complete coverage of the NP's surface with the label. Additionally, a (further) physical and chemical protection of the probes can be achieved by encapsulation with a silica layer as conceptually introduced by Mulvaney and co-workers [91]. This has been further advanced by Tian and co-workers through their synthesis method to achieve ultrathin silica shells [31, 92, 93], thus causing minimal loss of SERS signal which decreases with the distance away from the nanoparticle surface. Hollow AuNPs can also be protected by PEG capping, offering stabilisation and red shifting of their SPR from 700 nm to allow optimal excitation by 785 and 1064 nm irradiation [94]. Many SERS labels (also called reporter molecules) have been developed since the technique was established including enzymes, dyes, peptides and DNA which allow for multiplexed detection of numerous molecules using only a single laser wavelength. Multiplexing of SERS reporter functionalised nanoprobes allows for a greater number of analytes to be simultaneously studied when compared with fluorescence microscopy, due to spectral line-widths of Raman peaks of <2 nm which can enable access to between 10–100 unique optical signatures as provided by varying reporter molecules [56].

SERS reporter experiments benefit from the specificity of attaching a known organic molecule to the NP and detecting its known SERS signature, suitable for various intracellular sensing applications. Stevenson et al. presented the first use of SERS reporter methodology to monitor the intracellular activity of specific enzymes, a tool which could possibly be implemented to detect dysfunction underlying many diseases [95]. A recent example of this concept is the detection of carbon monoxide in normal human liver and HeLa cells [96]. This sensing is based upon the sensitivity of SERS signals from 40 nm AuNP-bound palladacycle reporter ligands to carbonylation, induced by the presence of carbon monoxide, whereby the palladium constituent is removed from the ligand to form a carboxylic acid group as presented in Fig. 10.3 [96]. Inside cells, the presence of carbon monoxide releasing molecules (CORMs) was thus evidenced by the appearance of spectral bands at 1032 cm^{-1} ($R\text{-}NH_2$ rocking), 1118 cm^{-1} (νCX stretch) and 1242 cm^{-1} (νCO stretch). The lowest detection limit of CORMs was found to be 0.5 μM in HeLa cells, attributed to the 1032 cm^{-1} vibrational peak [96]. This finding offers potential to utilise palladacycle based SERS reporters as an analytical technique to develop a much required knowledge of pathophysiological events involved with carbon monoxide. These examples demonstrate the typical sensing-styled applications of intracellular reporter SERS methodology as well as its high sensitivity.

As a well understood technique with high specificity, SERS-reporter methodologies have been developed into therapeutic applications, of particular interest to oncology. Much work has gone into the detection of biomarkers using the technique in conjunction with microfluidics [97–101]; however, this is mostly conducted at the cellular level. In diagnostic terms, cancer cells can be detected by the specific binding of SERS probes, functionalised with tumour cell antibodies and Raman reporter molecules at the Au surface [102, 103]. For instance, circulating tumour cells were detected in blood plasma by labelling of Ag-Au nanorods with a mixture of four SERS dyes, along with an 'antibody rainbow cocktail' of four breast cancer antibodies and a specific leukocyte CD-45 marker. This resulted in highly specific detection of a single cancer cell in a population of seven million blood cells when applied with multicolour and photothermal imaging. The presence of the SERS-reporter dyes could be effectively imaged on the membrane of such tumour cells [103]. Similarly, multicore

SERS-reporter labels functionalised with epithelial cell adhesion molecule antibodies have been shown to bind to the membrane of living MCF-7 tumour cells without cellular uptake following just 25 ms exposure time. Potential for enumeration and sorting of circulating tumour cells is therefore very promising with microfluidic chips [97].

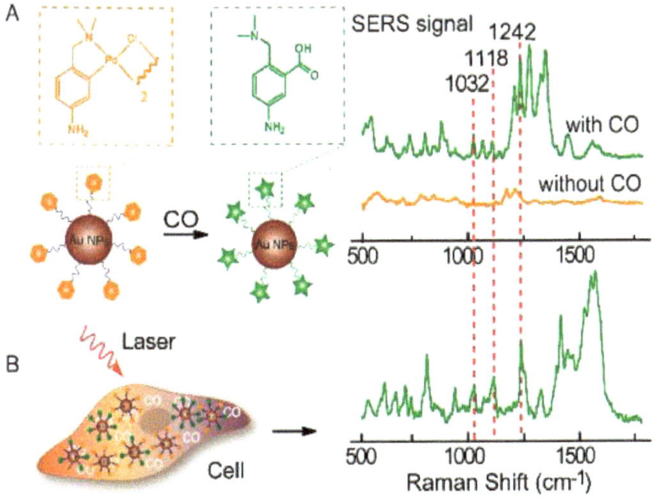

Figure 10.3 (A) SERS response and sensing mechanism of palladacycle carbonylation on AuNP/PC nanosensors for CO. (B) SERS detection of CO in living cells using AuNP/PC nanosensors. Reprinted with permission from Cao, Y., Li, D.-W., Zhao, Liu, L.-J., Cao, X.-M., and Long, Y.-T. (2015). *Anal. Chem.*, **87**, 9696–9701. Copyright (2015) *Anal. Chem.*

Of great relevance to subcellular events is utilising SERS-reporter NPs to achieve traceable intracellular delivery of therapeutics. AgNP-loaded, graphene oxide reporter-based nanoplatforms have been employed by Huang et al. to not only achieve pH-dependent delivery of anti-tumour drug DOX into live cells, but also to monitor its release by SERS [104]. Song et al. also performed successful delivery and monitoring of DOX in live SKBR-3 cells using hollow, amphiphilic SERS-reporter vehicles [105]. These particles consisted of 14 nm diameter AuNPs, coated with hydrophilic PEG, pH-sensitive hydrophobic PMMAVP grafts and the Raman reporter molecule BGLA (2-(4-(bis(4-(diethylamino)phenyl)(hydroxy)methyl)phenoxy)ethyl 5-(1,2-dithiolan-3-yl)pentanoate). The functionalised

AuNPs were then self-assembled into hollow 'plasmonic vesicles' tagged with HER2 antibodies for cancer cell targeting and loaded with DOX as shown in Fig. 10.4A–B. Upon cellular internalisation of the plasmonic vesicles, their disintegration is promoted by the pH decrease observed alongside the maturation of endosomes into lysosomes. Not only does this cause the release the drug cargo into the cell, but also diminishes the previously high SERS reporter signal intensity as the dye is no longer held in SERS hotspots of the assembled structure. However, as it is still bound to single AuNPs, signal remains sufficient to probe the chemical environment of the degraded vesicle. While the vesicle-labelled SKBR-3 cells displayed the strong Raman fingerprint of the BGLA probe, significantly weaker signals were detected from cells exposed to non-targeted vesicles and MCF-7 cells incubated with targeted vesicles (Fig. 10.4C–D). Those labelled with pH-sensitive vesicles also exhibited gradually reduced SERS intensity in the same timeline, as confirmed by the results of plasmonic imaging. Rather than merely detect the reporter compound employed, this approach monitors intracellular drug release based on the loss of hotspot-generated SERS reporter intensity to infer successful drug cargo delivery [105]. Uncontrolled AuNP aggregation often causes inhomogeneity in SERS measurements (discussed in the next section on label-free SERS), yet has been manipulated in this study to yield a novel analysis technique.

The lack of control surrounding AuNPs within the intracellular environment can be addressed with the creation of SERS superstructures. That is, depositing SERS-active NPs onto larger structures such that interparticle distances are controlled and relatively constant, offering a more reproducible level of SERS enhancement. Such superstructures can still be functionalised with therapeutics to produce drug delivery systems. Adopting this approach, DOX-bound AuNPs assembled around a silica nanocarrier were tracked through cancer cells by mapping of peaks of 4-MBA used as the Raman reporter [106]. Viability by MTT assay (cell metabolic activity) was assessed in HeLa and HEK-293 cells, demonstrating toxicity to the cervical cancer cells and biocompatibility to healthy embryonic kidney cells, respectively. Comparative results of tracing the particle by SERS and fluorescence was carried out and showed enhanced sensitivity and spectral resolution (narrow peaks) observed with SERS. Mapping of the distribution of particles (using 1078 cm^{-1} MBA vibration) across single HeLa cells also revealed intracellular details in SERS images, with a dark area of low signal representing the cell nucleus, which the sensors could not penetrate [106].

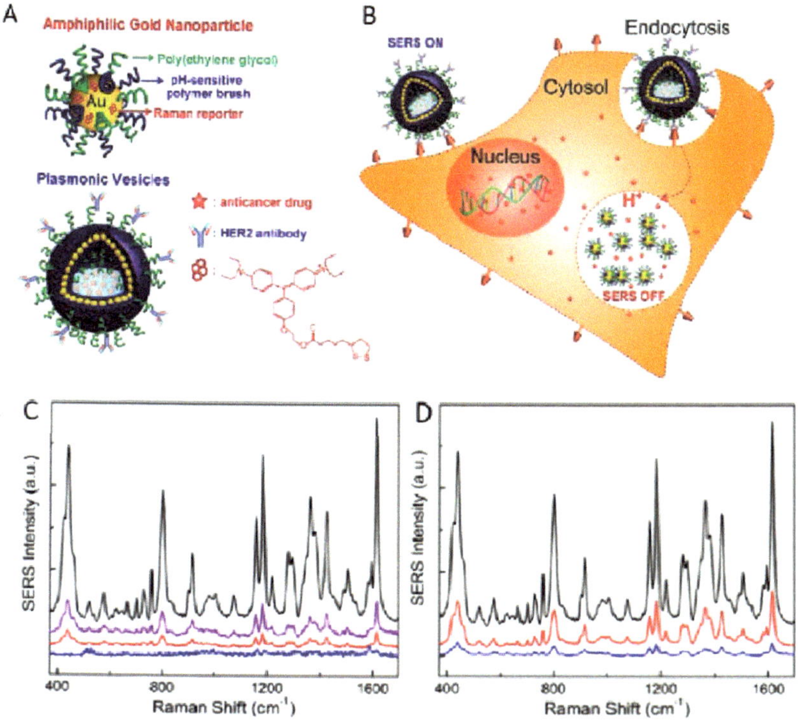

Figure 10.4 (A) Schematic illustration of the amphiphilic gold nanoparticle coated with Raman reporter BGLA and mixed polymer brushes of hydrophilic PEG and pH-sensitive hydrophobic PMMAVP grafts forming a drug-loaded 'plasmonic vesicle' tagged with HER2 antibody for cancer cell targeting. (B) The cellular binding, uptake, and intra-organelle disruption of the SERS-encoded pH-sensitive plasmonic vesicles. (C) Representative SERS spectra of SKBR-3 cells treated with HER2-targeted vesicles (black line) and untargeted vesicle (purple line), MCF-7 cells treated with targeted vesicles (red line), and SKBR-3 control cells without NPs (blue line). (D) Representative SERS spectra of SKBR-3 cells labelled with targeted vesicles after 30 min incubation (black line) and the post-incubation spectra of the cells at 60 min (red line) and 90 min (blue line). Reprinted with permission from Song, J., Zhou, J., and Duan, H. (2012). *J. Am. Chem. Soc.*, **134**, 13458–13469. Copyright 2012.

SERS reporter methodology may also be used to interrogate the intracellular distribution of various molecules by multiplexing the Raman signal of specific reporter molecules to produce SERS map images of single cells. The potential for this was demonstrated by Gregas et al., who targeted 4-MBA-AgNP probes to the cell nucleus with HIV-1 protein derived TAT sequence modification of NPs to characterise their location and produce two dimensional SERS images of PC-3 human prostate cells [107]. However, uncontrolled aggregation of spherical NPs not only causes locally heterogeneous enhancement of Raman signal by variable SERS hotspot generation, but also may prevent their internalisation or disturb cell function due to the size of agglomerates. Therefore Au nanostars of 60–70 nm diameter, labelled with toluidine blue (TB) and encapsulated in a 30 nm coating of silica have also been employed for probing of the intracellular matrix [108]. The performance of these particles was compared to spherical AuNPs with the same TB reporter and coating. Nanostars were shown by Transmission Electron Microscopy (TEM) to enter HeLa cells without any tendency to aggregate, albeit at a three-fold reduction in internalisation, which could be due to their larger size. Upon irradiation with an 830 nm laser, SERS signal intensity from the nanostars was mapped across single cells at intensities much higher than that generated by spherical NPs, which barely gave a signal. However, it must be noted when excited at 633 nm, a well-defined SERS signal could also be registered from the spherical NPs, demonstrating the presence of the label within the silica shell [108]. Therefore, the intensity differences could be due to the excitation wavelength and plasmon resonance dependence of enhancements. Apart from superstructures and nanostars, many new geometries of NPs such as nanoflowers [109] have been developed to improve sensitivity and circumvent aggregation.

The probing of the intracellular chemical environment is very much a key application for SERS given its high resolution and non-destructive nature. The ability to sensitively detect, map, monitor and quantify the localisation of both endogenous and extraneous molecules at the sub-cellular level would unlock a world of new information regarding the molecular changes undertaken by live cells during a range of cellular processes. The SERS reporter methods presented here offer promising steps towards visualisation and quantification of targeted molecules across individual cells. However, the effects of vast amounts and variety of biomolecules surrounding the nanoprobes go unaccounted for, limiting the amount of detailed information available as molecular events unfold. Despite the ability to multiplex data with numerous reporter molecules, this is a limitation of the technique when fully characterising cellular pathways or probing molecular interactions. The approach may also lack spectral resolution in some applications where characteristic reporter bands (such as the phenyl ring modes of 4-MBA) only exhibit small wavenumber shifts (few cm^{-1}) as a function of bound substituents [110–113].

10.3.2 Advances in Reporter-Free SERS

In contrast, in the reporter-free SERS approach the NPs probe their direct vicinity and thus can potentially sense the large complexity of the chemical environment around them. In reality, the affinity of molecules to the surface chemistry (charge) will determine the environment in the direct vicinity of plasmonic particles (e.g. AuNPs). Nevertheless, a more holistic intracellular view can be obtained compared to the SERS-reporter approach, playing to the multiplex advantage in SERS compared to fluorescence. Molecules near or adsorbed onto an NP or into a NP's protein corona determine the measured SERS spectrum and allow for (bio)chemical sensing and characterisation, to track cellular processes such as cellular differentiation by DNA/RNA ratios, for instance [57]. As discussed previously, single NPs can aggregate, giving rise to regions of intense electromagnetic field enhancements called 'hotspots' between close-proximity SERS probes. Molecules within hotspots experience greater SERS enhancements and consequently provide larger contributions to measured SERS spectra, albeit non-uniformly as the aggregation and therefore hotspot generation is largely uncontrolled inside cells. This phenomenon presents difficulty in quantifying both SERS reporter (unless the reporter is isolated within a shell) and reporter-free SERS data; however, it does allow for the qualitative, highly sensitive detection of the molecular environment inside cells [114]. While developing NPs which do not require aggregation for enhanced

SERS signals is an active area of research, most current work uses spherical AuNPs which become aggregated to produce high SERS signals.

Despite the simple preparation of reporter-free SERS NPs, selectively identifying molecular changes resulting from cellular processes in large and multidimensional data sets (arising from the large variety of biochemical molecules detected) can be challenging. Moreover, the lack of simple tools for extracting this information has so far resulted in limited characterisation of fundamental cellular processes, especially, by reporter-free SERS approaches. However, when applied effectively, this allows access to the vast wealth of information available from the raw SERS data acquired from biological samples such as cells [115].

Owing to the fact that this approach samples numerous constituents of the chemical environment around an NP, reporter-free SERS studies have been guided towards tracking and characterisation applications. The Kneipp group carried out an early study probing cellular compartments by reporter-free SERS in living macrophages and endothelial cells [11]. Differences in the SERS spectra were obtained in different cell lines and over time, in turn allowing the direct identification of physiologically relevant molecules. This study demonstrated that reporter-free SERS approaches are feasible for the characterisation of changing cellular environments and suitable for intracellular applications. Additionally, the tendency of internalised AuNPs to uncontrollably aggregate results from the changing chemical environment experienced by the probes, with dimer and trimer formation (revealed by TEM) leading to increased enhancement of the Raman signal compared to individual particles [9]. Thus, controlling or even preventing the aggregation of NPs inside the cell, in order to eliminate the incurred heterogeneity of acquired spectra from hotspots, remains one of the key challenges to reporter-free SERS experiments inside cells. Novel structures, which show high plasmonic enhancements at the single nanoparticle level compared to single spherical nanoparticles, have been employed to overcome this hurdle, as are discussed later in this section.

Reporter-free SERS for mapping the intracellular environment has nevertheless been used in many applications. In a study conducted by Zhang et al. a microfluidic chip was used for delivery and immobilisation of single cells [116]. They used ordered continuous flow to expose Chinese hamster ovary (CHO) K1 cells to intracellular flux agonist ionomycin. Following mapping to reveal the location of the cell nucleus, cytoplasm and membrane, the approach was found capable of monitoring the chemical changes occurring within the cell during ionomycin-evoked Ca^{2+} flux response. This was achieved by observing the time profile of changing amide I (1643 cm^{-1}) concentration, showing a similar trend to that of Ca^{2+} flux which is commonly used in biomedical research to stimulate the intracellular production of proteins such as interferons [117]. This demonstrates the potential to utilise SERS for in situ monitoring of molecules to characterise protein expression dynamics at subcellular levels. The continuous flow methodology developed also permits more time effective experimentation when a series of SERS measurements are required, relative to batch preparation and analysis [116].

Reporter-free SERS has also demonstrated the capability to differentiate between closely related cell phenotypes in progenitor and differentiated cell (DC) types [57]. The SV-40 large T nuclear localisation signal peptide, containing a fluorescein (flu) tag at its C terminus for confirmation of attachment, was bound to 40 nm AuNPs via a cysteine linker and used to probe human neuroblastoma (SH-SY5Y) cells with 633 nm excitation, producing cellular maps at a pixel resolution of 200 × 600 nm. Unambiguous characterisation of DC and undifferentiated cells (UDCs) was achieved through principal component analysis (PCA) across both the whole cell and the nuclear region only. Furthermore, characterisation of cell differentiation was carried out by examination of nuclear region, which undergo changes in ratios of molecular content as well as morphological and structural development [118]. The acquired SERS data revealed an increased ratio of DNA/RNA in DCs relative to UDCs. The number of protein peaks detected in DCs was also increased, indicating greater expression and variety of protein within the nuclear region and the whole cell [57]. This characterisation of cellular differentiation exhibits not only the ability of reporter-free SERS to distinguish between cell phenotypes but also how the careful application of statistical analysis methods such as PCA are utilised to effectively reduce the complex data matrices to unambiguously achieve classification.

More recent work by the same group further effectively demonstrated chemometric analysis in combination with tailored sample preparation to achieve detailed hyperspectral characterisation of endosomes and lysosomes in the endolysosomal pathway in the same cell line [115]. The novel method for creating a reference state depicted in Fig. 10.5 was derived from a conventional pulse-chase technique [9, 14, 119], whereby cells were subjected to a AuNP incubation pulse to ensure a wide distribution of the AuNPs within various types of vesicle and achieve high SERS signal. Cells analysed in this state were assigned as the *sample group*. A *reference state* was generated by carrying out the same incubation phase followed by a depletion phase in which the extracellular fluid was removed and replaced with fresh, AuNP-free media to ensure that the nanoprobes were processed through the endolysosomal system and only localised within lysosomes. Background subtraction of the generated SERS map data with principal component (linear summations of original data multiplied by a coefficient that describes the generated principle components, PCs) scores lower than 25% of the maximal PC1 value (Fig. 10.6A). PCA-LDA (Linear Discriminant Analysis) allows for the transformation of PC data to achieve maximum class segregation by hyperspectral peak positions, ratios and intensities. This was employed to assign PCA-LDA scores (Fig. 10.6C) as lysosomes (blue, scores common to both sets) or endosomes (green, scores exclusive to the *reference state*) with borderline case exclusion. Exemplarily, the back-projected pseudo colour map of the LD scores (Fig. 10.6D) clearly depicted the localisation of each vesicle type in a single cell, showing broad agreement with the supervised, multivariate analysis technique k-means clustering (Fig. 10.6E). Furthermore, characterisation of endosomes and lysosomes was achieved by observation of PC1 loadings, which were weighted with individual peak probabilities to reveal subtle differences in molecular content. These variations identified molecular processes occurring along the endolysosomal pathway as the breakdown of proteins and lipids, progressively raised acidity, and degradation of DNA/RNA in lysosomes [115].

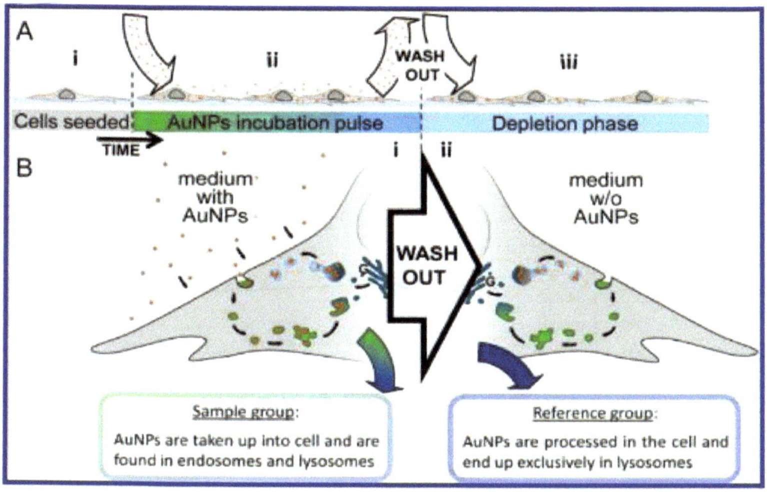

Figure 10.5 Schematic of experimental design. (A) Full experimental procedure involves AuNPs (red spheres) being added to the cell culture environment after (A.i) cells have sufficiently attached to the culture dish. (A.ii) Following uptake of the particles into the cell via endocytosis during the incubation pulse, extracellular particles were washed out. (A.iii) Fresh culture medium without AuNPs was added and the cells were left until incorporated particles were processed into lysosomes (depletion phase). (B.i) During the incubation phase, cells constantly internalize AuNPs, which accumulate inside endosomes (green vesicles) and lysosomes (blue vesicles), Their acquired SERS map data serves as the sample group for analysis. (B.ii) Following the wash-out, vesicular AuNPs are processed along the endolysosomal pathway and are eventually found exclusively in lysosomes. SERS maps of cells with only lysosomal AuNPs serve as the reference group for the data analysis. Reprinted with permission from Huefner, A., Kuan, W.-L., Müller, K. H., Skepper, J. N., Barker, R. A., and Mahajan, S. (2015). *ACS Nano.*, Copyright 2015 *ACS Nano.*

As mentioned earlier, the uniformity of SERS signals is extremely significant for reporter-free SERS, in particular for quantitative analysis. Considerable efforts have been made into counteracting the uncontrollable aggregation of internalised AuNPs by adaptation and modification of the particle structure. The production of gold lace nanoshells not only generates high SERS enhancements in the 1–3 nm hotspots between branch like structures without aggregation, but also offers further application

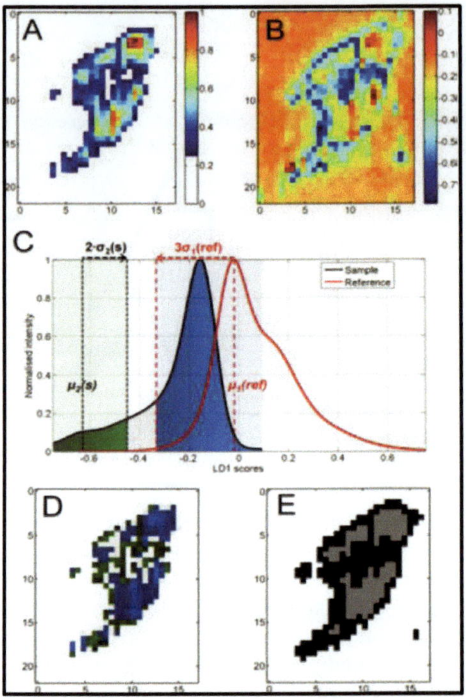

Figure 10.6 Demonstration of different intracellular SERS analysis steps and comparisons. Bright-field image (A) highlights the scanned area resulting in the corresponding absolute SERS intensity map (B). Pseudo colour PC1 intensity map (C) of the same cell after PC analysis shows the sample background as white for PC1 scores values smaller 0.25. LD1 scores map (D) of the sample cell prior to endosomes/lysosomes classification. (E) Assignment of components from the LD1 scores intensity distribution to either lysosomes (LD1(s) > −0.35) or endosomes (LD1(s) < -0.45) highlighted in blue and green, respectively. Assignments were made with respect to the modes $\mu1(ref)$ and $\mu2(s)$ as well as the standard deviations $\sigma1(ref)$ and $\sigma2(s)$ marked as dotted lines. The light grey region between the green and blue regions is a transition region to guarantee an accurate classification. (F) Based on scores for respective class assignments derived from (E), scores were projected back into a false-colour map. This generates a colour-coded reference-based PCA-LDA map showing endosomes (green) and lysosomes (blue). (G) A colour map was reconstructed after k-means clustering of the test data set for the three groups (white, grey, black) confirming the group segregation facilitated using our colour-coded reference-based PCA-LDA method. Scale bars: 10 μm, (C–D) and (F–G) pixel size: 600 nm × 600 nm. Reprinted with permission from Huefner, A., Kuan, W.-L., Müller, K. H., Skepper, J. N., Barker, R. A., and Mahajan, S. (2015). *ACS Nano*, Copyright 2015 *ACS Nano*.

as a model for the delivery and monitoring of hydrophobic drugs [34]. The topology generated by surfactant mediated production from polyurethanes (PUs) is similar to that of a Faberge egg, with the hydrophobic core (used to cargo drugs) provided by the amphiphilic structure of PU globules. SERS monitoring of loaded pyrene was carried out, the release of which was shown to plateau out after 30 h. The lace nanoshells offer tighter synthetic control of diameter (average 100 nm with 3–5 nm depth Au nanofinger layer) relative to thermosensitive polymer gold cages [34, 35] and liposome-gold nanocontainers [30] also used for hydrophobic drug delivery. This aids avoidance of first capture by macrophages and renal clearance when carriers are too large or small, respectively. The improved homogeneity of hotspots offers advantage in generating reproducible label-free SERS measurements over spherical NPs [34].

Similarly, the top-down construction of gold nanostructured microchips proposes an alternative means of removing the variability inherent to reporter-free SERS by aggregating spherical NPs. Chips with a uniform and reproducible nanodome gold pattern were generated by nanosphere lithography, depositing a gold nanofilm on a monolayer of polystyrene bead templates and thereafter released into solution [33]. Like spherical NPs, the microchips were taken up by normal human dermal fibroblast (NHDF) cells with no suggestion of reduced viability from observation of mitotic events. A proof-of-concept detection experiment was then carried out using the common extraneous SERS molecule, rhodamine 6G (R6G), serving as a stain. Following a 24 h incubation period, the intracellular location of the microchips and inhomogeneous distribution of R6G were confirmed in NHDF cells by confocal reflection and fluorescence microscopy (Fig. 10.7). The transmission image reveals the location of the cell which was mapped for the 1365 cm^{-1} R6G SERS peak (imaged in green).

This successful detection of a well-established SERS-active compound using nanostructured microchips paves the way for detection of extraneous molecules within live cells with improved reproducibility due to relative homogeneity of hotspot enhancements. This study also partially addressed a previous difficulty in determining of the exact location of internalised AuNPs, given that voluntarily internalised NPs exist in membrane-bound transport vesicles from which they may or may not escape to be free in the cytosol. A lack of co-localisation of the microchips with fluorescent dyes LysoTracker Red DND-99 and Vybrant Dil (accumulating in acidic cellular components and lipophilic membranes, respectively) suggested no localisation within vesicles of the endocytotic pathway, although further studies are required to verify exact location. This study also shows that the protein corona adsorption (an apparent barrier) on the gold plated chips still allows molecules to reach plasmonic hotspots. Overall this novel approach for reporter-free SERS takes positive steps towards homogenously enhanced, time-dependent and quantitative detection of intracellular molecules [33].

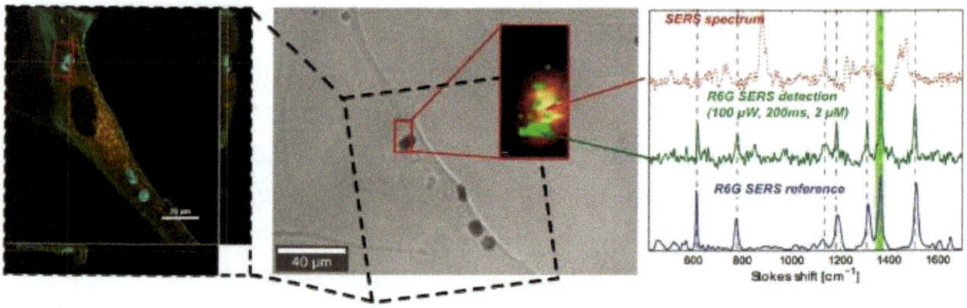

Figure 10.7 Reporter-free intracellular detection of Rhodamine-6G. (A) Confocal fluorescence scan of a single cell with several nanodome-patterned microchips show the intracellular localisation of the microchips, as well as the uptake and inhomogeneous distribution of R6G in the cell. (B) Transmission image of the same cell, where SERS spectra are mapped over the region marked with a red square. The inset shows the integrated number of counts in the 400–1700 cm^{-1} region (graded red-yellow), on top of which the presence of the 1365 cm^{-1} R6G-peak is highlighted in green, which corresponds to the green shaded area in (C). (C) SERS spectra on different positions of the microchip show the intracellular SERS detection of R6G (middle, green) and the R6G reference SERS spectrum (bottom, blue), while other positions show the presence of peaks related to other molecules adsorbed on the microchip (dotted red, top). Spectra are normalized and offset for clarity. Reprinted with permission from Wuytens, P. C., Subramanian, A. Z., De Vos, W. H., Skirtach, A. G., and Baets, R. (2015). *Analyst*, Published by The Royal Society of Chemistry.

Recently, in a powerful combination, reporter-free SERS and SERS reporter nanoprobes were employed by Chen et al. to simultaneously image the nucleus and plasma membrane of single HeLa cells in three dimensions, afforded by confocal SERS microscopy [56]. Characterisation of the nucleus and tracking of time dependent changes in molecular constitution during induced apoptosis was achieved using NLS-functionalised, reporter-free AuNPs. Consequently, slow condensation and degradation of chromatin, nuclear rupture and denaturation and degradation of proteins were all identified. AuNP-bound Raman dyes 4-MBA, Crystal Violet and Cresyl Violet acetate were used as SERS reporters to monitor the localisation and detachment of folate and luteinizing hormone-releasing hormone receptors in real time. The SERS reporter portion of this investigation provided verification that specific endogenous molecules are being monitored, which is sometimes an ambiguity in purely reporter-free studies. Although a fairly laborious process, the combination of reporter and reporter-free SERS techniques with application of suitable chemometric analysis shows immense promise for in situ characterisation of complex intracellular processes [120].

Along with fundamental studies, utilising reporter-free SERS to monitor the release and intracellular dynamics of drugs is emerging as a key application area. This is particularly relevant for delivering Raman-active, small molecule therapeutics which have a distinctive signature. While relevant to many diseases, most current examples are in oncology (cancer) and often exploit the pH-sensitive nature of NP-drug binding. Presenting both Raman activity and inherent fluorescence, DOX is often used as the drug molecule for such intracellular SERS studies to monitor delivery. An example of the reporter-free approach to tracking intracellular drug delivery was demonstrated with DOX-coated biohybrid AuNPs functionalised with anti-HER2 antibody, cell penetrating TAT peptide for increased efficacy and uptake along with a PEG cap [121]. DOX-NPs were administered to human cancer cells for 2 h, before washing and observation of drug release for a further 22 h period. SERS signals of DOX at 1275 cm^{-1} were observed in both SK-BR-3 and SH-SY5Y cells and allowed mapping of the intracellular distribution of the DOX-NPs across single cells. This SERS signature exhibited reduced intensity over time inside SK-BR-3 cells as the interaction with intracellular glutathione (GSH) triggered the release of DOX from NPs and consequently caused the loss of their SERS signal. The approach achieved quantification of the amount of drug delivered into the cell, observing a 90.23% DOX release after 24 h, which could be reduced to 2 h with application of additional GSH to cells. Examination of cell viability by MTT assay revealed that DOX-NPs were also slightly more effective than the free drug in preventing cancer cell proliferation [121]. Therefore, as in SERS-reporter studies, a model is offered demonstrating that SERS can be utilised inside cells to not only confirm and quantify the release of therapeutics but also that the conjugation to NPs can increase their efficacy by aiding delivery and increased bioavailability. While not shown in this study, further insight into intracellular drug dynamics will be beneficial and could potentially be carried out using drug's fluorescent signal, if any, to achieve dual mode delivery-monitoring. This has been achieved in the study by Zhuang et al. using AuNPs (30 nm) functionalised with the hyperbranched polymer HPVBEG (poly((S-(4-vinyl) benzyl S′-propyltrithiocarbonate)-co-(poly(ethylene glycol) methacrylate)) and use pH sensitive Schiff-base formation to bond DOX [122]. Thus, the bound state offers a strongly enhanced SERS-signal while the AuNP quenches fluorescence from the drug molecule. Following endocytotic uptake and lysosomal release, reduction of SERS intensity was accompanied by the appearance of fluorescence, which increased over time due to more release of DOX molecules. Another advantage of conjugating the drug onto AuNPs is its increased bioavailability inside cells compared to the free drug as also shown in this study [122]. A similar combined approach of using SERS and fluorescence has been demonstrated using asymmetric Janus particles (comprised of two-opposing faces of mesoporous silica and gold surfaces) functionalised with pH sensitively bound 6-mercaptopurine (6-MP, used in leukaemia chemotherapy) and the FRET pair DOX/CMR (7-hydroxycoumarin-3-carboxylate), respectively [123]. The two modalities were applied to observe 6-MP release in acidic lysosomes by loss of its SERS intensity in conjunction with the appearance of DOX fluorescence following its displacement from FRET donor CMR by cellular GSH. A similar enhanced efficiency of drug action was observed in HeLa cells by CCK-8 cytotoxicity assay. While the delivery of a drug into cells, monitoring its release by SERS and fluorescent mapping of the free drug distribution is highly useful, the major advantages of intracellular SERS, are not fully employed. That is, its increased spatial and molecular resolution, along with the wealth of molecular information contained within spectra. Moreover, most drug molecules are not fluorescent in the visible range which precludes use of such a combined approach.

An alternative approach may therefore be adopted in utilising reporter-free SERS methodologies to interrogate intracellular drug dynamics. Pre-incubation of cells with SERS NPs holds potential for direct detection of later administered drug molecules arriving inside cells (or at specific intracellular targets by AuNP-functionalisation) in an assay-based platform. In this vein, reporter-free SERS has been employed to observe the dynamics of non-fluorescent drug molecules 6-mercaptopurine (6-MP) and methimazole (MMI) inside living HeLa cells [124]. AgNPs were administered to cells in a 3/1 (v/v) media-NP dispersion before drug application. Inside the cell, both drugs localised to the immediate vicinity of the AgNPs (their structures having previously demonstrated adsorption to Ag electrodes) with each directly detected and identified by SERS following their co-incubation. Consequently, the internalisation and distribution of 6-MP and MMI was confirmed using reporter-free SERS by recognition of vibrational modes at 862 and 1360 cm^{-1}, respectively. SERS detection was highly sensitive, with a lower detection limit of 1 nM achieved for 6-MP. Furthermore, relative rates of metabolism were determined by ratiometric comparison of the disappearance of the drugs' respective vibrations revealed that 6-MP is metabolised faster than MMI [124]. Albeit using AgNPs and utilising drugs that were chosen due to their known affinity for silver, the study provides an excellent example of the detection power possessed by reporter-free intracellular SERS. In an assay-based format, this approach holds enormous potential to provide insights into drug delivery and cellular dynamics. This new knowledge in live, intact cells regarding compound bioavailability, distribution/localisation and metabolism would prove invaluable to pharmaceutical industry, identifying the pharmacodynamics of drug candidates for improved clinical outcomes.

10.4 Conclusions and Outlook

As a non-destructive, highly sensitive technique capable of sub-micron imaging resolutions, SERS provides a useful tool for interrogation of the intracellular environment across biomedical research. Acquisition times can be optimised down to the sub-millisecond range, sufficient for even single molecule detection [7, 8, 125]. Under appropriate laser powers, absence of signal deterioration effects such as photobleaching also potentiates live cell investigations over prolonged durations. Despite a requirement for standardisation of toxicity studies [68], the employed nanostructures, particularly those of gold, appear not to significantly reduce the viability of live cells. Intracellular SERS studies are still in their infancy; however, this is a fast-paced field attracting high levels of interest with an ever-growing body of research. Advances in nanotechnology have led to an increasingly varied library of SERS active structures, tailored to addressing the difficulties of *in cellulo* experiments such as cellular entry, pH dependent aggregation and targeting of specific compartments.

It is clear that the choice of intracellular SERS experiment performed is very dependent on not only the studied system but also the data required. SERS reporter methodology benefits greatly from simplicity of data analysis, and holds advantage over fluorescence microscopy in increased spectral resolution to allow multiplexing of a larger number of unique optical signatures. The discussed studies offer reporter-based SERS as an effective tool for the imaging of molecules and proteins with known vibrational signatures, as well as relatively simple quantification of simple cellular processes [88, 89, 96]. The approach has matured to therapeutic application with multipurpose nanostructures which not only detect and monitor but deliver a cargo of hydrophobic drug molecules into intracellular compartments [105].

Label-free SERS, on the other hand, despite being less widely developed is more suited as a powerful means of probing complex cellular processes due to the sheer wealth of information which can be extracted from a single measurement of the intracellular matrix about the vicinity of the NP. This, however, comes at the cost of throughput times, as careful processing and analysis of experimental data must be employed to extract desired information in order to accurately and unambiguously characterise processes at the molecular level. Such methodology was effectively achieved in the cases of endocytosis and apoptosis [115, 120]. Combination of label-free and SERS reporter techniques yields comprehensive findings, tracking both the localisation of specific proteins as well as monitoring changes in the diverse chemical environment. Tailoring of sample preparation to create a sample set upon which to train multivariate analysis methods such as PCA-LDA prove equally capable with the potentially complex synthesis of multiple functionalised SERS reporter probes. In either case, it remains that the development of statistical methodologies and data analysis toolkits are central to potentiating the use of reporter-free SERS to deconvolute the molecular dynamics of intracellular structures, pathways and processes.

Future developments in intracellular SERS will be of particular interest to the pharmaceutical industry, for elucidating new drug targets in both healthy and pathogenic cells. Additionally, there exists a requirement for determination of administered small molecules concentration and localisation at the sub cellular level, which no established technology can currently satisfy [126]. Without information on whether candidates are entering cells, let alone localising to and interacting with target organelles or biomolecules, pharmacodynamics cannot be fully understood which contributes to the lack of in vivo efficacy responsible for the failure of a large portion of clinical trials [126]. Microfluidics, for both nanoprobe synthesis and cellular interrogation, will also play a major role in the development of tightly controlled, reproducible and ultimately higher throughput techniques. Along with early-stage diagnostics by SERS-reporter and reporter-free SERS methodologies, these applications should form central goals for future proof-of-concept and process optimisation studies in order to develop and implement highly powerful intracellular SERS techniques for interrogation of the intracellular environment.

Acknowledgements

Support from the Institute for Life Sciences, University of Southampton, ERC grant NanoChemBioVision 638258 and

BBSRC iCase BB/N50404X/1 award with Astra Zeneca are kindly acknowledged.

References

1. Schlücker, S. (2014). Surface-enhanced Raman spectroscopy: Concepts and chemical applications, *Angew. Chem. Int. Ed.*, **53**, 4756–4795.
2. Le Ru, E. C. (2009). *Principles of Surface-Enhanced Raman Spectroscopy: And Related Plasmonic Effects*, 1st ed. Elsevier, Amsterdam; Boston.
3. Taylor, J., Huefner, A., Li, L., Wingfield, J., Mahajan, S. (2016). Nanoparticles and intracellular applications of surface-enhanced Raman spectroscopy, *Analyst*, **141**, 5037–5055.
4. Pallaoro, A., Braun, G. B., Moskovits, M. (2015). Biotags based on surface-enhanced Raman can be as bright as fluorescence tags, *Nano Lett.*, **15**, 6745–6750.
5. Sabatté, G., Keir, R., Lawlor, M., Black, M., Graham, D., Smith, W. E. (2008). Comparison of surface-enhanced resonance Raman scattering and fluorescence for detection of a labeled antibody, *Anal Chem.*, **80**, 2351–2356.
6. Kneipp, K., Wang, Y., Kneipp, H., Perelman, L. T., Itzkan, I., Dasari, R. R., Feld, M. S. (1997). Single molecule detection using surface-enhanced Raman scattering (SERS), *Phys. Rev. Lett.*, **78**, 1667–1670.
7. Le Ru, E. C., Meyer, M., Etchegoin, P. G. (2006). Proof of single-molecule sensitivity in surface enhanced Raman scattering (SERS) by means of a two-analyte technique, *J. Phys. Chem. B*, **110**, 1944–1948.
8. Nie, S. (1997). Probing single molecules and single nanoparticles by surface-enhanced Raman scattering, *Science*, **275**, 1102–1106.
9. Kneipp, J., Kneipp, H., McLaughlin, M., Brown. D., Kneipp, K. (2006). In vivo molecular probing of cellular compartments with gold nanoparticles and nanoaggregates, *Nano Lett.*, **6**, 2225–2231.
10. Dreaden, E. C., Alkilany, A. M., Huang, X., Murphy, C. J., El-Sayed, M. A. (2012). The golden age: Gold nanoparticles for biomedicine, *Chem. Soc. Rev.*, **41**, 2740–2779.
11. Bhardwaj, V., Srinivasan, S., McGoron, A. J. (2015). Efficient intracellular delivery and improved biocompatibility of colloidal silver nanoparticles towards intracellular SERS immuno-sensing, *Analyst*, **140**, 3929–3934.

12. Lévy, R., Shaheen, U., Cesbron, Y., Sée, V. (2010). Gold nanoparticles delivery in mammalian live cells: A critical review, *Nano Rev.*, doi: 10.3402/nano.v1i0.4889.

13. Willets, K. A. (2009). Surface-enhanced Raman scattering (SERS) for probing internal cellular structure and dynamics, *Anal. Bioanal. Chem.*, **394**, 85–94.

14. Alberts, B. (2008). *Molecular Biology of the Cell*. Garland Science.

15. Alberts, B. (2010). *Essential Cell Biology*. Garland Science.

16. Conner, S. D., Schmid, S. L. (2003). Regulated portals of entry into the cell, *Nature*, **422**, 37–44.

17. Drescher, D., Kneipp, J. (2012). Nanomaterials in complex biological systems: Insights from Raman spectroscopy, *Chem. Soc. Rev.*, **41**, 5780.

18. Huotari, J., Helenius, A. (2011). Endosome maturation: Endosome maturation, *EMBO J.*, **30**, 3481–3500.

19. Kneipp, J., Drescher, D. (2014). SERS in Cells: From Concepts to Practical Applications. In: Ozaki, Y., Kneipp, K., Aroca, R. (eds.), *Frontiers of Surface Enhanced Raman Scattering*, John Wiley & Sons, Ltd, Chichester, UK, pp. 285–308.

20. Stern, S. T., Adiseshaiah, P. P., Crist, R. M. (2012). Autophagy and lysosomal dysfunction as emerging mechanisms of nanomaterial toxicity, *Part Fibre Toxicol.*, **9**, 20.

21. Hillaireau, H., Couvreur, P. (2009). Nanocarriers' entry into the cell: relevance to drug delivery, *Cell Mol. Life Sci.*, **66**, 2873–2896.

22. Tkachenko, A. G., Xie, H., Coleman, D., Glomm, W., Ryan, J., Anderson, M. F., Franzen, S., Feldheim, D. L. (2003). Multifunctional gold nanoparticle–peptide complexes for nuclear targeting, *J. Am. Chem. Soc.*, **125**, 4700–4701.

23. Desnick, R. J., Schuchman, E. H. (2002). Enzyme replacement and enhancement therapies: lessons from lysosomal disorders, *Nat. Rev. Genet.*, **3**, 954–966.

24. del Pino, P., Pelaz, B., Zhang, Q., Maffre, P., Nienhaus, G. U., Parak, W. J. (2014). Protein corona formation around nanoparticles: From the past to the future, *Mater Horiz.*, **1**, 301–313.

25. Chithrani, B. D., Chan, W. C. W. (2007). Elucidating the mechanism of cellular uptake and removal of protein-coated gold nanoparticles of different sizes and shapes, *Nano Lett.*, **7**, 1542–1550.

26. Murphy, C. J., Gole, A. M., Stone, J. W., Sisco, P. N., Alkilany, A. M., Goldsmith, E. C., Baxter, S. C. (2008). Gold nanoparticles in biology: Beyond toxicity to cellular imaging, *Acc. Chem. Res.*, **41**, 1721–1730.

27. Zhang, S., Li, J., Lykotrafitis, G., Bao, G., Suresh, S. (2009). Size-dependent endocytosis of nanoparticles, *Adv. Mater*, **21**, 419–424.

28. Barbosa, S., Agrawal, A., Rodríguez-Lorenzo, L., Pastoriza-Santos, I., Alvarez-Puebla, R. A., Kornowski, A., Weller, H., Liz-Marzán, L. M. (2010). Tuning size and sensing properties in colloidal gold nanostars, *Langmuir*, **26**, 14943–14950.

29. Fu, Q., Ran, G., Xu, W. (2015). A microfluidic-based controllable synthesis of rolled or rigid ultrathin gold nanoplates, *RSC Adv.*, **5**, 37512–37516.

30. Jin, Y., Gao, X. (2009). Spectrally tunable leakage-free gold nanocontainers, *J. Am. Chem. Soc.*, **131**, 17774–17776.

31. Li, J.-F., Tian, Z.-Q. (2014). Shell-Isolated Nanoparticle-Enhanced Raman Spectroscopy (SHINERS). In: Ozaki, Y., Kneipp, K., Aroca, R. (eds.), *Frontiers of Surface Enhanced Raman Scattering*, John Wiley & Sons, Ltd, Chichester, UK, pp. 163–192.

32. Perezjuste, J., Pastorizasantos, I., Lizmarzan, L., Mulvaney, P. (2005). Gold nanorods: Synthesis, characterization and applications, *Coord. Chem. Rev.*, **249**, 1870–1901.

33. Wuytens, P. C., Subramanian, A. Z., De Vos, W. H., Skirtach, A. G., Baets, R. (2015). Gold nanodome-patterned microchips for intracellular surface-enhanced Raman spectroscopy, *Analyst*, doi: 10.1039/C5AN01782C.

34. Yang, M., Alvarez-Puebla, R., Kim, H.-S., Aldeanueva-Potel, P., Liz-Marzán, L. M., Kotov, N. A. (2010). SERS-active gold lace nanoshells with built-in hotspots, *Nano Lett.*, **10**, 4013–4019.

35. Yavuz, M. S., Cheng, Y., Chen, J., Cobley, C. M., Zhang, Q., Rycenga, M., Xie, J., Kim, C., Song, K. H., Schwartz, A. G., Wang, L. V., Xia, Y. (2009). Gold nanocages covered by smart polymers for controlled release with near-infrared light, *Nat. Mater.*, **8**, 935–939.

36. Xie, W., Schlücker, S. (2013). Medical applications of surface-enhanced Raman scattering, *Phys. Chem. Chem. Phys.*, **15**, 5329.

37. Gao, H., Shi, W., Freund, L. B. (2005). From the cover: Mechanics of receptor-mediated endocytosis, *Proc. Natl. Acad. Sci.*, **102**, 9469–9474.

38. Bartczak, D., Muskens, O. L., Nitti, S., Sanchez-Elsner, T., Millar, T. M., Kanaras, A. G. (2012). Interactions of human endothelial cells with gold nanoparticles of different morphologies, *Small*, **8**, 122–130.

39. Monopoli, M. P., Walczyk, D., Campbell, A., Elia, G., Lynch, I., Baldelli Bombelli, F., Dawson, K. A. (2011). Physical–chemical aspects of

protein corona: Relevance to in vitro and in vivo biological impacts of nanoparticles, *J. Am. Chem. Soc.*, **133**, 2525–2534.

40. Leroueil, P. R., Hong, S., Mecke, A., Baker, J. R., Orr, B. G., Banaszak Holl, M. M. (2007). Nanoparticle interaction with biological membranes: Does nanotechnology present a Janus face? *Acc. Chem. Res.*, **40**, 335–342.

41. de la Fuente, J. M., Berry, C. C. (2005). Tat peptide as an efficient molecule to translocate gold nanoparticles into the cell nucleus, *Bioconjug. Chem.*, **16**, 1176–1180.

42. Verma, A., Stellacci, F. (2010). Effect of surface properties on nanoparticle-cell interactions, *Small Weinh. Bergstr. Ger.*, **6**, 12–21.

43. Chou, L. Y. T., Ming, K., Chan, W. C. W. (2011). Strategies for the intracellular delivery of nanoparticles, *Chem. Soc. Rev.*, **40**, 233–245.

44. Chatterjee, D., Rufaihah, A., Zhang, Y. (2008). Upconversion fluorescence imaging of cells and small animals using lanthanide doped nanocrystals, *Biomaterials*, **29**, 937–943.

45. Herrero, M. A., Toma, F. M., Al-Jamal, K. T., Kostarelos, K., Bianco, A., Da Ros, T., Bano, F., Casalis, L., Scoles, G., Prato, M. (2009). Synthesis and characterization of a carbon nanotube–dendron series for efficient siRNA delivery, *J. Am. Chem. Soc.*, **131**, 9843–9848.

46. Lutz, G. J., Sirsi, S. R., Williams, J. H. (2008). PEG–PEI Copolymers for oligonucleotide delivery to cells and tissues. In: Doux, J. M. L. (ed.), *Gene Therapy Protocols: Production and In Vivo Applications of Gene Transfer Vectors*. Humana Press, Totowa, NJ, pp. 141–150.

47. Babič, M., Horák, D., Trchová, M., Jendelová, P., Glogarová, K., Lesný, P., Herynek, V., Hájek, M., Syková, E. (2008). Poly(L-lysine)-modified iron oxide nanoparticles for stem cell labeling, *Bioconjug. Chem.*, **19**, 740–750.

48. Al-Jamal, W. T., Al-Jamal, K. T., Bomans, P. H., Frederik, P. M., Kostarelos, K. (2008). Functionalized-quantum-dot-liposome hybrids as multimodal nanoparticles for cancer, *Small*, **4**, 1406–1415.

49. Al-Jamal, W. T., Al-Jamal, K. T., Tian, B., Lacerda, L., Bomans, P. H., Frederik, P. M., Kostarelos, K. (2008). Lipid–quantum dot bilayer vesicles enhance tumor cell uptake and retention in vitro and in vivo, *ACS Nano*, **2**, 408–418.

50. Rhim, W.-K., Kim, J.-S., Nam, J.-M. (2008). Lipid-gold-nanoparticle hybrid-based gene delivery, *Small*, **4**, 1651–1655.

51. Krpetić, Ž., Saleemi, S., Prior, I. A., Sée, V., Qureshi, R., Brust, M. (2011). Negotiation of intracellular membrane barriers by TAT-modified gold nanoparticles, *ACS Nano*, **5**, 5195–5201.

52. Todorova, N., Chiappini, C., Mager, M., Simona, B., Patel, II., Stevens, M. M., Yarovsky, I. (2014). Surface presentation of functional peptides in solution determines cell internalization efficiency of TAT conjugated nanoparticles, *Nano Lett.*, **14**, 5229–5237.

53. Torchilin, V. P. (2008). Tat peptide-mediated intracellular delivery of pharmaceutical nanocarriers, *Adv. Drug Deliv. Rev.*, **60**, 548–558.

54. Ciobanasu, C., Siebrasse, J. P., Kubitscheck, U. (2010). Cell-penetrating HIV1 TAT peptides can generate pores in model membranes, *Biophys. J.*, **99**, 153–162.

55. Fischer, R., Fotin-Mleczek, M., Hufnagel, H., Brock, R. (2005). Break on through to the other side-biophysics and cell biology shed light on cell-penetrating peptides, *ChemBioChem*, **6**, 2126–2142.

56. Chen, Y., Bai, X., Su, L., Du, Z., Shen, A., Materny, A., Hu, J. (2016). Combined labelled and label-free SERS probes for triplex three-dimensional cellular imaging, *Sci. Rep.*, **6**, 19173.

57. Huefner, A., Kuan, W.-L., Barker, R. A., Mahajan, S. (2013). Intracellular SERS nanoprobes for distinction of different neuronal cell types, *Nano Lett.*, **13**, 2463–2470.

58. Alkilany, A. M., Murphy, C. J. (2010). Toxicity and cellular uptake of gold nanoparticles: What we have learned so far? *J. Nanoparticle Res.*, **12**, 2313–2333.

59. Ozaki, Y., Kneipp, K., Aroca, R. (2014). *Frontiers of Surface-Enhanced Raman Scattering: Single Nanoparticles and Single Cells*, John Wiley & Sons, Ltd, Chichester, UK.

60. Sakhtianchi, R., Minchin, R. F., Lee, K.-B., Alkilany, A. M., Serpooshan, V., Mahmoudi, M. (2013). Exocytosis of nanoparticles from cells: Role in cellular retention and toxicity, *Adv. Colloid Interface Sci.*, **201–202**, 18–29.

61. Goodman, C. M., McCusker, C. D., Yilmaz, T., Rotello, V. M. (2004). Toxicity of gold nanoparticles functionalized with cationic and anionic side chains, *Bioconjug. Chem.*, **15**, 897–900.

62. Pan, Y., Leifert, A., Ruau, D., Neuss, S., Bornemann, J., Schmid, G., Brandau, W., Simon, U., Jahnen-Dechent, W. (2009). Gold nanoparticles of diameter 1.4 nm trigger necrosis by oxidative stress and mitochondrial damage, *Small*, **5**, 2067–2076.

63. Pan, Y., Neuss, S., Leifert, A., Fischler, M., Wen, F., Simon, U., Schmid, G., Brandau, W., Jahnen-Dechent, W. (2007). Size-dependent cytotoxicity of gold nanoparticles, *Small*, **3**, 1941–1949.

64. Hutter, E., Boridy, S., Labrecque, S., Lalancette-Hébert, M., Kriz, J., Winnik, F. M., Maysinger, D. (2010). Microglial response to gold nanoparticles, *ACS Nano*, **4**, 2595–2606.

65. Niidome, T., Akiyama, Y., Yamagata, M., Kawano, T., Mori, T., Niidome, Y., Katayama, Y. (2009). Poly(ethylene glycol)-modified gold nanorods as a photothermal nanodevice for hyperthermia, *J. Biomater. Sci. Polym. Ed.*, **20**, 1203–1215.

66. Alkilany, A. M., Nagaria, P. K., Hexel, C. R., Shaw, T. J., Murphy, C. J., Wyatt, M. D. (2009). Cellular uptake and cytotoxicity of gold nanorods: Molecular origin of cytotoxicity and surface effects, *Small*, **5**, 701–708.

67. Leonov, A. P., Zheng, J., Clogston, J. D., Stern, S. T., Patri, A. K., Wei, A. (2008). Detoxification of gold nanorods by treatment with polystyrenesulfonate, *ACS Nano*, **2**, 2481–2488.

68. Fratoddi, I., Venditti, I., Cametti, C., Russo, M. V. (2015). The puzzle of toxicity of gold nanoparticles. The case-study of HeLa cells, *Toxicol. Res.*, **4**, 796–800.

69. Nabiev, I. R., Morjani, H., Manfait, M. (1991). Selective analysis of antitumor drug interaction with living cancer cells as probed by surface-enhanced Raman spectroscopy, *Eur. Biophys. J.*, doi: 10.1007/BF00183320.

70. Auchinvole, C. A. R, Richardson, P., McGuinnes, C., Mallikarjun, V., Donaldson, K., McNab, H., Campbell, C. J. (2012). Monitoring intracellular redox potential changes using SERS nanosensors, *ACS Nano*, **6**, 888–896.

71. Huh, Y. S., Chung, A. J., Erickson, D. (2009). Surface enhanced Raman spectroscopy and its application to molecular and cellular analysis, *Microfluid Nanofluidics*, **6**, 285–297.

72. Ando, J., Fujita, K., Smith, N. I., Kawata, S. (2011). Dynamic SERS imaging of cellular transport pathways with endocytosed gold nanoparticles, *Nano Lett.*, **11**, 5344–5348.

73. Ando, J., Yano, T., Fujita, K., Kawata, S. (2013). Metal nanoparticles for nano-imaging and nano-analysis, *Phys. Chem. Chem. Phys.*, **15**, 13713.

74. Hollywood, K. A., Shadi, I. T., Goodacre, R. (2010). Monitoring the succinate dehydrogenase activity isolated from mitochondria by surface enhanced Raman scattering, *J. Phys. Chem. C*, **114**, 7308–7313.

75. Larmour, I. A., Graham, D. (2011). Surface enhanced optical spectroscopies for bioanalysis, *Analyst*, **136**, 3831.

76. Escoriza, M. F., VanBriesen, J. M., Stewart, S., Maier, J. (2007). Raman spectroscopic discrimination of cell response to chemical and physical inactivation, *Appl. Spectrosc.*, **61**, 812–823.

77. Kang, B., Austin, L. A., El-Sayed, M. A. (2012). Real-time molecular imaging throughout the entire cell cycle by targeted plasmonic-enhanced rayleigh/Raman spectroscopy, *Nano Lett.*, **12**, 5369–5375.

78. Ong, Y. H., Lim, M., Liu, Q. (2012). Comparison of principal component analysis and biochemical component analysis in Raman spectroscopy for the discrimination of apoptosis and necrosis in K562 leukemia cells: Errata, *Opt. Express*, **20**, 25041.

79. Karataþ, Ö. F., Sezgin, E., Aydýn, Ö., Çulha, M. (2009). Interaction of gold nanoparticles with mitochondria. *Colloids Surf B Biointerfaces*, **71**, 315–318.

80. Kang, B., Afifi, M. M., Austin, L. A., El-Sayed, M. A. (2013). Exploiting the nanoparticle plasmon effect: Observing drug delivery dynamics in single cells *via* Raman/fluorescence imaging spectroscopy, *ACS Nano*, **7**, 7420–7427.

81. Ock, K., Jeon, W. I., Ganbold, E. O., Kim, M., Park, J., Seo, J. H., Cho, K., Joo, S.-W., Lee, S. Y. (2012). Real-time monitoring of glutathione-triggered thiopurine anticancer drug release in live cells investigated by surface-enhanced Raman scattering, *Anal. Chem.*, **84**, 2172–2178.

82. Yang, J., Wang, Z., Zong, S., Chen, H., Zhang, R., Cui, Y. (2014). Dual-mode tracking of tumor-cell-specific drug delivery using fluorescence and label-free SERS techniques, *Biosens. Bioelectron*, **51**, 82–89.

83. Cialla, D., März, A., Böhme, R., Theil, F., Weber, K., Schmitt, M., Popp, J. (2012). Surface-enhanced Raman spectroscopy (SERS): Progress and trends, *Anal. Bioanal. Chem.*, **403**, 27–54.

84. Kneipp, J., Drescher, D. (2014). SERS in Cells: from Concepts to Practical Applications. In: Ozaki, Y., Kneipp, K., Aroca, R. (eds.) *Frontiers of Surface Enhanced Raman Scattering*, John Wiley & Sons, Ltd, Chichester, UK, pp. 285–308.

85. Pahlow, S., März, A., Seise, B., Hartmann, K., Freitag, I., Kämmer, E., Böhme, R., Deckert, V., Weber, K., Cialla, D., Popp, J. (2012). Bioanalytical application of surface- and tip-enhanced Raman spectroscopy: Bioanalytical application of SERS and TERS, *Eng. Life Sci.*, **12**, 131–143.

86. Peng, R., Si, Y., Deng, T., Zheng, J., Li, J., Yang, R., Tan, W. (2016). A novel SERS nanoprobe for the ratiometric imaging of hydrogen peroxide in living cells, *Chem. Commun.*, **52**, 8553–8556.

87. Wang, W., Zhang, L., Li, L., Tian, Y. (2016). A single nanoprobe for ratiometric imaging and biosensing of hypochlorite and glutathione

in live cells using surface-enhanced Raman scattering, *Anal. Chem.*, **88**, 9518–9523.

88. Jamieson, L. E., Jaworska, A., Jiang, J., Baranska, M., Harrison, D. J., Campbell, C. J. (2015). Simultaneous intracellular redox potential and pH measurements in live cells using SERS nanosensors, *Analyst*, **140**, 2330–2335.

89. Talley, C. E., Jusinski, L., Hollars, C. W., Lane, S. M., Huser, T. (2004). Intracellular pH sensors based on surface-enhanced Raman scattering, *Anal Chem.*, **76**, 7064–7068.

90. Williams, A., Flynn, K. J., Xia, Z., Dunstan, P. R. (2016). Multivariate spectral analysis of pH SERS probes for improved sensing capabilities: Spectral analysis of pH SERS probes, *J. Raman Spectrosc.*, **47**, 819–827.

91. Mulvaney, S. P., Musick, M. D., Keating, C. D., Natan, M. J. (2003). Glass-coated, analyte-tagged nanoparticles: A new tagging system based on detection with surface-enhanced Raman scattering, *Langmuir*, **19**, 4784–4790.

92. Li, J. F., Huang, Y. F., Ding, Y., Yang, Z. L., Li, S. B., Zhou, X. S., Fan, F. R., Zhang, W., Zhou, Z. Y., Wu, D. Y., Ren, B., Wang, Z. L., Tian, Z. Q. (2010). Shell-isolated nanoparticle-enhanced Raman spectroscopy, *Nature*, **464**, 392–395.

93. Tian, X.-D., Liu, B.-J., Li, J.-F., Yang, Z.-L., Ren, B., Tian, Z.-Q. (2013). SHINERS and plasmonic properties of Au Core SiO_2 shell nanoparticles with optimal core size and shell thickness: SHINERS and properties of Au Core SiO_2 shell nanoparticles, *J. Raman Spectrosc.*, **44**, 994–998.

94. Moreton, S., Faulds, K., Shand, N. C., Bedics, M. A., Detty, M. R., Graham, D. (2015). Functionalisation of hollow gold nanospheres for use as stable, red-shifted SERS nanotags, *Nanoscale*, **7**, 6075–6082.

95. Stevenson, R., McAughtrie, S., Senior, L., Stokes, R. J., McGachy, H., Tetley, L., Nativo, P., Brewer, J. M., Alexander, J., Faulds, K., Graham, D. (2013). Analysis of intracellular enzyme activity by surface enhanced Raman scattering, *Analyst*, **138**, 6331.

96. Cao, Y., Li, D.-W., Zhao, L.-J., Liu, X.-Y., Cao, X.-M., Long, Y.-T. (2015). Highly selective detection of carbon monoxide in living cells by palladacycle carbonylation-based surface enhanced Raman spectroscopy nanosensors, *Anal. Chem.*, **87**, 9696–9701.

97. Freitag, I., Matthäus, C., Csaki, A., Clement, J. H., Cialla-May, D., Weber, K., Krafft, C., Popp, J. (2015). Differentiation of MCF-7 tumor cells from leukocytes and fibroblast cells using epithelial cell adhesion

molecule targeted multicore surface-enhanced Raman spectroscopy labels, *J. Biomed. Opt.*, **20**, 055002.

98. Li, J., Skeete, Z., Shan, S., Yan, S., Kurzatkowska, K., Zhao, W., Ngo, Q. M., Holubovska, P., Luo, J., Hepel, M., Zhong, C.-J. (2015). Surface enhanced Raman scattering detection of cancer biomarkers with bifunctional nanocomposite probes, *Anal. Chem.*, **87**, 10698–10702.

99. Li, Q., Gao, Q., Zhang, G. (2014). Classification for breast cancer diagnosis with Raman spectroscopy, *Biomed. Opt. Express*, **5**, 2435.

100. Tang, L., Casas, J., Venkataramasubramani, M. (2013). Magnetic nanoparticle mediated enhancement of localized surface plasmon resonance for ultrasensitive bioanalytical assay in human blood plasma, *Anal. Chem.*, **85**, 1431–1439.

101. Wu, L., Qu, X. (2015). Cancer biomarker detection: Recent achievements and challenges, *Chem. Soc. Rev.*, **44**, 2963–2997.

102. MacLaughlin, C. M., Mullaithilaga, N., Yang, G., Ip, S. Y., Wang, C., Walker, G. C. (2013). Surface-enhanced Raman scattering dye-labeled Au nanoparticles for triplexed detection of leukemia and lymphoma cells and SERS flow cytometry, *Langmuir*, **29**, 1908–1919.

103. Nima, Z. A., Mahmood, M., Xu, Y., Mustafa, T., Watanabe, F., Nedosekin, D. A., Juratli, M. A., Fahmi, T., Galanzha, E. I., Nolan, J. P., Basnakian, A. G., Zharov, V. P., Biris, A. S. (2014). Circulating tumor cell identification by functionalized silver-gold nanorods with multicolor, super-enhanced SERS and photothermal resonances, *Sci. Rep.*, doi: 10.1038/srep04752.

104. Huang, J., Zong, C., Shen, H., Cao, Y., Ren, B., Zhang, Z. (2013). Tracking the intracellular drug release from graphene oxide using surface-enhanced Raman spectroscopy, *Nanoscale*, **5**, 10591.

105. Song, J., Zhou, J., Duan, H. (2012). Self-assembled plasmonic vesicles of SERS-encoded amphiphilic gold nanoparticles for cancer cell targeting and traceable intracellular drug delivery, *J. Am. Chem. Soc.*, **134**, 13458–13469.

106. Liu, L., Tang, Y., Dai, S., Kleitz, F., Qiao, S. Z. (2016). Smart surface-enhanced Raman scattering traceable drug delivery systems, *Nanoscale*, **8**, 12803–12811.

107. Gregas, M. K., Scaffidi, J. P., Lauly, B., Vo-Dinh, T. (2010). Surface-enhanced Raman scattering detection and tracking of nanoprobes: Enhanced uptake and nuclear targeting in single cells, *Appl. Spectrosc.*, **64**, 858–866.

108. Rodríguez-Lorenzo, L., Krpetic, Z., Barbosa, S., Alvarez-Puebla, R. A., Liz-Marzán, L. M., Prior, I. A., Brust, M. (2011). Intracellular mapping with SERS-encoded gold nanostars, *Integr. Biol.*, **3**, 922.

109. Song, C. Y., Yang, B. Y., Chen, W. Q., Dou, Y. X., Yang, Y. J., Zhou, N., Wang, L. H. (2016). Gold nanoflowers with tunable sheet-like petals: Facile synthesis, SERS performances and cell imaging, *J. Mater. Chem. B*, **4**, 7112–7118.

110. Dougan, J. A., Faulds, K. (2012). Surface enhanced Raman scattering for multiplexed detection, *Analyst,* **137**, 545–554.

111. Matschulat, A., Drescher, D., Kneipp, J. (2010). Surface-enhanced Raman scattering hybrid nanoprobe multiplexing and imaging in biological systems, *ACS Nano*, **4**, 3259–3269.

112. Schlücker, S. (2009). SERS microscopy: Nanoparticle probes and biomedical applications, *ChemPhysChem*, **10**, 1344–1354.

113. Schütz, M., Müller, C. I., Salehi, M., Lambert, C., Schlücker, S. (2011). Design and synthesis of Raman reporter molecules for tissue imaging by immuno-SERS microscopy, *J. Biophotonics*, **4**, 453–463.

114. Lim, D.-K., Jeon, K.-S., Hwang, J.-H., Kim, H., Kwon, S., Suh, Y. D., Nam, J.-M. (2011). Highly uniform and reproducible surface-enhanced Raman scattering from DNA-tailorable nanoparticles with 1 nm interior gap, *Nat. Nanotechnol.,* **6**, 452–460.

115. Huefner, A., Kuan, W.-L., Müller, K. H., Skepper, J. N., Barker, R. A., Mahajan, S. (2015). Characterization and visualization of vesicles in the endo-lysosomal pathway with surface-enhanced Raman spectroscopy and chemometrics, *ACS Nano*, **10**, 307–316.

116. Zhang, X., Yin, H., Cooper, J. M., Haswell, S. J. (2008). Characterization of cellular chemical dynamics using combined microfluidic and Raman techniques, *Anal. Bioanal. Chem.,* **390**, 833–840.

117. Osna, N., Elliott, K., Khan, M. M. (2001). The effects of histamine on interferon gamma production are dependent on the stimulatory signals, *Int. Immunopharmacol.,* **1**, 135–145.

118. Grigoryev, S. A., Bulynko, Y. A., Popova, E. Y. (2006). The end adjusts the means: Heterochromatin remodelling during terminal cell differentiation, *Chromosome Res.*, **14**, 53–69.

119. Büchner, T., Drescher, D., Traub, H., Schrade, P., Bachmann, S., Jakubowski, N., Kneipp, J. (2014). Relating surface-enhanced Raman scattering signals of cells to gold nanoparticle aggregation as determined by LA-ICP-MS micromapping, *Anal. Bioanal. Chem.,* **406**, 7003–7014.

120. Chen, Y., Bai, X., Su, L., Du, Z., Shen, A., Materny, A., Hu, J. (2016). Combined labelled and label-free SERS probes for triplex three-dimensional cellular imaging, *Sci. Rep.*, **6**, 19173.

121. Hossain, M. K., Cho, H.-Y., Kim, K.-J., Choi, J.-W. (2015). In situ monitoring of doxorubicin release from biohybrid nanoparticles modified with antibody and cell-penetrating peptides in breast cancer cells using surface-enhanced Raman spectroscopy, *Biosens. Bioelectron.*, **71**, 300–305.

122. Zhuang, Y., Wang, D., Yin, C., Deng, H., Sun, M., He, L., Su, Y., Zhu, X. (2016). Tracing drug release process with dual-modal hyperbranched polymer-gold nanoparticle complexes, *Sci. China Chem.*, **59**, 1600–1608.

123. Cao, H., Yang, Y., Chen, X., Shao, Z. (2016). Intelligent Janus nanoparticles for intracellular real-time monitoring of dual drug release, *Nanoscale*, **8**, 6754–6760.

124. Yang, J., Cui, Y., Zong, S., Zhang, R., Song, C., Wang, Z. (2012). Tracking multiplex drugs and their dynamics in living cells using the label-free surface-enhanced Raman scattering technique, *Mol. Pharm.*, **9**, 842–849.

125. Zhang, R., Zhang, Y., Dong, Z. C., Jiang, S., Zhang, C., Chen, L. G., Zhang, L., Liao, Y., Aizpurua, J., Luo, Y., Yang, J. L., Hou, J. G. (2013). Chemical mapping of a single molecule by plasmon-enhanced Raman scattering, *Nature*, **498**, 82–86.

126. Dollery, C. T. (2013). Intracellular drug concentrations, *Clin. Pharmacol. Ther.*, **93**, 263–266.

Chapter 11

SERS-Based Nanotechnology for Imaging of Cellular Properties

Ewelina Wiercigroch and Kamilla Malek

Faculty of Chemistry, Jagiellonian University, Gronostajowa 2, Kraków 30-387, Poland

kamilla.malek@uj.edu.pl

Noble metal nanoparticles are one of the most powerful tools in nanotechnology, and are used to tune novel, purpose-built plasmonic nanostructures capable for intracellular probing. In imaging techniques based on surface-enhanced Raman spectroscopy (SERS), precise and controllable surface modifications are carefully designed to detect the localization of biomolecules, delivery, and monitoring of therapeutics and for characterization of complex cellular processes at the molecular level. This Chapter is aimed at introducing the reader to the field of utilization of nanomaterials for SERS biosensing with special emphasis on monitoring cellular response upon disease and exposure to stress conditions.

Plasmonics in Chemistry and Biology
Edited by Marc Lamy de la Chapelle and Nordin Felidj

ISBN 978-981-4800-03-7 (Hardcover), 978-0-429-45875-0 (eBook)
www.jennystanford.com

11.1 Cellular Interaction and Uptake of SERS Nanosensors

To date, in vitro applications of SERS nanoprobes have been mainly directed into functionalization of gold nanoparticles (AuNPs) with a diameter between 20–100 nm. The wide interest in AuNPs results from their properties such as high chemical stability, easy control of size during synthesis, and various functionalizations with nanoprobes possessing usually the thiol and amine groups, and the most important issue is the high biocompatibility of gold. However, similarly to other biomedical applications of nanomaterials, in vitro SERS-based nanotechnology requires the careful evaluation of cell-NPs interactions which in turn need to be assessed individually because these interactions are affected by physicochemical properties of nanosensors (size, coating, functionalization) and cell type [50]. Unfortunately, this issue has not been standardized so far in nanomedicine and its solving relies on individual reports while nanotoxicity of designed nanoprobes is a crucial concern for any further clinical translation of proof-of-concept studies in the field of SERS nanotechnology [49]. Among the most important key points in exposing live cells to gold nanoparticles are size of NPs and their coating/functionalization. Too small particles can reach intracellular compartments like nucleus and then they induce cell death [49]. For SERS biosensing, citrate-stabilized gold nanoparticles are usually synthesized with a diameter in the range of 40–80 nm and of spherical shape [2, 5, 36]. Nanotoxicity is also related with the exposure time of cells to incubated nanoparticles. Therefore, SERS studies have included toxicity testing by using MTT and LDH assays; however, several experiments have just employed overnight or even 24 h incubation of NPs. So long incubation time can cause several secondary effects on cell functionality such as viability, alternation of the cytoskeleton network and actin-mediated signaling, cell proliferation and differentiation, induction of ROS (reactive oxygen species), and finally apoptosis and/or necrosis [49, 50]. The internalization of metal nanoparticles is cell-dependent. For instance, the 14 h incubation of endothelial cells (EA.hy 926) with 40 nm AuNPs of

10 pM concentration resulted in massive apoptosis with viability of ca. 40% only [22, 23] whereas 3T3 fibroblast cells were not affected by even 100 pM AuNPs (d = 20–50 nm) after the 24 h incubation (viability of ca. 95%) [37].

When the cells are exposed to nanoparticles, a time and concentration-dependent uptake is observed. Uptake level increases gradually up to 2–6 h [5, 14, 23, 29, 49]. Transmission electron microscopy (TEM) often shows that the metal particles are located in intracellular vesicular organelles and the uptake mechanism relies on cellular chemical-mechanical energy (active mechanism-endocytosis) and/or on a random diffusion process (passive mechanism). In the endocytosis, NPs are enclosed in endosomes which undergo maturation to the latest stage: lysosomes. A simply schematic of this process is shown in Fig. 11.1.

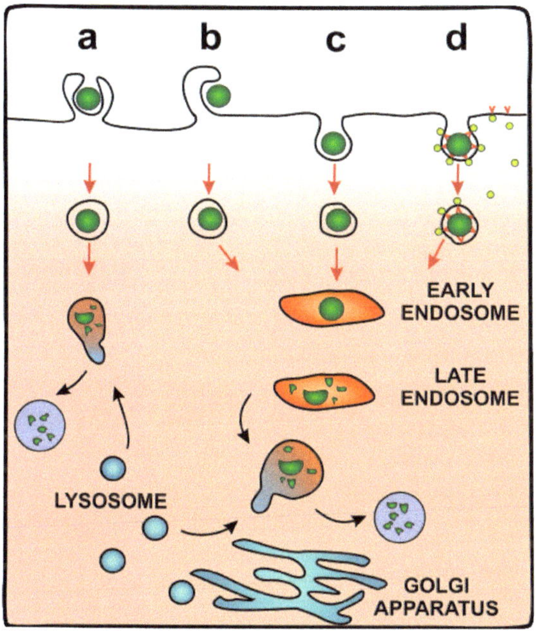

Figure 11.1 A schematic of different mechanisms of endocytosis: phagocytosis (a), macropinocytosis (b), non-specific absorptive endocytosis (c) and receptor-mediated endocytosis (RME) also called clathrin-mediated endocytosis. In (a), after ingesting, particles are internalized by lysosomes, while in (b), (c), (d), vesicles are converted from early endosomes to mature and late endosomes. Next, endosomes fuse to lyposomes located in Golgi apparatus. Based on [16, 38, 53].

Since AuNPs are usually prepared by using the reduction of chloroauric acid with citrate ions, the latter are adsorbed on the gold introducing negative surface charges. These nanoparticles are often transported through the cell membrane via the clathrin-mediated endocytosis. This process is rapid and takes several minutes [13, 39]. It is well-known that nanoparticles with a diameter less than 40 nm cannot be transported through the cell membrane via the endocytosis process because the docking the NPs does not produce enough energy to wrap the NPs by the cell membrane [23, 49]. Gold nanoparticles with a diameter larger than 70 nm are rarely internalized by cells, and they directly move to the apical surfaces of cells and are found bounded to the outer surface of the cell membrane [57]. Mechanisms of the uptake can be determined by tracking metal particles conjugated to a fluorescent label by fluorescence microscopy under optimal and energy depletion conditions [23], by plasmonic scattering [57], confocal Raman spectroscopy [5] or dark-filed microscopy [2]. Jaworska and coworkers have showed in their study that AuNPs with the diameter of ca. 40 nm and conjugated with rhodamine 6G (R6G) are taken by endothelial cell line (EA. hy 926) even at 4°C at the level only slightly lower than for the environment optimal for cell growing (37°C) [23]. It means that both active and passive mechanisms contribute into the uptake of SERS sensors. Nanoparticles are unevenly distributed in the cytosol as aggregates without entering to nucleus. Similarly, Wang et al. have showed by using plasmonic scattering imaging that the percentage of 45 nm AuNPs modified with single-stranded DNA inside two cancer cell lines (CL1-0 and HeLa) is ca. 60%. This amount of nanoparticles was accumulated within 1.5 h [57]. Taking into account several studies on uptake of metal nanoparticles, a particle diameter in the range of 40–60 nm is optimal for designing intracellular SERS nanoprobes since this size is compatible with membrane rigidity and ligand-receptor binding energy.

Chemical modification of nanoparticle surface by specific signal peptide sequences can deliver nanoparticles to a target position and then chemical information from SERS spectra is extracted to monitor biochemical events of interest. One must pay attention on the interpretation of SERS bands because of dynamic and fast motion of NPs within the cellular space and the small volume of interaction between particles and surrounding biomolecules. This temporal variation in SERS spectrum is widely observed on a large time scale (within 24 h) when AuNPs are conjugated with organic dyes as Raman reporters. The dynamic exchange between a Raman reporter and biomolecules near the metal surface are observed when the uptake of NPs is considered in investigations and therefore SERS spectra represent a large variability of Raman features typical for lipids, proteins, aromatic amino acids and nucleotides, see Fig. 11.2 [22, 23, 36]. Bando et al. have showed in their studies on the endocytic transportation of 80 nm AuNPs to HeLa cells how dual-focus dark-field imaging combined with SERS imaging technique can be employed to track motional trend of nanoparticles within seconds in cellular cytosol [2]. The motion of the single AuNP results from random and linear motion supported by motor proteins and/or simply from fluctuations in intracellular diffusion. The calculated speed of the nanoparticle changes during the acquisition reaching maximum of 500 nm s^{-1} typical for speeds of motor transport proteins such as kinesin or dynine. The simultaneously recorded SERS spectra and dark-field images showed that some peaks appear and disappear within seconds and this variation is weakly correlated with the particle trajectory and/or cellular transport [2].

An important aim in cellular SERS biosensing is designing nanosensors which are organelle-specific, in particular to recognize nucleus and mitochondria. A basic part of a nanoprobe is organelle localization signal. Thus, the transportation of AuNPs to the nucleus is supported by conjugation of NLS peptide (e.g., GGVKRKKKPGGE) to the gold and then the localization of nanosensors inside nucleus is monitored by an increase of SERS bands assigned to DNA and RNA [5, 19, 27]. However, some reports have shown that the nucleus-targeting strongly depends on cell line and the type of peptide [27, 47]. Moreover, the localizing peptide can be also a source of background SERS signal, which may disturb cellular signal [27]. Localization signal molecules with sequence of MLALLGWWWFFSRKKC or delocalized lipophilic cations (MLS) are often used for mitochondrial targeting, while cytoplasm can be exclusively filled out with AuNPs when their surface is decorated with RGD peptide

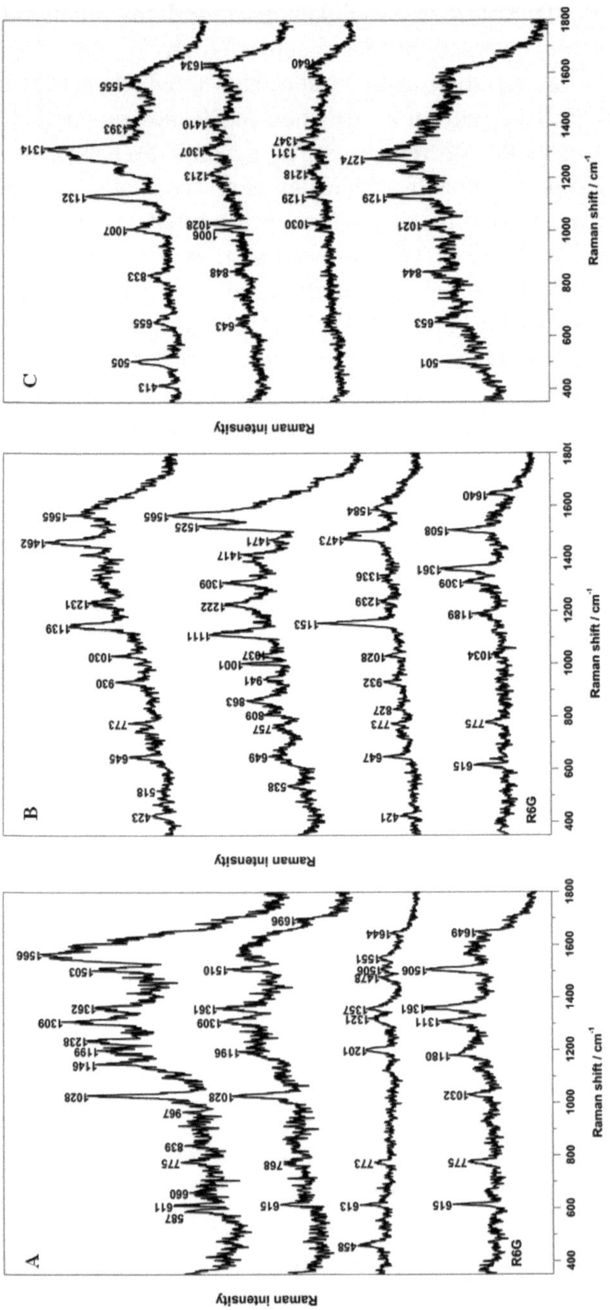

Figure 11.2 Variation of SERS spectra collected by internalization of R6G-AuNPs nanoprobes after 2 h (A), 6 h (B) and 16 h (C) uptake; adapted from [22]; IOS Press DOI: 10.3233/BSI-130047.

(e.g., RGDRGDRGDRGDPGC) [27, 47]. All together can be employed for targeting and multiplexed SERS-based cell imaging when each nanosensor is additionally immobilized by a specific Raman reporter, e.g., methyl blue, 4,4′-dipyridyl and 4,4′-azobis, see Fig. 11.3. The uptake of mitochondrial-targeting nanosensors is the fastest comparing to the uptake of the corresponding nanoprobes for cytoplasm and nucleus. On the other hand, the uptake of SERS probes by nucleus induces significant cytotoxic effects [27].

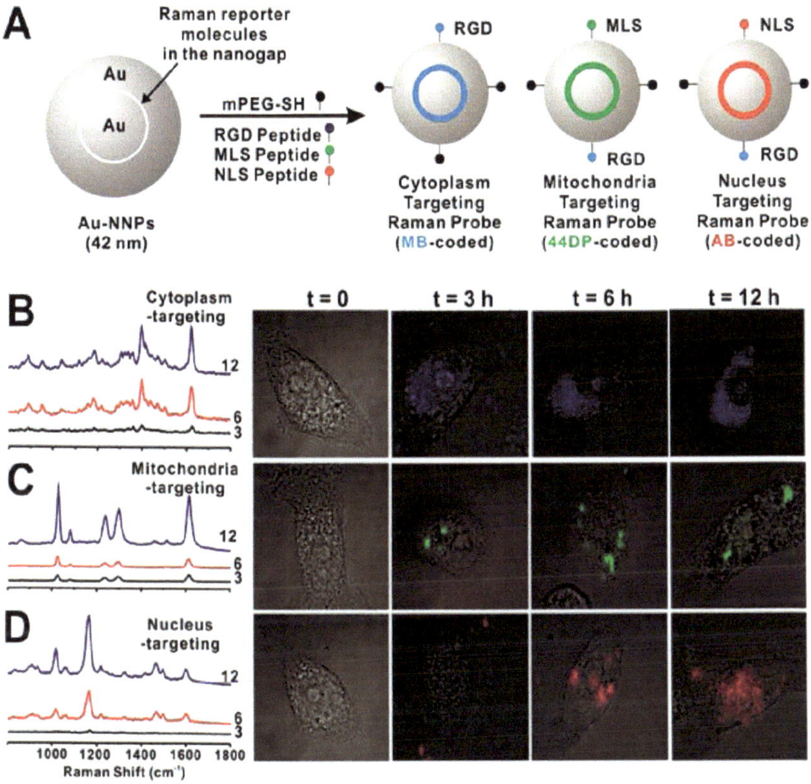

Figure 11.3 A. Scheme for the PEGylation and peptide modifications on the Raman dye-coded Au-NNPs, **B-D.** Time-dependent live cell Raman images after incubating with subcellular targeting NNPs and representative Raman spectra for cells incubated with: MB-coded NNPs for cytoplasm targeting (B), 44DP-coded NNPs for mitochondria targeting (C), AB-coded NNPs for nucleus targeting (D). Adapted from [27].

11.2 Designing of Cellular SERS Nanoprobes

An exciting aspect in the application of AuNPs for biosensing is designing multifunctional SERS nanosensors, which probe cellular environment to deliver chemical information and understand specific alternations in cellular machinery, including abnormalities in function of cells.

11.2.1 Simple Functionalization of the Metal Surface

Among several studies, the determination of intracellular pH, redox potential and quantification of exogenous chemicals have found a great interest in the development of SERS-based methodologies.

SERS-based monitoring of intracellular pH (pH_i) changes has been widely reported as a powerful tool for observation of cellular functions, endosomal trafficking altered due to action of environmental stimuli. Mostly, AuNPs are decorated with pH responsive Raman reporters such as 4-mercaptobenzoic acid (MBA) [21, 30, 42, 48, 51, 56], 4-mercaptopyridine [41] and aminothiophenols [4, 59, 64].

Intensities of SERS signals from the de/protonated carboxylic group of MBA at 1720 and 1400 cm^{-1}, respectively, correspond to changes in pH from 4 to 9 [44]. The determination of a Henderson-Hasselbach plot shows a linear response of ratio of these bands versus pH and helps constructing maps of pH_i distribution when Raman imaging technique is employed, *c.f.* Fig. 11.4. Sensitivity of the MBA-SERS nanoprobe to pH changes has been proven by stressing endothelial cells with acidic and alkaline environment. In addition, this sensor is also highly sensitive to detect changes in pH_i due to inflammation of endothelial cells induced by tumor necrosis factor α [44].

The ability of 4-mercaptobenzoic acid to response to subtle variation in pH (±0.1–0.2) has employed to design in situ SERS-active fiber-optic nanoprobes by deposition of a 6 nm silver layer on a nanoprobe and then functionalization of the silver by MBA. These nanoprobes are inserted into the cells using micromanipulators and do not induce apoptosis or strong lysosomal response in cells [48]. A more complex MBA-based nanoprobe was proposed by Pallaoro and coworkers. They have developed two-AgNPs clusters linked by a bifunctional diamine molecule and decorated with MBA. Additionally, the cluster is covered by PVP and dye labeled streptavidin with bovine serum albumin to quench aggregation of metal nanoparticles. Uptake of such large SERS-active clusters is supported by a cationic lipid designed for trafficking nucleic acid nanoparticles into cells. Tracking fluorescence and SERS probes indicated the internalization via the clathrin-mediated endocytosis [42]. SERS and fluorescence tags have been also used in the fabrication of a pH nanotracker aimed to monitor the dynamic change of local pH in living cancer HeLa cells during photothermal therapy (PTT) [33]. Gold nanorods are decorated with MBA and a cysteine residue-contained peptide conjugated with a fluorescent derivative of rhodamine. After 3-hour internalization of the sensor by nigericin-treated HeLa cells, the cells were exposed to white light up to 20 min. The distribution maps of pH_i indicated an increase of lysosomal pH after an increased death of illuminated cancer cells. In turn, Jung et al. have engineered "smart" AuNPs which were then used as a "turn-on" theragnostic agent suitable for simultaneous SERS diagnosis and photothermal therapy [26]. The design of nanoprobes relies on the conjugation of 10 nm spherical AuNPs with smart ligands containing hydrolysis-susceptible citraconic amide units and with MBA. Under mildly acidic conditions (pH of 5.5) "recognized" by MBA, the nanoprobes undergo rapid aggregation shifting the absorption to the NIR region. And this shift is next utilized for selective photothermal therapy. The proof of concept has been shown on B16 F10 cells mouse melanoma cell line in which high mortality was induced by a 785 nm laser excitation with a laser power ≥ 10.5 mW/cm^2 [26]. In turn, targeting cancer cells with pH-controllable drug carries has been reported in [10]. A three-layer nanoprobe consists of a AgNP as a SERS-active substrate functionalized with MBA. Then this layer is coated by a shell of mesaporous silica for doxorubicin (dox) loading and release and a pH-detectable layer of chitosan/polymethacryclic acid. The targeting ability of the drug carrier is improved by conjugation of the nanosensor with transferrin recognizing the Tf receptor on HeLa cells. The release ability of the drug carries was modeled in pH of 7.4 (specific for extracellular tumor environment), 6.8 (specific for normal tissues), and

5.5 (specific for endosomes). After 24 h, a release amount of dox reached 15 and 60% at pH 7.4 and 5.5, respectively, while maximal concentration of the drug was observed after 5 h [10]. A similar approach, based on a Au-dox-PEG system, has been developed by Lee and coworkers [31].

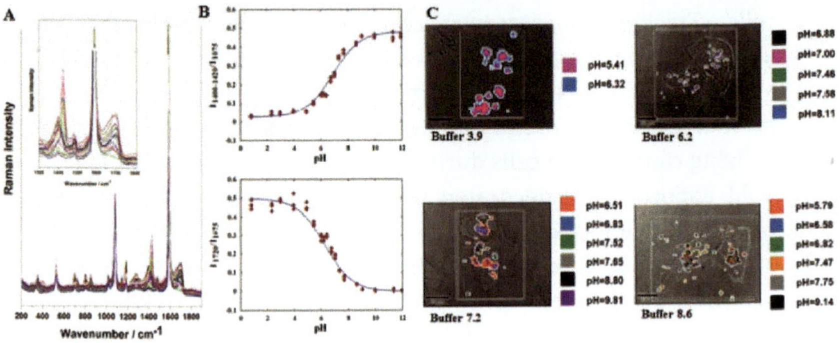

Figure 11.4 **A.** SERS spectra of MBA at different pH values. The inset shows the region of 1300–1800 cm^{-1} associated with pH-sensitive bands; **B.** Henderson-Hasselbalch plots for bands ratios: $I_{1400-1420}/I_{1070}$ (up) and I_{1720}/I_{1070} (down); **C.** Cluster maps (k-means cluster analysis) of SERS-based pH_i imaging for EA.hy926 cells incubated with MBA-AuNPs in buffers with pH 3.9, 6.2, 7.2, and 8.6. The colors decoded in the clusters maps correspond to pH values labeled by squares. Adapted from [21].

Talley and coworkers have pointed out that unwanted variation in pH values inside cells, when simple AuNPs-MBA probe is internalized by cells, can results from the nature of individual particles. Only 9% of nanosensors produced a significant SERS signal and only these AuNPs which were involved in the formation of clusters [30]. Another issue in the determination of pH_i is interaction between nanosensors and macromolecules in biological systems. To limit this interference as well as to avoid uncontrolled aggregation, Wang et al. have proposed to encapsulate the AgNPs-MBA sensor in a 30 nm thick silica layer with small pores [56]. Compared with this approach, Wei and others have showed that the use of protective HS-PEG layer provides a better coating of SERS pH probe because PEG has a long hydrophobic chain that does not inhibit proton diffusion to the SERS hot spot [60]. In other report, AuNP multimers covered by MBA were first produced in 50: 50 water : ethanol co-solvent. SERS spectra of the internalized pH probes by Pc-3 human prostate cancer cells showed no interference from either cells or the cultured medium [59]. The authors have also pointed out the necessity to determine a pH calibration curve in cell culture media since the shape of the curve and pH sensitive range differ from calibration parameters calculated for phosphate buffer. Another solution for uncontrolled aggregation of NPs and separation of sensors from biomolecules has reported by Cong and coworkers who fabricated sub-1-nm gap-separated linear chain AuNPs silica nanotube peapod (SNTP) in which MBA served as a pH sensor [6]. The encapsulation of pH-sensitive molecules in this way can prevent interactions with medium and cellular proteins and interference from monocations present in physiological fluids. This design also improves reliability and reproducibility of pH measurements.

Despite a wide application of MBA as a pH sensor, aminothiophenol (ATP) has similar sensitivity to intracellular pH. Wand and Cui have reported that this Raman reporter is a perfect nanoprobe to investigate live cells and is capable to measure pH over the range of 3.0–8.0 with high sensitivity in a more acidic region compared to MBA [4, 59, 64]. SERS bands of ATP at 1142, 1390 and 1432 cm^{-1} are strongly dependent on pH value. Functionalization of silver spherical nanoparticles with 2-ATP yield to exchange of the sensor molecules with cellular macromolecules after the 20 h incubation [59] that is not observed when p-ATP is conjugated with gold nanorods treated with HCl [64]. The authors have shown that the simple treatment of gold nanorods with HCl improves biocompatibility of these metal probes in HeLa cells [64]. The range of detected pH values can be extended when p-ATP decorates multi-walled carbon nanotubes (MWCNTs) coated with silver-coated gold NPs. This pH sensor retains its pH sensing ability after 3-hour incubation into HeLa cells and is found in the whole intracellular space detecting [4].

Next pH_i, redox potential is an equally important feature of cellular machinery regulating cell cycle, differentiation, inflammation and signaling. Campbell et al. have proposed compounds containing the quinone moiety (HQ and NQ) as redox sensitive SERS probes [1]. SERS spectra of the HQ and NQ reporters respond to changes in potential over the range of −12 ÷ +370 and −394 ÷ −194 mV, respectively. However, only NQ (1,8-diaza-4,5-dithia-1,8-di(2-chloro-[1,4]-naphthoquinone-3-yl)octane) is able to participate in electron transfer between typical cellular redox couples like GSH/GSSG, Cys/CySS, and $NADP^+$/NADPH. The authors have shown that uptake of the sensor is independent of energy, cholesterol and clathrin when cells are incubated in serum-free DMEM growth media. Nanoprobes are accumulated in this way in cytoplasm, not in vesicles [1]. Since redox potential of −315 ÷ −280 mV is found in healthy cells, the combination of both HQ and NQ reporters has been used to cover a wide range of potential and to study oxidative stress. Oxidative changes were measurable within minutes showing a great potential of this approach to monitor the dynamics of redox potential in cells under stress conditions. A similar experiment with the use of a dihydroanthracene derivative as a probe has been proposed to determine redox potential in hypoxic cells [25]. Whilst the use of this reporter in combination with MBA has been employed in multiplexing responsive of redox and pH in whole basis of live endothelial cells [20]. pH values slightly varied over the range of 6.9–7.1 in healthy cells while redox variation was more substantial (from −241 to −332 mV) and required correction based on determined pH values. Variation in redox potential may result from different stages of cell life-cycle.

Redox potential is a main biomarker for the detection of biochemical events associated with generation of ROS and accompanying oxidative stress and inflammatory state. All of them can be monitored by SERS nanotechnology when the gold surface is decorated with molecules sensitive to production of specific ROS molecules like hypochlorite (ClO^-), gluthatione (GSH) or hydrogen peroxide (H_2O_2) [45, 58]. A working principle of a nanosensor detecting ClO^- and GSH has been developed on the base on a single Raman reporter–4-mercaptophenol (4-MP) adsorbed on gold nanoflowers [58]. 4-MP specifically detects hypochlorite ion due to conversion of the hydroxybenzene group into the quinonyl group. As expected this reaction strongly affects SERS features of the Raman reporter. After subsequent addition of GSH, the reaction is reversed to 4-MP and intensity ratios of marker bands change linearly with concentrations of both ROS molecules; I_{635}/I_{822} for ClO^- and I_{1077}/I_{822} for GSH. Concentrations of ClO^- and GSH in macrophages determined by SERS agreed well with a reference fluorescent technique [58]. In turn, the detection of hydrogen peroxide is based on the oxidation reaction between H_2O_2 and 4-carboxyphenylboronate modifying gold nanospheres [45]. Due to this reaction the phenol is liberated and specific changes in SERS spectra of the nanoprobes are observed. The nanosensor exhibits a higher selectivity to H_2O_2 over other ROS, cellular thiols and other biological species with LOD of 80 nM. The proposed approach also estimates the rate production of hydrogen peroxide. For live tumor HeLa and normal liver cells, the rate was 1.08 and 0.85 nmol per 10^4 cells per hour, respectively [45].

Inflammation and the generation of ROS are also closely related with function of endothelial cells and signaling by NO messengers. SERS detection of nitric oxide was created by designing a novel probe recognizing specifically NO (3,4-diaminobenzene-thiol, DABT) [63]. DABT generates benzotriazole due to reaction with nitric oxide. A strong enhancement factor of ca. 8×10^6 has been achieved by immobilizing DABT onto trisoctahedral gold nanocrystals. Intensity ratio of SERS signals at 698 and 974 cm^{-1} shows a good linearity with the NO concentration in the range of 54–918 nM and no interference with other reactive species and metal ions is observed [63].

11.2.2 Antibody- and Peptide-Functionalized SERS Tags

Despite many obvious advantages, using colloidal nanoparticles for the cell analysis shows crucial obstacles. Aggregation, irreversible uptake and uncontrollable localization of nanoparticles inside the cell and variation in SERS signal have still remained unsolved [28, 61]. One of the methods to overcome these limitations is functionalization of the metal nanoparticle with target specific molecules [55]. Different SERS tags, such as peptides, antibodies or aptamers, have been used for this purpose. This approach provides both identification and localization information about target molecules. The localization of a target molecule is indirect because SERS signal of the Raman reporter is recorded, not from an analyte [9, 43].

Antibody is one of the widely used SERS tag. To stabilize this nanosensors, polymer molecules (spacers) are introduced directly into a Raman reporter molecule, and they additionally conjugate nanoparticles with antibodies. For example monoethylene glycol (MEG) with the terminal OH groups and triethylene glycol (TEG) with the terminal COOH groups bind antibody with the SERS probe [24]. The nanolabel prepared in this way can recognize an unique antigen or receptor present in cells. This technique is referred as an immunoSERS assay. A schematic of the design of an antibody-based SERS tag is shown in Fig. 11.5. It should be emphasized here that immunoSERS technology has several advantages over other immunochemical techniques. Conventional immunofluorescence staining, which is the most common technique for cell investigation, is limited by several factors such as autofluorescence of a biosample, photobleaching of fluorescent labels, or overlapping of absorption/emission bands in the multi-target analysis. ImmunoSERS overcomes these limitations since sharp and easily distinguishable SERS bands are highly photostable and offer multiplexed quantification [43, 62]. Antibody-labeled SERS tags have been successfully used for the detection of proteins overexpressed on membrane of cancer cells in oral [8], lung [3] and breast tissues [32]. Lee at al. have developed a method to identify growth factor receptors on breast cancer cell membranes by multiplexed SERS imaging [32]. For this purpose, expression of three receptors (epidermal growth factor EGF, insulin-like growth factor-1 IGF-1 and ErbB2) in three cell lines of human breast cancer has been determined. SERS nanotags are fabricated by decoration of hollow gold nanosphere (HGNs) with three Raman reporters and conjugation to antibodies against each of growth factor receptors (see Fig. 11.6). Additionally, protein overexpression has been quantified from pixels showing the presence of Raman signal. It is worthy to stress here that detection and quantification of cancer cells are currently performed by two techniques separately, i.e., fluorescence cytometry and densitometry in Western blots, respectively. Other reports have confirmed potential of multiplex SERS imaging for triple detection of cancerous cells and also demonstrated the capability of immunoSERS as flow cytometry technique [35].

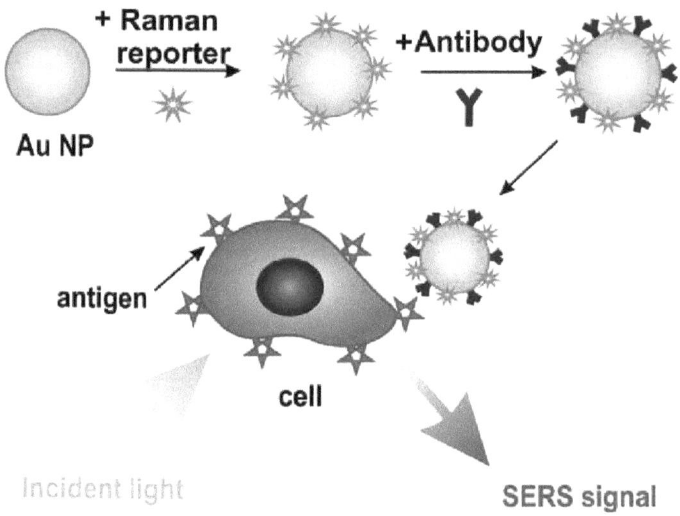

Figure 11.5 Schematic illustration of fabrication of SERS probes for targeting antigens or receptors. Based on [3, 44].

SERS immunoassay has been employed as identification tag and drug carrier [12, 43]. Parchur et al. have designed hybrid Au nanorods functionalized with Prussian blue (PB) that is the approved clinically anticancer drug. AuNPs are additionally attached to anti-HER2 antibody, which selectively recognizes HER2/neu glycoproteins amplified and overexpressed on the membrane of invasive breast cancer cell. PB plays a double role of the Raman reporter and the agent of PTT therapy since it efficiently transforms irradiation of laser into heat in cancer treatment [12, 43].

SERS nanoprobes can be also indirectly labeled by antibody as shown by Nithipatikom and coworkers [40]. They have conjugated crystal violet (CV) and dimethylaminoazobenzene (DAB) to AgNPs functionalized with biotin. The latter identifies biotinylated secondary antibodies and consequently target proteins, c.f. Fig. 11.7. This SERS probe has been successfully employed to localize of 12-lipoxygenase (12-LO) and cyclooxygenase-1 (COX-1) in single bovine coronary artery endothelial cells. 12-LO was mainly detected by SERS in the cytosolic region of cells but not in nucleus or on cell membrane while COX-1 was found in both endoplasmic reticulum and cytoplasm as expected [40].

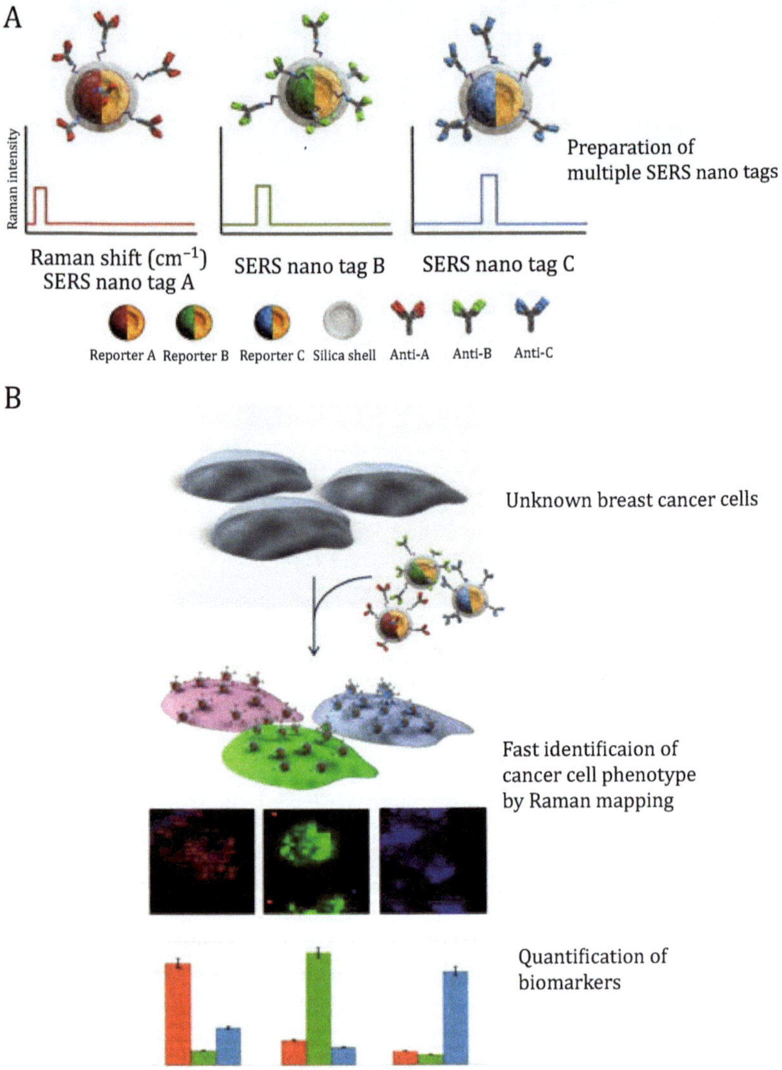

Figure 11.6 Schematic illustration of SERS-based cell imaging for rapid detection and quantification of growth factor receptors on breast cancer cell membranes. (A) Preparation of Si-encapsulated SERS nanotags bearing three different functional structures. Three different Raman reporter-antibody sets were conjugated to SEHGNs to target specific breast cancer phenotypic markers. (B) Rapid detection and quantitative evaluation of phenotypic markers on cell surface membranes using the SERS imaging technique. Adapted from [32].

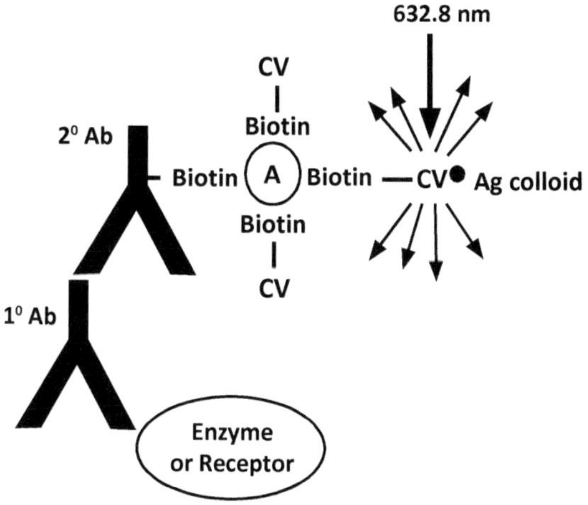

Figure 11.7 A schematic diagram depicting the immunocomplexation of enzyme or receptor with primary antibody, biotin-labeled secondary antibody, avidin (A), and Raman-labeled biotin (with crystal violet, CV) for detection of enzymes or receptors in single cells. Adapted from [40].

As mentioned in the previous section, peptides have been widely used as SERS tags. This type of nanoprobes has been employed to distinguish differentiation of lung resident mesenchymal stem cells (LR-MSCs) from bone marrow (BM)-derived MSCs The former is a crucial regulator of tissue repair and regeneration, angiogenesis, inflammation or tumor formation [11]. LR-MSCs differentiate into alveolar epithelial type II (ATII) cells, thus they represent an endogenous population of lung progenitor cells. However, recent studies have indicated that in altered pulmonary micro-environment these cells can inhibit their capacity of regeneration and are transformed into myofibroblasts responsible for several pulmonary diseases [11]. Because of morphological similarities, the distinction between the undifferentiated and differentiated LR-MSCs has still remained a great challenge. Shi and coworkers have shown that this issue can be solved by using transactivator of transcription (TAT) peptide conjugated to Au nanostars as cellular SERS probes. An evaluation of SERS spectra by principal component analysis (PCA) enables to identify differentiated subtypes of cells and confirms a pathway of differentiation of ATII cells into alveolar epithelial cells and fibroblasts [3].

Protein-labeled SERS probes are as effective for recognition of cancer cells as antibody-conjugated tags as showed by Ramya et al. [46]. A series of tetraphenylethylene (TPE)-appended fluorogens has been utilized as Raman reporters to produce a multiplexing signal pattern. One of the TPE analogues with indoline (TPE-In) as an electrons donor can be used for the identification of prostate specific antigen (PSA) expression when it is conjugated with specific peptide sequence (Cys-Ser-Lys-Leu-Gln-OH). This approach offers detection of PSA up to 0.5 ng, which is a great achievement in prostate cancer diagnosis [46].

11.2.3 Other Solutions for Cellular SERS-Sensing

Using solid Ag and Au nanostructures, which can be prepared by dispersing a layer of nanoparticles on nanotemplate, is not a new idea in the development of diagnostic platforms in clinics. This direct method of the detection provides molecular information after interaction of a SERS substrate with living cells. A compact layer of nanoparticles exhibits a high SERS enhancement and reduces the optical transparency of the substrate. For example, Huang and coworkers have proposed the 3-aminopropyltrimethoxysilane (APTMS)-functionalized surfaces of indium tin oxide (ITO) and immobilized by Au nanoparticles as a SERS substrate to study live cells [18], while gold-coated anodic porous alumina (APA) templates have been developed by Toccafondi et al. for similar purpose [54]. In turn, the deposition of the gold on Si plate has been used to record signal of carotenoids and lipids from algal cells (*Chlorella vulgaris*) [7]. Similarly to gold colloidal nanoparticles, solid nanostructures show high biocompatibility to living organisms.

To improve reproducibility and specificity of SERS substrates sandwich immunoassays have been introduced in the field of SERS nanotechnology [15, 52]. First, this approach has been employed for counting cells of *Escherichia coli* (*E. coli*) by combining SERS spectroscopy with immunomagnetic separation (IMS) [15]. Biotin-labeled antibodies against *E. coli* are conjugated with avidin-coated magnetic nanoparticles and then used to separate bacteria cells in water samples. Next, this complex interacts with AuNPs functionalized with 5,5-dithiobis-2-nitrobenzoic acid (DTNB) as a molecular recognizer. The calibration curve is established by calculating intensity of the DTNB Raman band at 1325 cm^{-1} against a number of cells. The effectiveness of the SERS-based sandwich assay has been proven by comparing with a classical counting method [15]. A multiplex sandwich SERS immunoassay has been also developed by Tang and coworkers to detect liver cancer biomarkers at a subpicomolar level [52].

Probing the intracellular environment gives direct information about biological processes appeared due to various pathologies. However, in many of them, signaling pathways are important for fundamental understanding of disease and its diagnosis at an early stage. Several methodologies rely on the development of nanopipettes decorated with gold nanostructures and alternatively with Raman reporters sensitive to a biomarker of a pathological process. For instance, Lussier and coworkers have recently proposed dynamic-SERS optophysiology to monitor secretion events. A gold-decorated nanopipette detects simultaneously a few metabolites like pyruvate, lactate, ATP and urea when the nanopipette is present in the close distance to cells [34]. Due to action of lyses on kidney epithelial cells, a semiquantitative analysis has indicted an increase in metabolite concentration and dynamic changes in relative concentrations of lactate to pyruvate. A specific design of nanopipettes has been constructed by Hanif et al. who prepared the nanoprobe by covering a glass capillary tip with a gold layer functionalized by 4-mercaptobenzonitrile (MBN) [17]. The cyanide group of MBN easily binds Fe^{3+} ions and oxy-Hb and this binding is manifested by the appearance of bands assigned to the stretching modes of the C≡N and S-H groups. LOD was 0.1 pM and 0.08 μM for Fe^{3+} ions in both free and protein-bound states, respectively. The insertion of the nanoprobe with the penetration depth of 2–3 μm does not cause damage of cells [17].

11.3 Conclusion

In this chapter, we aimed to review the single-cells analysis with surface-enhanced Raman spectroscopy by stressing the behavior of gold nanoparticles in the intracellular space and an enormous potential of decorating the metal surface by selectivity elements which exhibit specific response to cellular machinery. The current literature review indicates new trends in SERS nanotechnology toward designing multifunctional nanoprobes monitoring specific biological processes and serving as drug carries. Although the state of the art of SERS-based nanotechnology has been undoubtedly proven in several reports, there are still challenges which must be resolved before standardization of this methodology for clinical applications. Current efforts aim to improve reproducibility of SERS activity of metal NPs and designed nanoprobes and their quantitation capability. Even though we can synthesize nanoparticles with various morphology (from nanospheres, nanostars to nanoflowers or nanocubes, etc.) and tunable sizes (from 10 to 250 nm), still the synthetic roots strongly depend on even small variabilities in reaction parameters yielding to different products than desirable. Thus, designing automated and continuous-flow mass production of AuNPs is required to develop and translate SERS-based in vitro nanotechnologies into standard use in research and into clinics. This will undoubtedly provide multiplexed and multifunctional targeted imaging of single cells with high lateral resolution up to 300 nm.

Acknowledgments

The authors thank the Polish National Centre of Science for financial support (Grant Opus 11 No. 2016/21/B/ST4/02151).

References

1. Auchinvole, C. A. R., Richardson, P., McGuinnes, C., Mallikarjun, V., Donaldson, K., McNab, H., Campbell, C. J., Auchinvole, D. O., Document, E., Statement, P. R., and Society, A. C. (2012). Monitoring intracellular redox potential changes using SERS nanosensors, *ACS Nano*, 6, 888–896.
2. Bando, K., Smith, N. I., Ando, J., Fujita, K., and Kawata, S. (2015). Analysis of dynamic SERS spectra measured with a nanoparticle during intracellular transportation in 3D, *J. Opt.*, 17, 114023.
3. Cao, X., Shi, C., Lu, W., Zhao, H., Wang, M., Zhang, M., Chen, X., Dong, J., Han, X., and Qian, W. (2016). Surface enhanced Raman scattering probes based on antibody conjugated Au nanostars for distinguishing

lung cancer cells from normal cells, *J. Nanosci. Nanotechnol.*, 16, 12161–12171.

4. Chen, P., Wang, Z., Zong, S., Chen, H., Zhu, D., Zhong, Y., and Cui, Y. (2014). A wide range optical pH sensor for living cells using Au@Ag nanoparticles functionalized carbon nanotubes based on SERS signals, *Anal. Bioanal. Chem.*, 406, 6337–6346.
5. Chen, Y., Bai, X., Su, L., Du, Z., Shen, A., Materny, A., and Hu, J. (2016). Combined labelled and label-free SERS probes for triplex three-dimensional cellular imaging, *Sci. Rep.*, 1–12, doi:10.1038/srep19173.
6. Cong, V. T., Ganbold, E.-O., Saha, J. K., Jang, J., Min, J., Choo, J., Kim, S., Song, N. W., Son, S. J., Lee, S. B., and Joo, S.-W. (2014). Gold nanoparticle silica nanopeapods, *J. Am. Chem. Soc.*, 136, 3833–3841.
7. Deng, Y. L., and Juang, Y. J. (2014). Black silicon SERS substrate: Effect of surface morphology on SERS detection and application of single algal cell analysis, *Biosens. Bioelectron.*, 53, 37–42.
8. El-Sayed, I. H., Huang, X., and El-Sayed, M. A. (2005). Surface plasmon resonance scattering and absorption of anti-EGFR antibody conjugated gold nanoparticles in cancer diagnostics: Applications in oral cancer, *Nano Lett.*, 5, 829–834.
9. Fabris, L. (2015). Gold-based SERS tags for biomedical imaging, *J. Opt.*, 17, 114002.
10. Fang, W., Wang, Z., Zong, S., Chen, H., Zhu, D., Zhong, Y., and Cui, Y. (2014). pH-controllable drug carrier with SERS activity for targeting cancer cells, *Biosens. Bioelectron.*, 57, 10–15.
11. Foronjy, R. F., and Majka, S. M. (2012). The potential for resident lung mesenchymal stem cells to promote functional tissue regeneration: Understanding microenvironmental cues, *Cells*, 1, 874–885.
12. Fu, G., Liu, W., Feng, S., and Yue, X. (2012). Prussian blue nanoparticles operate as a new generation of photothermal ablation agents for cancer therapy, *Chem. Commun.*, 48, 11567–11569.
13. Gu, Y., Di, X., Sun, W., Wang, G., and Fang, N. (2012). Three-dimensional super-localization and tracking of single gold nanoparticles in cells, *Anal. Chem.*, 84, 4111–4117.
14. Guo, X., Guo, Z., Jin, Y., Liu, Z., Zhang, W., and Huang, D. (2012). Silver–gold core-shell nanoparticles containing methylene blue as SERS labels for probing and imaging of live cells, *Microchim. Acta*, 178, 229–236.

15. Basaran-Akgul, N., Temur, E., Tamer, U., and Boyaci, I. H. (2011). SERS-based sandwich immunoassay using antibody coated magnetic nanoparticles for *Escherichia coli* enumeration., *Analyst*, 136, 740–748.

16. Haglund, K., and Dikic, I. (2012). The role of ubiquitylation in receptor endocytosis and endosomal sorting, *J. Cell Sci.*, 125, 265 lp-275.

17. Hanif, S., H. Liu, H., Chen, M., Muhammad, P., Zhou, Y., Cao, J., Ahmed, S. A., Xu, J., Xia, X., Chen, H., and Wang, K. (2017). Organic cyanide decorated SERS active nanopipettes for quantitative detection of hemeproteins and Fe^{3+} in single cells, *Anal. Chem.*, 89, 2522–2530.

18. Huang, J.-Y., Zong, C., Xu, L., Cui, J. Y., and Ren, B. (2011). Clean and modified substrates for direct detection of living cells by surface-enhanced Raman spectroscopy, *Chem. Commun. (Camb).*, 47, 5738–5740.

19. Huefner, A., Kuan, W., Barker, R., and Mahajan, S. (2013). Intracellular SERS nanoprobes for distinction of different neuronal cell types, *Nano Lett.*, 13, 2463–2470.

20. Jamieson, L. E., Jaworska, A., Jiang, J., Baranska, M., Harrison, D. J., and Campbell, C. J. (2015). Simultaneous intracellular redox potential and pH measurements in live cells using SERS nanosensors, *Analyst*, 140, 2330–2335.

21. Jaworska, A., Jamieson, L. E., Malek, K., Campbell, C. J., Choo, J., Chlopicki, S., and Baranska, M. (2014). SERS-based monitoring of the intracellular pH in endothelial cells: The influence of the extracellular environment and tumour necrosis factor-α, *Analyst*, 0, 1–9.

22. Jaworska, A., Małek, K., Chłopicki, S., Barańska, M., and Kachamakova-Trojanowska, N. (2013). The uptake of gold nanoparticles by endothelial cells studied by surface-enhanced Raman spectroscopy, *Biomed. Spectrosc. Imaging*, 2, 183–189.

23. Jaworska, A., Wojcik, T., Malek, K., Kwolek, U., Kepczynski, M., Ansary, A. A., Chlopicki, S., and Baranska, M. (2015). Rhodamine 6G conjugated to gold nanoparticles as labels for both SERS and fluorescence studies on live endothelial cells, *Mikrochim. Acta*, 182, 119–127.

24. Jehn, C., Kustner, B., Adam, P., Marx, A., Strobel, P., Schmuck, C., and Schlucker, S. (2009). Water soluble SERS labels comprising a SAM with dual spacers for controlled bioconjugation, *Phys. Chem. Chem. Phys.*, 11, 7499–7504.

25. Jiang, J., Auchinvole, C., Fisher, K., and Campbell, C. J. (2014). Quantitative measurement of redox potential in hypoxic cells using SERS nanosensors, *Nanoscale*, 6, 12104–12110.

26. Jung, S., Nam, J., Hwang, S., Park, J., Hur, J., Im, K., Park, N., and Kim, S. (2013). Theragnostic pH-sensitive gold nanoparticles for the selective surface enhanced Raman scattering and photothermal cancer therapy, *Anal. Chem.*, 85, 7674–7681.

27. Kang, J. So, P. T. C., Dasari, R. R., and Lim, D. K. (2015). High resolution live cell Raman imaging using subcellular organelle-targeting SERS-sensitive gold nanoparticles with highly narrow intra-nanogap, *Nano Lett.*, 15, 1766–1772.

28. Kneipp, J., Kneipp, H., McLaughlin, M., Brown, D., and Kneipp, K. (2006). In vivo molecular probing of cellular compartments with gold nanoparticles and nanoaggregates, *Nano Lett.*, 6, 2225–2231.

29. Kneipp, J., Kneipp, H., Rajadurai, A., Redmond, R. W., and Kneipp, K. (2009). Optical probing and imaging of live cells using SERS labels, *J. Raman Spectrosc.*, 40, 1–5.

30. Kneipp, J., Kneipp, H., Wittig, B., and Kneipp, K. (2010). Following the dynamics of pH in endosomes of live cells with SERS nanosensors, *J. Phys. Chem. C*, 114, 7421–7426.

31. Lee, K. Y. J., Wang, Y., and Nie, S. (2015). In vitro study of a pH-sensitive multifunctional doxorubicin-gold nanoparticle system: Therapeutic effect and surface enhanced Raman scattering, *RSC Adv.*, 5, 65651–65659.

32. Lee, S., Chon, H., Lee, J., Ko, J., Chung, B. H., Lim, D. W., and Choo, J. (2014). Rapid and sensitive phenotypic marker detection on breast cancer cells using surface-enhanced Raman scattering (SERS) imaging, *Biosens. Bioelectron.*, 51, 238–243.

33. Luo, R., Li, Y., Zhou, Q., Zheng, J., Ma, D., Tang, P., Yang, S., Qing, Z., and Yang, R. (2016). SERS monitoring the dynamics of local pH in lysosome of living cells during photothermal therapy, *Analyst*, 141, 1–4.

34. Lussier, Li, Y., Zhou, Q., Zheng, J., Ma, D., Tang, P., Yang, S., Qing, Z., and Yang, R. (2016). Dynamic-SERS optophysiology: A nanosensor for monitoring cell secretion events, *Nano Lett.*, 16, 3866–3871.

35. MacLaughlin, C. M., Mullaithilaga, N., Yang, G., Ip, S. Y., Wang, C., and Walker, G. C. (2013). Surface-enhanced Raman scattering dye-labeled Au nanoparticles for triplexed detection of leukemia and lymphoma cells and SERS flow cytometry, *Langmuir*, 29, 1908–1919.

36. Małek, K., Jaworska, A., Krala, P., Kachamakova-Trojanowska, N., and Barańska, M. (2013). Imaging of macrophages by surface enhanced Raman spectroscopy (SERS), *Biomed. Spectrosc. Imaging,* 2, 349–357.

37. Matschulat, A., Drescher, D., and Kneipp, J. (2010). Surface-enhanced Raman scattering hybrid nanoprobe multiplexing and imaging in biological systems, *ACS Nano*, 4, 3259–3269.

38. Mayor, S., Pagano, R. E. (2007). Pathways of clathrin-independent endocytosis, *Nat. Rev. Mol. Cell Biol.*, 8, 603–612.

39. Ng, C. T., Tang, F. M. A., Li, J. J., Ong, C., Yung, L. L. Y., and Bay, B. H. (2015). Clathrin-mediated endocytosis of gold nanoparticles in vitro, *Anat. Rec.*, 298, 418–427.

40. Nithipatikom, K., McCoy, M. J., Hawi, S. R., Nakamoto, K., Adar, F., and Campbell, W. B. (2003). Characterization and application of Raman labels for confocal Raman microspectroscopic detection of cellular proteins in single cells, *Anal. Biochem.*, 322, 198–207.

41. Nowak-Lovato, Wilson, B. S., and Rector, K. D. (2010). SERS nanosensors that report pH of endocytic compartments during FcεRI transit, *Anal. Bioanal. Chem.*, 398, 2019–2029.

42. Pallaoro, A., Braun, G. B., Reich, N. O., and Moskovits, M. (2010). Mapping local pH in live cells using encapsulated fluorescent SERS nanotags, *Small*, 6, 618–622.

43. Parchur, A. K., Li, Q., and Zhou, A. (2016). Near-infrared photothermal therapy of Prussian-blue-functionalized lanthanide-ion-doped inorganic/plasmonic multifunctional nanostructures for the selective targeting of HER2-expressing breast cancer cells, *Biomater. Sci.*, 4, 1781–1791.

44. Porter, M. D., Lipert, R. J., Siperko, L. M., Wang, G., and Narayanan, R. (2008). SERS as a bioassay platform: Fundamentals, design, and applications., *Chem. Soc. Rev.*, 37, 1001–1011.

45. Qu, L.-L., Liu, Y.-Y., He, S.-H., Chen, J.-Q., Liang, Y., and Li, H.-T. (2016). Highly selective and sensitive surface enhanced Raman scattering nanosensors for detection of hydrogen peroxide in living cells, *Biosens. Bioelectron.*, 77, 292–298.

46. Ramya, A. N., Joseph, M. M., Nair, J. B., Karunakaran, V., Narayanan, N., and Maiti, K. K. (2016). New insight of tetraphenylethylene-based Raman signatures for targeted SERS nanoprobe construction toward prostate cancer cell detection, *ACS Appl. Mater. Interfaces*, 8, 10220–10225.

47. Sakhrani, N. M., and Padh, H. (2013). Organelle targeting: Third level of drug targeting, *Drug Des. Devel. Ther.*, 7, 585–599.

48. Scaffidi, J. P., Gregas, M. K., Seewaldt, V., and Vo-Dinh, T. (2009). SERS-based plasmonic nanobiosensing in single living cells, *Anal. Bioanal. Chem.*, 393, 1135–1141.

49. Soenen, S. J., Manshian, B., Montenegro, J. M., Amin, F., Meermann, B., Thiron, T., Cornelissen, M., Vanhaecke, F., Doak, S., Parak, W. J., De Smedt, S., and Braeckmans, K. (2012). Cytotoxic effects of gold nanoparticles: A multiparametric study, *ACS Nano*, 6, 5767–5783.

50. Soenen, S. J., Rivera-Gil, P., Montenegro, J.-M., Parak, W. J., De Smedt, S. C., and Braeckmans, K. (2011). Cellular toxicity of inorganic nanoparticles: Common aspects and guidelines for improved nanotoxicity evaluation, *Nano Today*, 6, 446–465.

51. Talley, C. E., Jusinski, L., Hollars, C. W., Lane, S. M., and Huser, T. (2004). Intracellular pH sensors based on surface-enhanced Raman scattering, *Anal. Chem.*, 76, 7064–7068.

52. Tang, B., Wang, J., Hutchison, J. A., Ma, L., Zhang, N., Guo, H., Hu, Z., Li, M., and Zhao, Y. (2016). Ultrasensitive, multiplex Raman frequency shift immunoassay of liver cancer biomarkers in physiological media, *ACS Nano*, 10, 871–879.

53. Taylor, J., Huefner, A, Li, L., Wingfield, J., and Mahajan, S. (2016). Nanoparticles and intracellular applications of surface-enhanced Raman spectroscopy, *Analyst*, 141, 5037–5055.

54. Toccafondi, C., Thorat, S., La Rocca, R., Scarpellini, A., Salerno, M., Dante, S., and Das, G. (2014). Multifunctional substrates of thin porous alumina for cell biosensors, *J. Mater. Sci. Mater. Med.*, 25, 2411–2420.

55. Vitol, E. A., Orynbayeva, Z., Friedman, G., and Gogotsi, Y. (2012). Nanoprobes for intracellular and single cell surface-enhanced Raman spectroscopy (SERS), *J. Raman Spectrosc.*, 43, 817–827.

56. Wang, F., Widejko, R. G., Yang, Z., Nguyen, K. T., Chen, H., Fernando, L. P., Christensen, K. A., and Anker, J. N. (2012). Surface-enhanced Raman scattering detection of pH with silica-encapsulated 4-mercaptobenzoic acid-functionalized silver nanoparticles, *Anal. Chem.*, 84, 8013–8019.

57. Wang, S.-H., Lee, C.-W., Chiou, A., and Wei, P.-K. (2010). Size-dependent endocytosis of gold nanoparticles studied by three-dimensional mapping of plasmonic scattering images, *J. Nanobiotechnol.*, 8, 33.

58. Wang, W., Zhang, L., Li, L., and Tian, Y. (2016). A single nanoprobe for ratiometric imaging and biosensing of hypochlorite and glutathione in live cells using surface-enhanced Raman scattering, *Anal. Chem.*, 88, 9518–9523.

59. Wang, Z., Bonoiu, A., Samoc, M., Cui, Y., and Prasad, P. N. (2008). Biological pH sensing based on surface enhanced Raman scattering through a 2-aminothiophenol-silver probe, *Biosens. Bioelectron.*, 23, 886–891.

60. Wei, H., Willner, M. R., Marr, L. C., and Vikesland, P. J. (2016). Highly stable SERS pH nanoprobes produced by co-solvent controlled AuNP aggregation, *Analyst*, 141, 5159–5169.

61. Willets, K. A. (2009). Surface-enhanced Raman scattering (SERS) for probing internal cellular structure and dynamics, *Anal. Bioanal. Chem.*, 394, 85–94.

62. Woo, M., Lee, S., Kim, G., Baek, J., Noh, M. S., Kim, J. E., Park, S. J., Minai-tehrani, A., Park, S., Tai, Y., Kim, Y., Lee, Y., Jeong, D. H., Cho, M., and Seo, Y. T. (2009). Multiplex immunoassay using fluorescent-surface enhanced Raman spectroscopic dots for the detection of bronchioalveolar stem cells in murine lung multiplex immunoassay using fluorescent-surface enhanced Raman spectroscopic dots for the detection of bronch, *Anal. Chem.*, 81, 1008–1015.

63. Xu, Q., Liu, W., Li, L., Zhou, F., Zhou, J., and Tian, Y. (2017). Ratiometric SERS imaging and selective biosensing of nitric oxide in live cells based on trisoctahedral gold nanostructures, *Chem. Commun.*, 53, 1880–1883.

64. Zong, S., Wang, Z., Yang, J., and Cui, Y. (2011). Intracellular pH sensing using p-aminothiophenol functionalized gold nanorods with low cytotoxicity, *Anal. Chem.*, 83, 4178–4183.

Index

AB, *see* azobenzene
AB photoisomerization 44, 52–53
AB photoswitching 52
adenine 94–95, 97
AFM, *see* atomic force microscopy
AgNPs, *see* silver nanoparticles
amino acids 82, 89–91, 98, 219–220, 224–225, 227
anticancer drugs 162, 170, 172, 175–176, 179, 291
APV, *see* avipoxvirus
aryl diazonium salts 2, 15–16, 21
aryl film thickness 3–5, 11
aryl films 2–6, 8–11
atom transfer, surface-initiated 38–39
atomic force microscopy (AFM) 3–4, 10, 14, 17–18, 39, 54, 61–62, 72, 77, 86–88, 97, 222
atomic transfer radical polymerization (ATRP) 54, 60–61
ATRP, *see* atomic transfer radical polymerization
AuNPs, *see* gold nanoparticles
  functionalised 244
  internalised 254, 257, 259
AuNRs, *see* gold nanorods
avipoxvirus (APV) 88
azobenzene (AB) 35–37, 40, 42, 44, 48–52, 54
azobenzenes, alkyne 37–38

biomedical applications 168, 278
biomolecules/markers 245
Bovine Serum Albumin (BSA) 59, 64–67, 227, 285
2-bromoisobutyryl bromide 37, 39, 60, 62
BSA, *see* Bovine Serum Albumin
BSA adsorption 65–67
BSA molecules 66–67

cancer 159–160, 169, 171–172, 177, 184, 188, 247, 261
cancer cells 162, 165, 172, 177–180, 187–188, 248, 250–251, 290, 294
  breast 173, 176
  lung 168
cancer chemotherapy, image-guided 175
cancer diagnostic 160, 162, 189
cancer theranostics 160, 162, 169, 171, 173, 175, 177–179, 181, 183, 185, 187
carbon monoxide 248
catalysts 110–113, 115–117, 119, 121, 125, 130, 134
CDH, *see* cellobiose dehydrogenase
cell biology 150
cell membranes 83–85, 280, 291
cell penetrating peptides (CPPs) 241–242
cellobiose dehydrogenase (CDH) 116, 119, 124–125
cells
  bacterial 87

biofuel 112, 125, 127
blood 86, 248
C-26 185–186
endothelial 161, 254, 278, 284, 288–289
fibroblast 279
hypoxic 288
mammalian 172
spectro-electrochemical 121
cellular interaction and uptake of SERS nanosensors 278–279, 281, 283
cellular SERS nanoprobes 284–285, 287, 289, 291, 293
chemometrics 88–89
chemotherapeutic drug molecules 189
chemotherapeutic drugs 160, 162, 174, 189
chemotherapy 172, 177
CPPs, *see* cell penetrating peptides
crystal violet (CV) 115–116, 164, 261, 291, 293
CV, *see* crystal violet
cytochromes 85–86, 113–114, 119, 127, 129–131, 201–202, 206, 208–211
cytosine 94–95

DDA, *see* discrete dipole approximation
density-functional theory (DFT) 94
DFT, *see* density-functional theory
di-mercaptoazobenzene (DMAB) 75, 77, 79–80
diarylethenes 34, 52–53
diazonium salts 2–5, 7–9, 11, 13, 15, 17–19, 21, 54
discrete dipole approximation (DDA) 3, 6, 13, 17, 21
DLS, *see* dynamic light scattering
DMAB, *see* di-mercaptoazobenzene
DNA 82, 94–95, 98, 147, 149, 247, 281
DNA strands, single 94–96
double strand breaks (DSBs) 97
DOX 174, 244, 249–250, 261, 285–286
DOX-AuNPs@gelatin 174
DOX nanoparticles 261–262
doxorubicin 172–173, 176–179, 285
doxorubicin molecules 172–173
drug delivery 2, 58, 173–175, 189, 263
dynamic process of 162, 172–173, 177
drug release 171, 173, 177–178, 261
drugs, photosensitizing 170, 189
DSBs, *see* double strand breaks
Dynamic light scattering (DLS) 227

E-field, *see* electromagnetic field
E-field distribution 204–205
EBL, *see* electron beam lithography
EF, *see* enhancement factor
electrocatalysis 110–111, 113, 130
electrocatalytic reactions 111
electrochemical sensors 110, 112
electrochemistry 113, 115, 120
electrodes 111–113, 115–117, 119–121, 123–126, 128, 130–131, 133–134
electromagnetic field (E-field) 21, 147, 200–201, 205, 210, 229

electron beam lithography (EBL) 11, 15, 22, 37, 54, 59, 61, 145
electron transfer 2, 8, 77, 111–112, 125, 130, 288
electron transfer pathways 124–125, 129
emitters, fluorescent 142–143, 149–150
endocytosis 236–238, 242, 257, 265, 279
endolysosomal pathway 241, 256–257
endosomes 237, 250, 256–258, 279, 286
early 237–238, 279
late 237–238, 279
enhancement, local excitation intensity 145–146
enhancement factor (EF) 8, 46–47, 54, 143–145, 148–149, 183, 225, 228, 289
enzymes 110, 112–113, 124–125, 127–128, 130, 140, 181, 238, 247–248, 293
single polymerase 149
excitation 10, 19, 45, 49, 78, 141, 146, 163, 165, 168, 183, 185–186, 199–201, 206, 211
excitation intensity 141–143

FCS, *see* fluorescence correlation spectroscopy
fluorescence correlation spectroscopy (FCS) 147, 227
fluorescence emission 144, 148, 182
fluorescence enhancement 140, 142, 144–147, 150
fluorescence enhancement factors 143, 145
fluorescence imaging 172, 182, 187, 233
fluorescence signal 150, 173, 185
fluorescence spectroscopy enhancement on photonic nanoantennas 140, 142, 144, 146, 148, 150
Förster resonance energy transfer (FRET) 150–151, 176, 220
Fourier polar analysis (FPA) 224–225
FPA, *see* Fourier polar analysis
FRET, *see* Förster resonance energy transfer
fuel cells 110–111, 134

galvanic replacement 200–201, 204, 212
gold nanodisks 15–16, 18, 21
gold nanoparticles (AuNPs) 1–4, 33, 35–37, 39–42, 54, 164–165, 172–174, 234, 239, 241–243, 255–257, 278, 280–281, 284, 294–296
gold nanorods (AuNRs) 11–14, 146, 166–167, 178–179, 239, 285, 287
gold nanostars 174–175, 183
gold nanostripes 4, 6, 8, 11
gold nanostructures 66, 74, 295
Golgi apparatus 237–238, 279
grafting, aryl film 8, 10–13
grafting of PNIPAM 39, 62–63
growth factor receptors 290, 292

hangman complexes 130–131
HCA, *see* hierarchical clustering analysis

HEBDT, *see* 4-(2-hydroxyethyl)-benzene diazonium tetrafluoroborate
hemozoin crystals 86
hierarchical clustering analysis (HCA) 224–225
hot electron chemistry 72–73
hot electron–induced reaction 74–75
4-(2-hydroxyethyl)-benzene diazonium tetrafluoroborate (HEBDT) 3–4, 61

imaging, chemical 84
intracellular surface-enhanced Raman spectroscopy 164, 166, 234–236, 238–240, 242, 244–266
isomerization 35, 42, 44, 48
isopropylacrylamide 33, 58–59

label-free SERS detection of heme-proteins 199–200, 202, 204, 206, 208, 210, 212
laser excitation 6, 45, 51, 53, 121, 165–166, 183–184, 207, 209, 285
laser irradiation 3–8, 10, 15, 21, 183, 189, 291
LCST, *see* lower critical solution temperature
LDA, *see* linear discriminant analysis
light irradiation 1, 34, 42, 44, 49–51
linear discriminant analysis (LDA) 224, 256
localized surface plasmon (LSP) 1, 8, 10–11, 42, 54, 73, 160, 200
localized surface plasmon resonance (LSPR) 16, 26, 28, 35, 41, 44, 51, 54, 160, 163, 166, 200, 204
lower critical solution temperature (LCST) 36, 40–41, 45, 48, 51–52, 54, 58, 62, 64–65
lowest unoccupied molecular orbital (LUMO) 74, 77
LSP, *see* localized surface plasmon
LSP excitation 6, 8–11, 13–14
LSP resonance 4, 6, 8, 14
LSP wavelength 3–4, 7, 64, 66–67
LSPR, *see* localized surface plasmon resonance
LSPR excitation 16, 34–35, 45
LSPR wavelength 63–65
LUMO, *see* lowest unoccupied molecular orbital
lysosomes 250, 256–258, 279
  primary 237–238

macromolecules 58–59, 229, 236, 238, 240, 286
MEF, *see* metal-enhanced fluorescence
metal-enhanced fluorescence (MEF) 161
metal nanoparticles 147, 163, 181, 278, 280, 285, 289
microchips 259–260
molecular catalysts 110, 112, 130, 134
myoglobin 114, 201–203, 206–208, 210–211

nano-cylinder 22–28

nanoantennas 139, 141–146, 148, 150–151
nanocubes 200–204, 207, 209, 296
nanodisks 16–18
nanoparticle–cell interactions 235, 237, 239, 241, 243
nanoparticles
  anisotropic 166
  biohybrid 172
nanopores 202, 205, 210
nanoprobes 161, 253, 256, 278, 281–285, 288–289, 293, 295
nanorods 11–14, 146–147, 200
nanosensors 278, 281, 283, 285–286, 288–290

optical antennas 140, 147
optical nanoantennas 142–143, 150–151

PB, *see* Prussian blue
PCA, *see* principal component analysis
phosphodiester vibrations 96–97
photodynamic therapy 179
photoirradiation 35, 49, 51–52
photoisomerization 35, 45, 49, 53
plasma membrane 236–237, 243, 260
plasmon decay 72, 74, 78
plasmon excitation 3, 6, 12–16, 161
plasmon-induced grafting 6, 12–14, 16–17, 19–20
plasmon-induced surface functionalization 2–3, 5, 7, 9, 11, 13, 15, 17, 19, 21
plasmonic nanoantennas 143, 150–151
plasmonic nanoparticles 1, 159–160, 181, 188–190
plasmonic vesicles 176, 250–251
porphyrins 114, 117, 119, 131, 180
primary lysosome 237–238
principal component analysis (PCA) 83, 87, 224–225, 247, 255, 293
proteins 64–65, 83–84, 90–92, 117, 119, 125–126, 128, 130–131, 201–202, 208–212, 219–227, 229, 240–241, 255–256, 264–265
Prussian blue (PB) 291

Raman bands 45, 75, 84, 224, 227, 229
Raman reporter molecules 161, 163, 168, 170, 248, 290
Raman scattering, tip-enhanced 72–73
Raman shift 225, 292
Raman signal 72, 78, 86, 117, 199, 206, 252, 254, 290
Raman spectroscopy 72–73, 117, 172, 174, 176, 184, 220–221, 233
  surface-enhanced 22
reaction time constant 26
reactive oxidative species (ROS) 243, 246, 278, 289
receptor-mediated endocytosis (RME) 236–237, 242, 279
redox chemistry 128, 131
redox reactions 115
redox transition 115, 121, 128
reporter-free SERS 245, 247, 253–255, 257, 259–261, 263, 265

RME, *see* receptor-mediated endocytosis
RNA 82, 94–95, 281
ROS, *see* reactive oxidative species

SAMs, *see* self-assembled monolayer
scanning electron microscopy (SEM) 3, 16–17, 19–20, 23
scanning probe microscopy (SPM) 72, 77
secondary lysosome (SL) 237–238
SEF, *see* surface-enhanced fluorescence
self-assembled monolayer (SAMs) 26, 52
SEM, *see* scanning electron microscopy
sensing applications 22, 246, 248
sensitive detection of proteins 201
sensitivity 84–85, 87, 94–95, 117, 131, 134, 140, 150, 163, 166, 201, 221–222, 234, 248, 251–252
SERS, *see* surface-enhanced Raman spectroscopy
SERS-traceable drug nanocarriers 160, 162
SERS-traceable plasmonic nanoparticles 171, 189
silica, mesoporous 176–177, 262
silver electrodes 76, 199, 263
silver nanocubes 147, 204–205
silver nanoparticles (AgNPs) 164, 168, 234–236, 263, 285, 291
silver porous nanocubes 209, 211–212
singlet oxygen generation 181, 183, 189
SL, *see* secondary lysosome
spexels 226–227
SPM, *see* scanning probe microscopy
surface-enhanced fluorescence (SEF) 45, 53–54
surface-enhanced Raman scattering 22, 72, 160, 199
surface-enhanced Raman spectroscopy (SERS) 20–22, 47–48, 72–73, 75–78, 160–161, 163, 176–177, 179, 182–185, 199–200, 233–235, 244–245, 249–253, 262–264, 285–286
surface plasmons 72–75
synergistic chemo-photothermal therapy 177–178

TEM, *see* transmission electron microscopy
TERS, *see* tip-enhanced Raman spectroscopy
theranostic agents 162, 169–170, 178, 185, 189
theranostic nanoparticles 162, 170, 184–185
theranostic platforms 178–179
theranostics 162, 175
thiol-ene "click" chemistry 2, 23, 25, 27
thiol-ene reaction 27–28
thiophenol 24, 26–28, 77
thymine 94–95
TMV, *see* tobacco mosaic virus
tobacco mosaic virus (TMV) 88
toxicity 166, 234–235, 242–243, 245, 250, 278
transmission electron microscopy (TEM) 203–204, 252, 254, 279

tumor 160–162, 170–171, 174
tumor cells, circulating 248–249

UV light 22–24, 28, 43, 52

Varicella-zoster virus (VZV) 88–89
VZV, *see* Varicella-zoster virus

zero-mode waveguides 149